The
Slate Roof Bible
Second Edition

Joseph C. Jenkins

Joseph Jenkins, Inc.
143 Forest Lane, Grove City, PA 16127 USA
Ph: 814-786-9085; Fax: 814-786-8209
joseph-jenkins.com

The Slate Roof Bible

Second Edition — Completely Revised, Expanded and Updated

© 2003 by Joseph C. Jenkins — All rights reserved.
ISBN 0-9644258-1-5; Library of Congress Control Number: 2003096286
Second Printing, 2006 • Printed in China

Published by Joseph Jenkins, Inc., 143 Forest Lane, Grove City, PA 16127 USA
Toll free: 866-641-7141 or 814-786-9085

Distributed to the book trade by Chelsea Green Publishing, PO Box 428, White River Junction, VT 05001 (1-800-639-4099)

Cover art by Tom Griffin of the Otter Creek Store, Mercer, PA 16137

Please visit our websites for online sales, message boards, tools, materials and informative articles:

SLATEROOFCENTRAL.COM
(online sales, slate roofing tools, materials and supplies, slate industry
source lists, public message board — see ad in back of this book)

TRADITIONALROOFING.COM
(informative articles about slate and tile roofs available free online)

JENKINSPUBLISHING.COM
(other books by Joseph Jenkins, public message board)

SLATEROOFERS.ORG
(Slate Roofing Contractors Association of North America)

JOSEPH-JENKINS.COM
Joseph Jenkins, Inc. web site.

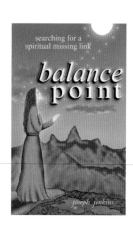

Have you seen a slate roof glisten as the sun rounds the ridge?

THE
SLATE
ROOF
POEM

Philip Terman

Have you seen it shimmer in a splash of a sky full of stars?
Have you ever seen a close-up of slates interwoven like leaves
on a tree or the way slats of stones will seem to quiver in a creekbed,

or how from a distance they can look like steps ascending into the sky,
or how they float in mid-air, the way they hang loose on their nails,
or how they make a joyful noise when they are tapped
and when rain patters like a waterfall as you lie under them…

Because the rock was here before we came and will be here
after we leave. Our first dwelling and our last, the veins we tap,
the secrets we extract. Have you driven by a farmhouse
and noticed the roof with the year of its construction slated in,

or a chicken coop with a heart-shape or a gravestone carved
with a poem? The slate on my study in Pennsylvania is from
a farmhouse in New England. Would we recognize each other,
we who were protected underneath the same substance?

Consider, as you sit shaded from the sun or dry from the rain,
all that geological time, the silt depositing and collecting
and compacting-shimmying and hopping and bopping
to the geologic boogie-into what has now become your roof.

Consider all that mineralogical pressure. Consider how
it must have been created on the first day, with all the other stuff
of the earth. Consider how its origins are a mystery, like love
and death and our desire to use a material that will last

beyond our lifetimes, we of the long-term, we who have faith
in the future, we who play our minor role in the eternal drama
of the significant stone, we who want to know what our part
of forever feels like. Because each square of slate is really

a petal of flowering sediments, a small piece of earth
from within the earth. A slate roof looks like a book before
it is bound. All of the pages are written by a collaborative
effort of weather and time, Hold one slate in your hands.

It is smooth and in places wrinkled like handmade paper.
You can't see yourself in the surface but it is beautiful,
it makes any building-pigsty or majestic dome-
look like a quaint hut, a cottage in a fairy tale.

Because it is so of-the-earth it is otherworldly.
Something mysterious about the way it is always shining,
something perplexing about the way it absorbs history
and stays its place. One cannot imagine a slate roof

without farms or woods or flowers around it, someone
baking bread beneath it. In one dream we walk on slate
but we slide because it is slick like trying to stand
on water. In another we are in a burning building

but we are safe inside our walls of slate.
and when we're effaced, these pieces of hard earth
above us for a short time will be beheld by others, interlocked
like our own and offering like our own their securities,

making again their music and their light.

Philip Terman is a Professor of English at Clarion University of Pennsylvania.
He is the author of *The House of Sages* and *Book of the Unbroken Days*, both from Mammoth Books (mammothbooks.com).
He lives under a slate roof in Barkeyville, PA.

THE SLATE ROOF BIBLE
• CONTENTS •

PART I — UNDERSTANDING SLATE ROOFS

PART II — INSTALLING SLATE ROOFS

PART III — REPAIRING AND RESTORING SLATE ROOFS

Opposite page: Stone roof (not slate) in northern Italy. Photo by author.

DEATH OF A BARN

I watched a hundred-year-old barn burn to the ground one day. As the crackling flames shot in the air and rapidly engulfed the old building, a large raccoon suddenly appeared on the ridge of the roof. The raccoon had been roused from its daytime sleep in the hayloft by the smoke and flames and shouts of the firemen, and had quickly scampered up the oak barn beams to find a safe place to hide. As the flames spread and the heat grew more intense, the raccoon began to suspect that it had made a wrong move and started pacing nervously back and forth along the ridge. Those of us on the ground who were gaping at the spectacle were horrified by the thought that the raccoon was about to become broiled alive, right before our eyes. The firefighters tried to knock the animal off the roof using the spray of the firehose, but the raccoon, not as dumb as it appeared, simply dodged the water by ducking behind the roof ridge. Soon the situation grew unbearable for the animal as the flames licked over the barn roof, and the raccoon, in an obvious act of desperation, suddenly made a swan dive off the gable end into a clump of bushes forty feet below. It looked like a suicide dive to me, but the animal surprised everyone by bouncing off the ground with a thud, then running away. Perhaps raccoons, like cats, have nine lives. Not true for barns.

In this case, the owner wanted the old, hand-hewn barn destroyed and had called the fire department to burn it down. Why? It was, after all, a beautiful, century-old structure, a memento of the agrarian days of rural Pennsylvania, of a time when everyone in the country had farm animals and lifestyles that centered around animal husbandry. When the barn was built just before the beginning of the 20th century, people still got together and helped each other in their communities. They didn't hop in a car and drive to each other's farms, because there weren't any cars. There were hardly any roads, and what roads existed tended to be rough and impassable in bad weather. There wasn't any electricity out on the farms either. People relied on their horses for transportation, and if the roads were too muddy for horse and buggy, they just stayed home, or they walked.

A cooperative group had raised this barn, probably the same group who got together at the local church on Sunday to pray for good crops, or a safe birth for a young couple, or life renewed for a sick elder. On Saturdays they congregated at the local grange, or in this very barn to plan political strategies or just to have a good time square dancing by the light of kerosene lanterns, if not the moon.

When the barn was built, it symbolized all the hope and promise of a developing country. People were settling the land, clearing the forest and planting the fields. Their survival rested upon the constructive relationships they maintained with their neighbors and their cows, horses, sheep, pigs, goats and chickens. The barn would house these animals, and thereby provide a livelihood for the country people whose new homesteads sprang up in the Pennsylvania countryside like mushrooms after a rain in the years following the Civil War.

The barn bore testimony to the use of local trees for construction — the beams were wrought from white oak, chestnut, or yellow poplar, while the siding was cut from pine or hemlock. The post and beam frame required no nails, but was fastened together with elegant mortise and tenon joints locked into place with stout wooden pegs called trunnels or "treenails." The beams were shaped by hand with broad axes, adzes, and chisels. They were cut with hand saws of various types, and drilled with breast drills, braces, bits, and augers. No power tools were used to build this barn, which remained square and erect after a century of use. The proud builders no doubt stood back and admired their work when the construction was completed a hundred years ago; then they communally feasted and together celebrated their accomplishments, their communities, their country, and their lives.

The elderly widow who eventually owned this farm called me one day and asked me if I wanted the slate roof that was on the barn. The barn had to go, she said, as it was beginning to collapse and was presenting a hazard. I could have the roof if I wanted it, as well as any other parts of the barn, but I only had two weeks to get it. I went over to the farm and inspected the barn to see if salvaging the slate was possible. Some barns are so far gone that climbing on the roof would pose an unacceptable danger. This barn only had one corner rotted beyond reasonable repair; the rest of it was in good condition, so I agreed to remove the slates.

Later, when I climbed on the barn roof, I saw why that bad corner had rotted: three slates were missing on a spot that was, coincidentally, directly over a main support post. Aside from those three slates, the rest of the sea green slate roof was in good condition. Of the 3,000 original slates on the roof, three of them, or one tenth of one percent, had fallen off and never been replaced. The resulting hole in the roof was on the back of the barn where it was not easily seen, and so it was ignored. The owners had allowed the rain water to leak through the roof decade after decade until the barn was crippled, then destroyed. Three slates take about one hour to replace, including setting up and taking down ladders. For want of one hour of professional maintenance, another hand-hewn barn, a symbol of America's rural heritage, died.

The barn wasn't a complete loss, though, because the slate roof was salvaged and used on a new building. And I'm willing to bet the clever raccoon found itself a new home too, but probably not in a barn.

REBIRTH OF A BARN

Not all barns are neglected and then destroyed. This century-old barn near Zelienople, PA, suffered storm damage to its original side-lapped slate roof. The oak barn was structurally sound, however, and a new Vermont "sea green" slate roof was installed in the more durable standard-lap method.The slates have been removed (top, left) and the wide spaces between the existing oak boards are being filled with one-inch-thick rough-sawn hemlock lumber. Orion Jenkins and Brent Ulisky put the finishing touches on the lightning rod re-installations (top right). Below is the newly slated barn and the slaters who installed it, with a view from the rear (center photo).

Photos by author.

Chapter One

UNDERSTANDING SLATE ROOFS

"Beneath the moss covering the roof of the rugged Saxon Chapel, erected during the eighth century, at Bradford-on-Avon in Wiltshire, England, is a slate shingle roof in good condition after twelve centuries of exposure to the elements in one of the most severe climates in the world, a monument to the enduring qualities of slate." H.O. Eisenberg

Slate roofs are the world's finest roofs. They're fireproof, waterproof, natural, will last centuries, and have a track record that goes back thousands of years and spans the entire Earth. They are beautiful, simple roofs made of rock on wood — ingenious, effective, and fabulously successful. Yet, in the United States, they have been under intense and relentless attack for decades, destroyed by the thousands, cast away and forgotten. The culprit, in a word, is ignorance. The following conversation illustrates this point:

The telephone rang. "Hello," a lady said, "Are you the guy who buys used roof slate?"

"Yes."

"Well, I have enough for a whole roof, and you can have them all if you want them."

"Where are they?"

"They're on my house!"

"They're still on your roof?"

"Yes! And you can have them for nothing if you'll take them off."

I paused for a moment, then asked, "Why do you want your slate roof taken off?"

"Because it leaks. We've already bought fiberglass shingles to replace the roof; they're sitting in the driveway. We just need somebody to take the old roof off."

"Well, ma'am, the reason I buy roof slate is because I repair and restore slate roofs professionally. Maybe I should have a look at your roof before you take it off."

There was a pause at the other end of the line. I could almost hear the thoughts racing through the lady's head: What? I can *repair* my slate roof?

"But we already bought the shingles."

"You can return them and get your money back *if* you don't need them, which you may not. What kind of slates do you have on your roof?"

"What kind? I didn't know there were different kinds."

"Slates *generally* fall into two categories—

hard and soft. If you have hard slates, they should last the life of your house and you won't need to replace them. If you have soft slates, then you may not have a choice — you may have to replace the roof. I can tell at a glance what kind of slate you have and whether the roof needs replaced or not."

The next day I stopped to look at the roof. The slates were hard Vermont "sea green" slates with a general life expectancy of 150 or more years. The house was about eighty years old. There was one slate missing from the roof — *one slate!* — and the roof had a small leak at the spot where the slate had broken off. Otherwise, the roof was beautiful. So I offered to repair the roof for a small fee, explaining that the roof should never have to be replaced in her lifetime. The lady accepted, I did the hour-long repair job, she saved both her roof and several thousand dollars, and I haven't seen her since. This is a true and typical story.

I can go on and on with these kinds of stories. One young lady who had a beautiful old Victorian house with a hard, Vermont slate roof in very good condition told me she was also considering having the slates taken off the roof. "Why?" I asked. "Well, I thought you were just *supposed* to replace slate roofs when they got old," she replied. We were standing in her front yard when she said that. The summer sun was glistening off her stone roof. Her impeccable white house stood like a majestic tribute to a time gone by when things were built with quality in mind — built to last. The roof had been cared for by her father, a spry 85-year-old man who recognized something of value when he saw it. Now the daughter owned the house and her first thought was to rip off the slate roof, which didn't even leak. I stared at the house, at the perfectly good, hard slate roof, wondering how many thousands of dollars it would cost her to remove her stone roof and put on a cheap, artificial roof. I tried to imagine how temporary asphalt shingles would look on that proud home.

"No, you don't have to replace slate roofs

Opposite Page: Castle near Mayen, Germany with German slate roof.
Photo by author.

▲ This 1860 roof in Fair Haven, Vermont, was 143 years old when photographed — still in good condition. The roof is made of Vermont purple slate with a Vermont green slate inscription. The 1862 roof located in western Vermont is also made from Vermont unfading green slate lettering on a purple Vermont slate roof, 141 years old at the time of photographing and still going strong. These roofs are built of stone (slate) nailed to a board roof deck — a roofing system easily replicable today. Slate shingles on a wood board roof deck is a tried and proven water-tight roofing system guaranteed to last centuries.

Photos by author.

when they get old," I finally said. "If the slate is hard, like yours is, you can expect it to last as long as your house lasts. It will certainly last your lifetime, and probably your children's."

"Really!?"

Recently, a man called me about the roof on his church. He said he had good, hard, black Pennsylvania "Peach Bottom" slate on the church roof, which is some of the best slate in the world, but he thought it should be replaced. Why? Because the valleys and flashings (metal joints) on the roof had deteriorated and were leaking.

"Well, replace the metal!" I advised. But my advice went in one ear and flew out the other as the man explained to me how much it would cost to replace the entire roof with new slates.

"We'll have about $150,000 to $200,000 in replacing that roof," he said.

"Don't replace the whole slate roof, just replace what *needs* to be replaced," I replied. "It's not that hard to replace roof *metal*. By keeping the *slate* but replacing the *flashing* you'll cut your costs to a fraction of what you're talking about. Probably down to *less* than 10%. The Peach Bottom slates you have on your church might last another hundred years. Why take them off? It's a lot of work to take off a slate roof, and it doesn't make a whole lot of sense to do so when the slate is still good. Besides, you can't even get Peach Bottom slate anymore — it hasn't been quarried in decades, even though it was once

internationally recognized as the best slate in the world."

The man was in a hurry, though, and he didn't seem to hear what I was saying. He had to call someone *else* about removing the roof (since I wasn't interested). End of conversation. End of roof.

Then there was the guy who had the beautiful Peach Bottom slate roof on his old Victorian house. Some Peach Bottom slate that's 250 years old is still good, and no one knows how long it can last. Four hundred years is a fair estimate. This fellow decided to have a professional roofing contractor look at his 90-year-old roof and estimate its remaining life. He was careful to find the oldest, most established roofer in the city. The roofer had only to take one look at the roof to state with certainty the number of years the roof had left: "Ten," he announced, with the air and authority of a professional. And of course that declaration was followed by the next, obvious one: "You'll have to tear it off. Would you like me to give you an estimate?" The homeowner was cautious and intelligent enough not to take the contractor up on the offer. Instead he saved his roof by repairing it at a fraction of the cost of replacement, and he'll never have to put another roof on that big old house as long as he lives. Otherwise, if the slate had been removed, he would have been destined to replace the roof about every twenty years at increasing cost. Which, by the way, is what many roofing contractors live for.

▲ A picture is worth a thousand words. Photographed by the author in 2002, this roof, covered with Vermont purple slate with a Vermont unfading green inscription, was 151 years old at the time and still quite functional. In contrast, the average American asphalt roof looks rather unpleasant and only lasts between 15 and 20 years. You just can't beat a good slate roof for beauty, utility, or even cost — when the entire life span of the roof is taken into consideration.

▲ Barn near Townville, PA, in poor repair. Main roof is Lehigh-Northampton (black PA) slate; designs are made of Vermont green slate. Inscription reads, "L. Jones - 1902."

Photos by author.

Slate roofs originated in Europe, and some of the earliest Europeans to develop a proficiency in slate quarrying and slate roofing were the Welsh. Slate roofs are common in Europe where they're valued enough that some are still in good repair after several centuries. It is against cultural mores to replace a slate roof with an asphalt roof in Europe, and the people there are so adamant about it that they have passed laws preventing people from replacing slate roofs with cheap substitutes. There are photos of European slate roofs in this book to help illustrate the contrast between American values and European values. There, they know a good thing when they see it. Here, we rip off our slate roofs at the first sign of a leak and replace them with substitutes that are miserable by comparison. If you own a slate roof, you will likely hear every conceivable excuse to tear it off. The strongest urging may come from contractors who stand to gain financially from the destruction of your stone roof.

I hear it all. One common myth is that slate roofs have to be replaced because their nails go bad. Usually, this is not true. At the time of this writing, I personally have worked on well over 1,000 slate roofs over the past 35 years. *Old* slate roofs. That means I've climbed on a thousand slate roofs, torn them apart and put them back together, examined them, assessed them, removed and recycled them, repaired, restored and installed them. Their average age has been about 100 years (not counting the new

installations). Most of the nails on these roofs are still good. In fact, they're so good that I could use some of them over again on a new roof. *Even after a century!* And 99% of these nails are hot-dipped galvanized or cut steel, *not* copper. A photo of these old nails can be found in Chapter 13.

In some cases, on some parts of roofs, especially leaking and neglected roofs, nails can and do go bad. This is especially true when the *wrong* nails were used *in the first place*, which happens, but not very often. As a rule, bad nails are rarely a reason to replace a slate roof. If the nails are going bad, it's either because the slates themselves are going bad, meaning they're worn out soft slates, or the wrong nails were used when the roof was installed. However, in Europe, when a roof goes bad because of rusting nails or rotting wood underlayment, the slates are simply removed, the wood replaced, and then the same slates, or new slates, can be nailed to the roof with new nails. A slate roof can be kept alive for centuries in this manner.

Another dubious reason to want to replace a slate roof is because the roof paper or *underlayment* (the felt underneath the slate) has deteriorated. After a hundred years, or less, the roofing felt paper turns to dust. People peer through the gaps in the roof boards in their attics and see that the paper has turned to dust, and they conclude, *"Time to replace the roof!"* Wrong! *Roofing felt paper is not a necessary part of a slate roof.* It is typically used as a temporary cover to protect the building in the event of rain until the finished roof is installed. Roof paper *does* help temporarily to insulate the roof, but its usefulness in shedding water is greatly reduced when thousands of holes are poked through the paper as the slates are nailed on. Barns typically do not have *any* roof paper under their slates, and they're as waterproof as any other slate roof, even after a century. It is foolish to think that the paper underneath the slate is an important part of the roof. It is not. But believe me, there are hordes of Cro-Magnon types out there who will bend over backward to convince you otherwise.

Then there's the *curtain* theory. One middle-aged couple called me to look at their roof. They were considering removing it and replacing it with fiberglass shingles — so-called "premium" asphalt shingles that are guaranteed to fail in twenty years, after which they curl up like potato chips and blow away in the wind. Anyway, I stood out in front of the house looking at the beautiful, decoratively-cut Vermont purple slate roof, one of the best slate roofs a house can have, with the homeowners standing by my side.

"That's where it leaks," said the lady as she pointed to a section of roof on the front of the house where a Neanderthal had applied tar.

"Not hard to fix," I replied, then I walked around the house to get a good look at the rest of the roof. I could tell that the husband wanted to keep the slate roof, but the wife didn't. Every now and then I'd overhear bits of conversation as they continued to stand out in their front yard gesturing toward the house.

"The slate roof doesn't match our new curtains!"

"I know, dear."

"We need a brown roof if we're going to match the curtains."

"I know, dear."

When I returned to the couple, I briefly explained to them that their purple slates would last the life of their house, and could last hundreds of years, nobody really knows. Purple Vermont slates that have seen a century of wear look pretty much the same as new purple Vermont slates, so who knows how long they'll last? I told them it would cost a few hundred dollars to do the routine maintenance on their 90-year-old roof, as opposed to ten times as much to replace the slate with asphalt shingles that were guaranteed to quickly fail. The lady was adamant about not liking the color, though, and I sensed that this roof was doomed. "If you do decide to remove the roof," I offered, "I'd be willing to remove the slates free of charge, for salvage purposes."

"Well, we'll call," they said. A month or two later I drove by the house. A brand new, *brown* asphalt shingle roof sat on top of the home. I knew what happened to the beautiful slates—they were carelessly destroyed by the roofers who tore them off. I have seen it many times—the roof slates are rudely ripped off, thrown on the ground, hauled off in the back of a dump truck, and unceremoniously dumped in a landfill. The homeowners are several thousand dollars poorer, they've permanently lost their wonderful slate roof, and now they're faced with having to replace the cheap asphalt roof every twenty years at a constantly increasing cost. But the roof matched the curtains! At least I think it did—I didn't notice the curtains—who ever does?

Every now and then I see a slate roof where one side of the roof has been replaced with asphalt or fiberglass shingles. Picture this if you can: you're balancing on the ridge of the roof and to one side of the ridge you see a slate roof, the other side a shingle roof. The shingle roof is curling up and falling apart, worn thin, brittle and leaking. The slate roof, already 90 years old, looks nearly the same as it did when first installed, and exactly the same as it did when the shingle roof was installed 20 years ago. You wonder why the homeowner ever took the slate off the one side in the first place. Now look into the future: the decrepit shingle roof will soon be

---

OLD ROOFING MATERIAL IS FILLING UP LANDFILL SPACE

When old roofing is removed from buildings, it is trucked to landfills. Non-biodegradable, petrochemical roofing made of asphalt and fiberglass typically lasts about 20 years, then is discarded. Roofs such as these that are guaranteed (by the manufacturer) to *fail* in two decades, are guaranteed to clog landfills year after year. According to scientific studies, construction and demolition debris make up 28% of the weight and 28% of the volume of the mixed waste in landfills, and is much more significant to landfill management policy than Styrofoam, fast-food packaging, disposable diapers and the total of all plastic packaging *combined*! About one fifth of all construction debris is roofing waste. Environmental concern is only one reason to have a slate roof, but it's an important one.

[Source: Garbage Magazine, Mar/Apr 1992, p. 67, and Jul/Aug 1992, pp. 20-21]

replaced and the slate roof will, once again, be left alone. In another 20 years the same scene will repeat itself. Then again. Then again. Each time, the shingle roof will need to be *completely* replaced — the old shingles hauled to a landfill, *if there is any landfill space left by then*, and the homeowner will needlessly fork out more money to eager roofing contractors.

The slate side, if left alone, and repaired occasionally as needed, will smugly sit there, stone quiet, stoically oblivious to the vagaries and vicissitudes of the human race. It will continue to do what it's supposed to be doing: protecting the dwelling and sheltering the people who live there. And it will do so with a level of aesthetics and a richness of history that a fiberglass shingle couldn't hold a candle to. The folly of replacing such a roof with a cheap substitute becomes clearly evident when one roof displays both roofing materials at the same time, and their performance can be seen and compared side by side.

A common reason why people feel they need to remove their slate roofs is because *"We can't find anyone to fix our roof."* If they do find someone, they run the risk of hiring a contractor who wants to "gouge" them (i.e. charge a small fortune for repairing the roof). Most contractors don't specialize in slate, or even roofing, and when they're not building decks or installing bathrooms, they'll "repair" your roof. Most of the contractors I've met who repair slate roofs do it improperly. And because they're not equipped and experienced to work on high, steep roofs, they charge an exorbitant fee, knowing that if they get paid enough, they'll do the work. So *if* a homeowner can find someone to fix the slate roof, and *if* she can afford him, then she will probably still have to pay *to have the job done wrong!* Which means she'll have to pay again to eventually have the job done right, and re-doing someone else's lousy work is always harder than doing the job right in the first place. This makes for a frustrating situation for owners of slate roofs, and it's one of the main reasons this book was written.

I recently met a contractor who advertised himself as a specialist in slate roof repair, and who told me he goes through about 15,000 pieces of slate every year doing roof repairs. Then he proudly informed me with a grin and a nod, *"I face-nail every one of 'em."* What roof owners don't know is that every face-nailed slate (nailed through the face of the slate so that the nail-head remains exposed to the weather after the repair is completed) is an improperly installed slate, one that will leak, and one that has been ruined by the roofer who punctured it. So here's a guy whose roofing company face-nails fifteen thousand slates every year. Every single one of those ruined slates will have to be taken off and replaced someday. Before they do, though, they'll probably leak. This roofer gets paid good money to slowly destroy people's slate roofs. And I'm not singling him out intentionally. He's *typical!* (Although he's since gone out of business.)

Why do contractors face-nail slates? Because

DECIDES TOO LATE TO REROOF WITH SLATE

$75,000 FIREPLACE BLAZE

"York, Me., Oct. 1 (A. P.).—The home of John C. Breckenbridge, on the York River, burned tonight with $75,000 loss. Sparks from the fireplace set fire to the roof."
—From clipping Boston Advertiser. Sent us by E. A. Bullard.

A few days later we learned that these people were considering reroofing with slate. Had actually wired Rising & Nelson Slate Company a few days before about it. Their home could have been saved had they considered earlier and put on slate, the sheltering stone, before they started up the first fire of the Fall in the fireplace. Buildings far away from community fire apparatus need protection of household fire extinguishers and slate roofs.

Capitalize such cases when you see them in the papers to hasten decisions of those in your city whose homes need to be reroofed to avoid such fire hazards.

W. L. Hassenplug, Philadelphia office, Rising & Nelson Slate Company, recalls a similar case in Maryland three years ago. A home was destroyed from sparks from fireplace chimney. It was rebuilt on original foundations, but first thing the owner insisted on was a slate roof.

From the *Slate News Bulletin*, May, 1926

THE BURNING OF THE "MYSTERY HOUSE"

Though staged primarily to prove the fire-resisting qualities of certain building materials, the "Mystery House" of Indianapolis was an advertising feature that aroused public curiosity, attention and approval. The house was built by Williams Creek Development Company for the purpose of a fire test. The house was a two-story building made of fire-proofed material and roofed with slate.

To provide a severe test, the lower floor was filled with kindling and lumber scraps saturated with kerosene. The oil-saturated wood, highly combustible, was ignited, and for three quarters of an hour (until the fire was extinguished) the building confined the flames like a huge furnace. To extinguish the flames, the fire department deluged the building within and without. A careful and critical examination after this severe treatment showed the slate roof intact without a single slate cracked from the fire and water. Publicity of the test was heightened by the presence of the mayor and the city fire chief of Indianapolis. The mayor started the fire and the chief put it out.

they don't know any better, and/or because it's a little bit easier and quicker than doing it the right way. Some contractors focus more on how much money they're making than on how well they're doing the job. There *are* instances when face-nailing *is* appropriate, but they're few and far between. We'll get into all of that when we cover slate roof repairs and restoration step-by-step in the third part of this book.

If you don't want to fix your roof yourself, but do want to hire a competent contractor, make sure the contractor you hire is either experienced with slate or has read this book *before* he's set a foot on your roof! If your contractor tries to tell you he already knows everything there is to know about repairing or installing slate roofs (a common line), then ask him a few questions to see whether he's

telling the truth. Ask him where your slates came from and how long they'll last; ask him to explain how he replaces a single slate; ask to see his slate cutter, slate hammer, slate ripper and ladder hooks. You may know a lot more about your slate roof than he does by the time you've read through this book, and if the guy is an imposter, you'll know that, too!

There are numerous reasons why old slate roofs should be preserved, in addition to longevity. One is *aesthetics*. In this plastic world a stone roof is a symbol of durability, of quality and craftsmanship. Once slate roofs become old they fall into the category of *antique* roofs, plain and simple. They're rich in history. Imagine the human effort and struggle that enabled people to bring many tons of massively heavy stones up from hundreds of feet underground, split them into shingles by hand, and do it all in the

Official inspection of the "Mystery House" after the fire

horse and buggy days, before electricity was available. People will give their eye teeth for a hundred-year-old chest of drawers, yet they'll discard at whim a perfectly useful, irreplaceable, antique stone roof worth $10,000, or much more. Why?

Let's also not overlook the fact that the purpose of a roof is protection. Stone roofs are *fireproof* and withstand the elements like no other roof. If you live under a properly maintained stone roof, then you know exactly what I mean. There's no feeling like the feeling of security you get when the wind is howling and torrential rain or sleet or ice is beating down upon the earth and pummeling your house, and a stone roof stands guard overhead, deflecting mother nature's blows. Living under a stone roof is kind of like living in a cave. It's almost a primal experience, and with a little imagination, we can see why such a roofing material appeals to us instinctively. There is no shelter quite as secure from the harsh elements as a cave, and any roof made of pure rock comes close to providing the same level of protection as the earth itself.

Speaking of caves, it's time to take a look at the *Neanderthal Syndrome*:

THE NEANDERTHAL SYNDROME

The Neanderthal Syndrome was discovered by the author, who has had the opportunity to study the traces of human impact left on slate roofs over the years, sort of like an anthropologist studying the campsites of cavepeople.

Many times I have positioned myself atop a slate roof and marveled at the utter and absolute folly with which the roof had been treated. I have shaken my head in disbelief, theorizing in vain as to

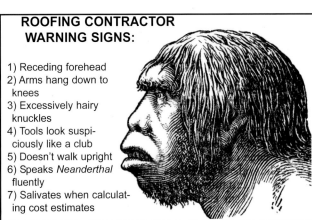

ROOFING CONTRACTOR WARNING SIGNS:

1) Receding forehead
2) Arms hang down to knees
3) Excessively hairy knuckles
4) Tools look suspiciously like a club
5) Doesn't walk upright
6) Speaks *Neanderthal* fluently
7) Salivates when calculating cost estimates

▲ A slate roof enhances the beauty of the lowliest building, such as the author's chicken coop (above), as well as the grandest of castles.
▼ The date on this Vermont "sea green" slate roof was made by using red slate found in the Granville, NY, area.
This roof was 113 years old when photographed and, *if properly cared for*, will outlast the house.

▲ A Vermont "sea green" slate roof located in western Pennsylvania with a PA black slate date, probably 1887.

(Left) This purple and green Vermont slate roof, dated 1862, will last for generations if properly cared for, despite its age of 131 years at the time of this photo. Age does not destroy slate roofs as quickly as improper maintenance.

Photos by author.

what could possibly have caused the prior "roofers" to put replacement slates in upside down and/or backwards, to tar an entire roof in order to cover a pinhole, to drive steel spikes through the slate roof willy-nilly, like a rampaging lunatic who just escaped from an insane asylum and decided, on a whim, to climb up on someone's roof and start pounding nails into it. No doubt I have seen every imaginable assault and insult to a slate roof, much of which is beyond human comprehension, without any reasonable explanation, and, at times, infuriating enough to become a laughing matter simply for the sake of sanity (mine).

The explanation is simple: Neanderthals never became extinct, they simply evolved into roofing contractors. The traces of their work are on almost every old American slate roof I have ever seen. I call these traces "Neanderthal tracks."

It appears that there was once a one-armed, blind roofer. The slates he used to replace broken slates didn't match the roof at all. When he repaired a nice green roof, he used black slate. When he repaired a black roof, he used green slate. When he replaced square slate, he used slates with cut corners. Nothing matched and it all looked bad when he was done, but it didn't bother this fellow — he was blind!

He had a crew who each had his own style of roof "repair." This crew must have worked very hard for many years because their work can be found across the United States. One of the hardest workers kept his nail bag full of 3½-inch-long *spikes*, and he

drove a spike into the roof wherever a slate looked loose. Sometimes he'd drive a spike into the roof for no apparent reason, perhaps just to look busy when the boss was around. Of course, a sixteen penny nail driven into a slate roof cannot be removed without breaking a few slates in the process, and if it's left in the roof, it will leak.

Another crew member loved tar, otherwise known as roof cement or mastic, and he used it for all his slate roof repairs. He kept a bucket in hand at all times as he spread the tar liberally over the surfaces of the roof. If he was in a really good mood, he'd use a thin tar that he could brush on and then he'd really go to town painting everything in sight. If a roof had a tiny hole in it, he would tar the whole thing. No pinhole was going to get the better of him!

Such legendary roofers worked on slate roofs for decades, and it's speculated that they grew old and died, but I'm not so sure about that. They might still be out there working away, reincarnated as younger Neanderthals. Ironically, these guys are slowly putting themselves out of business, because they deface and destroy slate roofs to the extent that the roof owners can't stand their own roofs anymore. When a slate roof has so many metal patches, tar blotches and leaks that never stop, it looks ugly, gets frustrating and drives the roof owner to throw up his hands and shout, "Tear the damn thing off!"

Fortunately, virtually any Neanderthal track can be erased, no matter how bad or how hopeless the roof looks to the roof owner (see Chapter 17).

There is one exception to this rule. Old slate

The above Peach Bottom slate gravestone, a scant one inch thick, is dated 1743. The inscription is not worn at all after 250 years of exposure to the elements. It reads, "Here lys y body of Jamer Roger who departed this life Novembry 9, 1743 aged 33 years." This stone bears testimony to the longevity of slate from the Peach Bottom, Pennsylvania/Maryland slate region.

Photo taken by the author at the Chestnut Level Cemetery, Quarryville, PA.

roofs that are made of the *softer ("S2") slates* cannot be restored. Soft slate will flake, crumble and fall apart sometime between 55 and 125 years. If the roof slate is otherwise hard, the roof can usually be restored and may still last for generations, even if it's already a century old. If the slate is old and soft, it may still last for years, but will need to be replaced before long. Of course, any worn out slate roof should be replaced with new or recycled roofing slate.

Unquestionably, historic slate roofs can have an air of mystery about them. The men who installed them have long since died and the old roofs (in the United States in particular) have been virtually ignored by almost everyone since. The origins of these old roofs and their remaining longevity and value are difficult to determine for laypeople and professionals alike.

The author of this book, a professional slate roofing contractor and consultant, set out on a quest determined to clear up this confusion. In a ten year process, he located the original sources of various old American roof slates and visited the actual quarry sites, which range from Georgia, up the eastern seaboard of the U.S. through Virginia, Maryland, Pennsylvania, Vermont, New Jersey, New York and Maine. Old roofs at each American quarry site were examined to determine their conditions after a cen-

tury or more of wear, in order to help ascertain the longevity of the slates from each region. Museums and libraries were scoured for historical information in order to understand the conditions and techniques involved in producing the slates still gracing older homes and buildings today. Personal interviews with quarry workers, roofers (there are some good ones, by the way), architects, homeowners and other history buffs, along with a multitude of photos, added to the investigation.

Not content only with information about *American* slate, the author journeyed to northern Quebec and on to the far coastal reaches of Newfoundland, Canada, then across the Atlantic Ocean to Ireland, Wales, Scotland, England, France, Germany, Spain and Italy, looking at slate roofs, hiking through slate quarries, descending into slate mines, and gathering information from slate distributors, slate roofing masters, slate geologists, and anyone else who had something to say about stone roofs.

The information gathered during these voyages, combined with decades of on-the-job experience and observation, have led to the creation of this much-needed book. I, the author, will be the first to admit that there are many different opinions about slate roofs, many ways to install them, numerous ways to repair, restore, appraise, assess and evaluate them — some quite sound, but others quite mistaken.

Let the reader be warned that there is a strong trend in U. S. roofing these days to largely abandon traditional roofing styles, methods and materials in exchange for those that are convenient, expeditious and cheap. You will find that a gross lack of understanding about slate roofs is firmly entrenched at the highest levels of the roofing trades, with misinformation running rampant. Not only roofers, but also architects can be notoriously bad sources of information. Luckily, slate roof systems are incredibly simple, and when their simplicity is understood and respected, these roofs are also unbelievably successful.

For an on-going discussion on slate roofing with the latest information about tools, techniques, sources of materials, supplies, and even contractors, visit the author's web site at jenkinsslate.com.

Chapter Two

FROM ROCK TO ROOF – WHAT IS SLATE?

"Slate tombs high in the Alps near Oisans, France (which, from money and jewels found in them, archeologists have concluded were constructed about 500 B.C.), are still in good condition." Oliver Bowles

It will help to understand slate roofs if one first understands what slate *is*. Yes, it is a rock, it comes from the ground, sometimes it's black, sometimes it isn't. Most people, believe it or not, don't even know that much. A newspaper reporter once did a story on a local slate roof restoration business, and the first thing she asked was, "What's a slate roof?" The roofer replied, "You are joking, aren't you?" and she said, "No, I've never seen a slate roof. I didn't know there was such a thing." "Next time you drive through town look up," he replied. "The slate roofs are the ones made of stone. They usually have a sheen to them, they look natural, they tend to be on older homes, you can't miss them. There are thousands of them right in our small town, and millions throughout the United States and the world."

For those of you who want to know more than that about slate roofs, let's start with some fundamentals. There are three kinds of slate: mica, clay and igneous. Mica slate is the only kind we're concerned with because that's what roof slates are made from.

Mica slate is considered to have formed from clay-containing silts originally deposited under water in horizontal beds, such as on the floor of ancient river beds or seas. These clay sediments compacted over many millions of years under the pressure of sedimentary deposits above them so that the final clay content became very small through metamorphosis. Some beds were subjected to horizontal geologic forces which caused the beds to fold, or heave vertically, or even turn upside down in the earth.

This geologic pressure forced the material to undergo fundamental changes in its chemical composition, eventually to become what we now call slate, made up primarily of mica in the form of fine flakes arranged in parallel order. *Mica* is a generic term for any group of minerals that crystallizes in thin, easily separated layers. In slate, the mineral is primarily silicon dioxide, often in the form of crystalline quartz.

The age of slate ranges from *Cambrian* (from the word *Cambria,* which is a name for Wales) a time 600 million years ago when life dawned on earth and the first abundant marine life appeared, to *Silurian* (another Welsh name), a time 425 million years ago when mountains were forming in Europe and the first small land plants appeared.

Slate is a finely layered stone that can be easily split, somewhat like a deck of cards. If you lay a deck of cards on a table, then "cut the deck," it's obvious that it can be split in two horizontal halves quite readily. However, if you try to split the deck in two halves *vertically,* forget it. The horizontal layering of the deck lengthwise, or of the slate, constitutes the *cleavage* plane of the slate, as it is the plane on which the slate readily splits. This plane is determined *not* by the sedimentary layering of the slate over the eons as one would expect (that's the *bedding* plane), but by the geological forces that squeezed the clay deposits together. The cleavage plane may be entirely independent of the bedding plane, which is, by definition, a distinct peculiarity of slate. Some

MINERAL COMPOSITION OF AVERAGE SLATE		
Quartz .... 31-45 %		
Mica (sericite) .... 38-40 %		
Chlorite .... 6-18 %		
Hematite .... 3-6 %		
Rutile .... 1-1.5 %		

Mica is here also known as secondary muscovite, or white mica, chemically composed of potash and aluminum. Chlorite is a mica-like mineral usually containing aluminum, iron and/or magnesium. [Source: Bowles, The Stone Industries, 1934]

Mineral%	Vermont	NY	PA	VA
Quartz	59-68	56-68	55-65	54-62
Al_2O_3	14-19	10-13	15-22	17-25
Fe_2O_3	0.8-5.2	1.5-5.6	1.4-4.5	7.0-7.
FeO	2.5-6.8	1.2-3.8	2.3-9.0	
CaO	0.3-2.2	0.1-5.1	0.2-4.2	0.4-1.9
MgO	2.2-3.4	3.2-6.4	1.5-3.8	1.5-3.9
K_2O	3.5-5.5	2.8-4.4	1.1-3.7	
Na_2O	1.1-1.9	0.2-0.8	0.5-3.5	
CO_2	0.1-3.0	0-7.4	1.6-3.7	0.2-2.0

US Dept. of Commerce, Bureau of Standards, Journal of Research, V9, No.3, 9/32 roof slates have darker or lighter bands, known as *ribbons*, running across their face showing where the bedding plane was intersected when the slate was split along the cleavage plane.

Slate can also be split along the *grain*, on a plane perpendicular to the cleavage plane. The "grain" of the slate is very important when quarrying, as roofing slate is usually split so that the longer sides of the roofing slates are in the direction of the grain. This helps to reduce breakage. Or, as one writer put it, *"In splitting and dressing the roofing slate, it is always done so that the grain runs parallel to the longer side of the rectangle. This grain, although never so marked as that in the timber, has a similar effect upon the strength in different directions [Peach Bottom Roofing Slate, 1898]."*

Slate can also be split across the grain, by drilling a hole in the block of slate and wedging a "plug and feathers" into the hole, forcing the block to pop in two (don't try this on a deck of cards).

All of this, obviously, can be quite confusing to someone not familiar with quarry geology. If you would like your eyeballs to glass over further with incomprehension, then read the following industry definitions:

"The term 'cleavage' has been applied to three different structures in slate. It has been used to designate parting along numerous, close-spaced and parallel planes that bear no relation to bedding; this is true 'cleavage' and the use of the term should be restricted to this feature. The term has also been applied to a tendency to part along small, joint-like openings, more closely spaced than joints, but less so than true cleavage planes, and generally inclined to the cleavage; this may be called 'false cleavage'. A third set of fracture planes of which no trace is visible to the naked eye until the slate is actually broken and which is generally at right angles to the true cleavage, is the 'grain', 'sculp' or 'scallop' of quarrymen. The cleavage plane in slate is the direction of easiest parting and is due chiefly to the arrangement of the individual crystals in the rock." [Slate in Pennsylvania, p. 29]

I vote we leave the splitting of the slate to the professional quarrymen. Generally, though, you get the picture (I hope).

Slate is, as we were saying, a very dense, heavy, finely grained rock, with an average particle size of 0.1 to 0.01 millimeters. This fine particle size aids slate in its ability to be perfectly split, and this cleavability of slate gives it its practical value. A true slate can be split into thin sheets with smooth, even surfaces. Some slates can be split as thin as one thirty-second of an inch. In the manufacture of blackboard slate, a 4'x6' piece can be split to a uniform thickness of three-eighths inch or one-half inch, depending on the slate source.

A scientific theory first published in 1912 by German meteorologist Alfred Wegener stated that all of the earth's land masses were once joined together as a single land mass, called *"Pangaea."* Pangaea is theorized to have split apart approximately 300 million years ago into two continents, one of which eventually became Africa and Europe, while another became the Americas and Asia. The basis for this theory is apparent when one looks at maps of the continents today and sees how they seem to fit together, as if they've broken apart from each other. Interestingly, the slate deposits that run up the east coast of the US, across the eastern edge of Canada and into Wales evidently formed before Pangaea broke apart, when Europe and the Americas were still joined. This is why Canadian and Welsh slate are nearly chemically identical, despite the fact that they're separated by thousands of miles of Atlantic Ocean. They're also both very similar to Vermont slate, which is not far from Canada.

Since slate formed from sedimentary deposits, an ancient river or an inland sea must have existed on Pangaea where we now have slate beds.

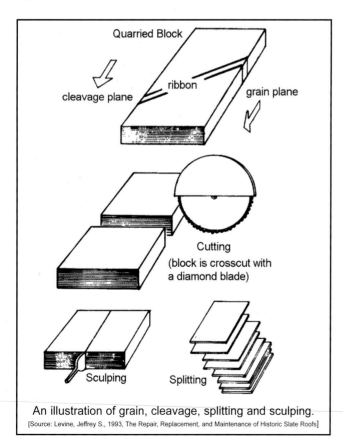

An illustration of grain, cleavage, splitting and sculping.
[Source: Levine, Jeffrey S., 1993, The Repair, Replacement, and Maintenance of Historic Slate Roofs]

When one considers that the slate deposits in eastern Pennsylvania are about a half mile thick, and that slate is an intensely compressed material, then one can speculate that an immensely deep body of water must have existed for eons and collected silt for many millennia of Earth's time. This scenario becomes even more fascinating when one understands that during the time of earlier slate formation, no animal or plant life existed on Earth, as the planet was too young, and life in the seas was just beginning to blossom. This may explain why fossils are only very rarely found in slate deposits — the only fossils the author has seen in slate (from Germany) were Cambrian Trilobites.

Due to unknown reasons, the slate that formed on Pangaea developed different characteristics depending on the location of the slate bed. For example, the slate in Wales has many color varieties, including black, purple, blue and green. Purple and green slate are also abundant in Newfoundland, Canada. Continue south into Maine and you find a black slate, but carry on a little farther into Vermont and New York and the slate is either purple, green, gray, red or black, and may be streaked, mottled or layered with iron-containing minerals that change color with exposure to the weather. Travel a bit further south again into eastern Pennsylvania and you find huge deposits of black slate, but a few miles further south in Chapman, PA, the black slate contains gray bands. Down in Maryland you find both a black and a purple slate that were once quarried there, but continue into Virginia and the slate is black again, and on close examination contains a sparkly quartz that glistens in the sunlight. Black, red and green slate deposits exist in Arkansas; both green slate and black slate can be found in Georgia, and green slate in Tennessee.

The period of slate formation ranged from 600 million years ago or more, to approximately 450 million years ago, which allows for a span of about 150 million years during which slate was formed. One hundred fifty million years is too long to comprehend, but if sedimentary deposits occurred over millions of years, one can understand why slate would vary not only according to the location of the slate in the earth's sub-surface, but also according to the depth level of the slate in the thick layers of the beds. In eastern Pennsylvania, the main commercial slate deposits are 2,800 feet thick, thirty miles long and two to five miles wide. That's a big chunk of solid slate lying just below the surface of the earth. 2,800 feet thick means more than a half mile *thick* layer of slate, and in that half mile are billions of thin layers, each representing a period of time so far in the distant past as to be quite unfathomable to us short-lived humans. So as quarrymen dig down through the slate, they come upon layer after layer with different characteristics, some softer, some harder, some useless rock.

For example, in the Pen Argyl region of eastern Pennsylvania, one of the world's most prolific sources of black slate, the layering of the slate is particularly evident. Beginning with the topmost beds, slate *runs* appear in the following succession: *Pennsylvania Run, United States Run, Diamond Run, Albion Run, Acme Run and Phoenix Run,* each run bearing a slate of somewhat different quality. The runs aren't in direct contact with each other but are separated by intervening beds 75 to 280 feet thick of unworkable slate-like rock.

Each run is itself then separated into individual *beds.* The Albion Run, for example, consists of 12 beds combined together to form a total run thickness of 184 feet. Some of the beds are "big beds," some are "ribbon" slate, and some are unworkable rock. The Albion Gray bed is known for its exceptionally high quality.

Some slate deposits have thin (one or two-inch) layers of minerals that will leave a visible strip across the finished roof slate. You may see green streaks across purple slate, and various colored streaks or mottled spots on different slate varieties. Normally, these do no harm but instead add some color and character to the slate. However, the most notorious "ribbons" occur in some black Pennsylvania slates, where an almost invisible band of gray/black material, high in carbon content, leaves a vulnerable spot in the slate. As carbon deteriorates more readily than the more common slate minerals such as quartz and mica, these ribbons will turn soft over time and the slate will then break apart. A carbon ribbon such as is common in Pennsylvania ribbon slate will turn soft enough over time (50 - 100 years) that you will be able to push your finger through it.

Other types of slate, most notably the Vermont "sea green" that is so common in the U.S., change colors with exposure to the weather. The result is an interesting mottled red and gray appearance which the roof develops over the years. This phenomenon is known as the "weathering" or "fading" of roofing slate by people in the slate industry. Many people have "sea green" slates on their roofs, also known as "semi-weathering gray-green" slate, as these slate deposits were abundant, easy to work, high quality and therefore popular for roofing.

Slates that are harder to work because, for example, the grain is not straight or uniform enough or the slate is too hard and brittle, will not be so widespread on roofs. The same goes for slate that is found only in small deposits.

Many people say, "I didn't realize that there were so many different kinds of slate. I thought all slate was the same!" As you can now see, all slate is not the same. In our younger years, the only slate we may have been aware of was the slate blackboard, and unless we grew up in a slate quarrying area, we may have developed the idea that all slate is black. Instead, we actually have a great variety of slate with a great variety of qualities, colors and characteristics. Roofs have been covered with all types of slate, as well as other stone. Some black slate from Italy even turns *white* with exposure to the weather!

One of the most important things to ascertain about any slate roof is what *type* of slate is on the roof, because then you will know how long the roof *should* last. We'll look at how to identify the slate on individual roofs in the next chapter of this book.

DETERMINING THE QUALITY OF ROOFING SLATE

We have established that some slate is harder, some softer, depending on such factors as geological age, the degree of metamorphosis and the mineral composition. For the sake of simplicity, soft slate is here being generally defined as *slate that more readily softens with prolonged exposure to the weather*, while hard slate is *slate that resists softening and stays hard with prolonged exposure to the weather*. Hard slate will obviously outlast soft slate on a roof.

I have seen soft slate (Pennsylvania ribbon slate) that had to be replaced in 55 years, and I have seen hard slate 250 years old (Pennsylvania Peach Bottom slate) that had no apparent deterioration. I keep a piece of hard "stone slate" in my shop that I brought back from a 16th century abbey in Wales, an original piece, which shows virtually no wear. It had been hung on the roof by a single wooden peg, two inches long and the thickness of a pencil. The peg had been driven through a hole in the stone and hooked on a thin piece of roof lath, which was the way Welsh slate was originally installed.

Measurements of slates from various regions of the United States show that the strength of a slate when newly quarried is not necessarily an indication of how long it will last on a roof. Some slates that last much longer than others will not measure much stronger than the others when newly quarried, and

may even measure weaker. After many years of exposure to the seasons, some slate will deteriorate more rapidly than other slate, regardless of how strong the slate is when new. So the longevity of slate cannot be judged by the apparent hardness of the slate when newly quarried, and instead, a variety of factors are considered.

The U.S. slate industry uses several highly technical standards to determine the strength, elasticity (bendability), abrasive hardness, toughness, acid resistance and porosity of slate. A variety of contraptions may be used to determine these measurements, thereby allowing for the slate to be categorized according to *expected* durability.

One indication of the potential durability of slate on a roof is the absorption characteristics of the slate. Some slate absorbs more water than others, although no slate absorbs very much. The chart on page 18 shows some average absorption levels for various roofing slates after immersion in water for 48 hours. Those slates that have a very low absorption ability have proven to have a very high rooftop longevity.

To simplify the complicated process of rating the durability of slate, the industry has adopted a rating scale which rates slate as S1, S2 and S3. This rating scale combines various factors pertinent to the longevity of the slate such as acid resistance, strength and ability to absorb water. Slates that have a high acid resistance, low water absorption and high strength are listed as S1 slate, which is considered the best slate to use for roofing. Generally speaking, these are the "hard" slates to which we have been referring — the roofing slates that will last several human lifetimes, and perhaps several centuries, on a roof.

S2 slates are not as acid resistant as S1 slate, and they absorb slightly more water than S1 slate, and although they are not expected to last as long on a roof as S1 slate, they can still make an excellent roofing material. Much of the S2 slate found on roofs today originated in eastern PA in Lehigh and Northampton Counties, and appear as black slate, or brownish-black slate (also called "blue-black," and gray slates). These are the most common "soft" slates in the US, and they may last only as long as 50 years, or maybe as long as 150 years, depending on the place of origin. Many original soft slate roofs have already deteriorated beyond repair and have been replaced, and many more will be replaced in the next few decades. Although soft slates don't last as long as hard slates, they'll still outlast just about any other roofing material on the market and are well worth buying for roofing material.

Soft slates appear crumbly or flaky when old and will give a dull thud when tapped with a hammer, while hard slates will appear smooth and will ring when tapped. But *some* hard slates *will* appear a little flaky too after a century or so, particularly some varieties of sea green slates. In addition, although *many* black slates *are* soft, some of the most durable roof slates in the world are black. So it is important to try to pinpoint the exact place of origin of the slates on old roofs in order to understand their qualities and to judge the remaining life expectancy of the roof.

Despite the ASTM testing standards that allow a slate to be rated S1, S2 or S3, this rating system has generated a lot of controversy in the slate roofing industry and may be abandoned in the future for a more accurate system. It seems that a single piece of roofing slate, when divided into four equal pieces and each piece tested separately, can yield four different test results, even when tested by the same laboratory! This problem seems to cause quite a bit of consternation among slate quarriers. This rating system is further undermined by the fact that any slate producer can simply provide the results of their very best test score even if it's twenty years old and came from a piece of slate from a section of the quarry no longer being worked. In short, the ASTM S1-S3 ratings may not be the most reliable way to determine the quality of a piece of slate.

To add to the confusion, in Europe, for example, a slate may be poorly rated if it is considered "fading," which means it changes color with age. However, some of the best slates in the United States, and probably the most popular, namely the "sea green" or semi-weathering green or gray green slates from Vermont, are the weathering or "fading" type of slates. If these slates were to be rejected out-of-hand as they may be in Europe, the world would be deprived of an incredibly valuable resource with the loss of millions of beautiful roofs. Roofs that change color with age are not bad roofs and many people even prefer them for their unusual character, oblivious to the unreasonable rejection of these sorts of roofing slates in some roofing/architectural circles. Some Europeans also turn their noses up at roof slates that contain a high carbonate content, such as the Italian slates that start out black and turn white with exposure to the weather. The local Italian people who mine these slates and use them for roofing have learned that they make charming and long lasting roofs, despite the carbonate content, by simply splitting the slates thicker and installing them with increased headlap.

The truest test of a roofing slate is the test of time. Most existing types of roofing slates have a historical element that can be observed and measured, although this has never been considered a measurement of quality by the testing laboratories. Perhaps, in the future, roof slate varieties will be judged according to their historical record of performance as well as according to mechanical or chemical laboratory test results. In the meantime, we can observe century-old slate roofs and see how well they have held up, if we want the truest indication of the longevity of any particular type of roof slate.

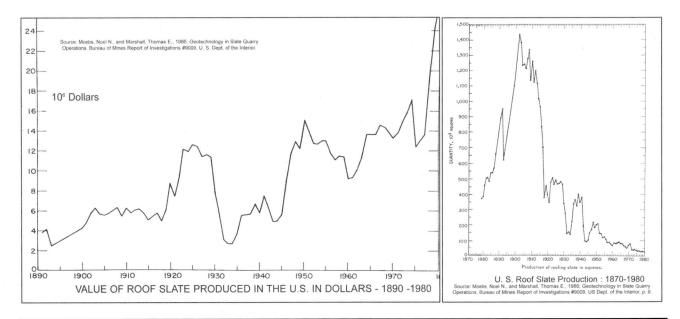

Source: Moebs, Noel N., and Marshall, Thomas E., 1986; Geotechnology in Slate Quarry Operations, Bureau of Mines Report of Investigations #9009, U. S. Dept. of the Interior.

10⁶ Dollars

VALUE OF ROOF SLATE PRODUCED IN THE U.S. IN DOLLARS - 1890 -1980

U. S. Roof Slate Production : 1870-1980
Source: Moebs, Noel N., and Marshall, Thomas E., 1986; Geotechnology in Slate Quarry Operations, Bureau of Mines Report of Investigations #9009, US Dept. of the Interior, p. 9.

Production of roofing slate in squares.

▲ Stone roofs in northern Italy are made of rock taken from nearby mountains, pointed out by the roofer doing the installation. Although this rock is not slate, it creates a roof with inimitable character and longevity, nevertheless. The worker (opposite) drills the stone in preparation for nailing in place. A single large nail fastens the massive stone to the roof.

Photos by author.

Chapter Three

IDENTIFYING ROOF SLATE

(Understanding What's Overhead)

In order to understand an existing American slate roof and its historical significance, as well as its expected remaining longevity and overall value, one must first identify the type of slate on the roof. That simply means one must determine where the slate was quarried. Slate varies significantly from quarry to quarry and from slate region to slate region. In general, slate can be expected to last from 55 years to centuries on a roof in the United States, depending on the *type* of slate. Some slates may last significantly longer if made from a high grade of slate rock and split thicker than the standard 3/16". Some of the better grades of slate rock can be expected to last 200 years even at 3/16" thickness.

A good slate roofer can tell what type of slate you have at a glance, but you'll have to figure this out on your own if an experienced slater is not readily available. Don't be disappointed if you can't find anyone to identify the slate on your roof. Remember that the men who installed the old slate roofs are almost all dead and gone, and most younger roofers today can't be bothered with trying to find out where roof slate came from, especially as most of them don't work on slate roofs anyway.

The easiest way to identify roof slate is visually. Look at the color and smoothness. Next, handle the slate, break it apart and judge its strength and density. Old soft slate may almost fall apart in your hands and will be visibly flaking. Hard slate, on the other hand, generally remains smooth on the surface and stays hard, even after a century of wear.

After breaking open a piece of slate, look at its interior. The interior of the slate is the same color as when it was originally quarried. The exterior of some slate changes color over time with exposure to weather or pollution, making it hard to identify the slate with certainty without breaking a piece open. Some slate roofs have been completely tarred over by well-meaning but misguided roofers and the only way to identify these slates is by breaking one open (use a pair of pliers and break a small piece off an edge of the roof if you have to). It may also be helpful to remove a slate from the roof and look at the back side if the front is tarred.

In any case, you can look at a slate till you're blue in the face, but if you don't know what to look for, it won't do you a bit of good. On the next page is a map of the eastern seaboard of the United States indicating the five main slate producing regions. There's nearly a 100% chance that older American roof slates came from one of these regions. Otherwise, the slates may have been imported from a foreign source. This is not likely on older roofs, except on very expensive establishments or perhaps at very old historic coastal sites where slate may have been brought over as ballast in ocean-going vessels. There are a few other areas of the United States that have produced quantities of roof slate, and if you live near one of them, you may have that type of slate on your roof.

For those who live in any of the main slate producing regions, your identification problems are more than likely solved. If you live near Monson, Maine, you can be pretty certain you have Maine slate on your roof. Same goes for Peach Bottom, Vermont-New York, Buckingham and Eastern Pennsylvania. In visiting all of these slate quarrying regions I wasn't surprised to see that in each region they used their own slate on their own roofs. If you don't live in any of these regions and have a slate roof, then you'll have to familiarize yourself with the types of slate each region produces.

Let's start at the beginning: slate has several basic colors — namely black, "blue-black," gray, purple, green, "sea green" and red. You know what black looks like (like a blackboard), although some black slates turn brownish with age, and some develop white overtones; blue-black looks black with a slight hint of blue; green slate does have a greenish tint to it when quarried, but often turns light gray with age; sea green looks blue-green when quarried, but turns light gray and develops mottled hues of

Opposite page: A mix of Vermont slates has created a beautiful, rounded valley on a graduated slate roof in Grove City, Pennsylvania.

Photo by author.

▲ VT purple slates with a sea green band and NY red slate florets. ▼ PA black slates and VT unfading green slates at 120 years. The black is getting soft, the green is still hard. Photos by author.

▲ PA black slates that are just about worn out. ▼ VT sea green slates that have weathered brown due to the quarry of origin and/or to environmental conditions. Roof is still quite good.

WHERE DO AMERICAN ROOF SLATES COME FROM?

★ = Inactive roof slate production today

✪ = Active roof slate production today

All regions were active at the time most older American roofs were installed.

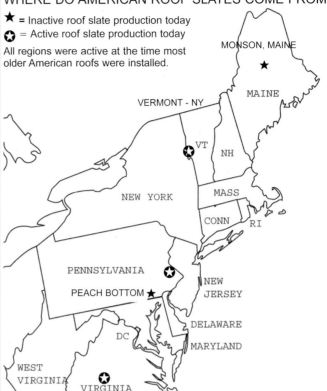

There are five main roof slate producing regions in the United States, all on the eastern seaboard. Only three of these regions are producing roof slate today, although there are many slate roofs still in existence bearing slates from all regions. The regions that are no longer producing roof slates have not stopped because they've run out of slate. Quite the contrary, the quantity of slate that has been removed from the earth has so far, in all regions, only scratched the surface of the immense deposits. Roof slate production has stopped in the Monson, Maine region and the Peach Bottom, Pennsylvania region due to economic factors: the roof slates have simply become too expensive to produce.

WHERE U.S. ROOFING SLATES WERE USED (1929)

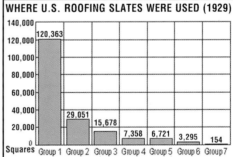

	Squares
Group 1	120,363
Group 2	29,051
Group 3	15,678
Group 4	7,358
Group 5	6,721
Group 6	3,295
Group 7	154

Distribution of Roofing Slate by Destination

GROUP 1: Delaware, District of Columbia, Maryland, New Jersey, New York, Pennsylvania, West Virginia

GROUP 2: Connecticut, Maine, Massachusetts, New Hampshire, Rhode Island, Vermont

GROUP 3: Illinois, Indiana, Michigan, Ohio

GROUP 4: Alabama, Kentucky, Louisiana, Mississippi, Tennessee

GROUP 5: Arkansas, Iowa, Kansas, Minnesota, Missouri, Nebraska, Oklahoma, Texas, Wisconsin

GROUP 6: Arizona, California, Colorado, Idaho, Montana, Nevada, New Mexico, North Dakota, Oregon, South Dakota, Utah, Washington, Wyoming

GROUP 7: Florida, Georgia, North Carolina, South Carolina, Virginia

The graph at left shows which states in the U.S. used the most roofing slates in 1929 (group 1), and which used the least (group 7). This provides a rough estimate of where the majority of American slate roofs are located today. Note that the southern states seem to have the least number of slate roofs. This may be due to the heat of the south, which can be absorbed by the stone roofs, or to a lack of proximity to prolific quarries.

[Source: Bowles, (1930), Slate in 1929].

The Smithsonian Institution Building, known as the "Castle," is covered in a mix of fairly new and also recycled Buckingham, Virginia slates.

Photo by author.

red, tan, pink or orange (depending on your eyesight) with age; purple slate is dark purple when quarried and stays that way with age, but may look black or dark gray to an untrained eye (some purple slates have green streaks in them, variegated purple has lots of green streaks); red slate is red like terra cotta ceramic tile, and stays red with age; gray slate is somewhat light gray when quarried, may have black streaks, and retains its original color with age. The industry term for slate that changes color with age and weather is "fading" or "weathering" slate, and slate that does not change color is called "unfading." Each region produces colors of slate peculiar to that region, as follows:

MONSON, MAINE: Solid black slate with slight luster, hard, durable. One of the best slates available. May last hundreds of years. These slates are no longer quarried, so recycling old roofs is imperative.

VERMONT/NEW YORK BORDER: Green, sea green, gray, red, purple, black. Many of the *colored* slates come from this area. Most of these slates are of exceptional quality and very durable and can be expected to last between 150 to 200 years, or longer. These slates are still being quarried and are readily available commercially.

EASTERN PENNSYLVANIA: Black, blueblack, dark gray and various other shades of black. Virtually all of the soft slates come from this region, with lifespans of as little as 50 years or less, although some of the black and gray slate from this area can be quite long-lasting (100-150 years). There is a considerable variety of blackish slate from this region, including CHAPMAN slate (a striated, durable black slate), and "Cathedral Gray," a lighter, durable black. "Ribbon" slate also come from here, having a carbon band across the face that can cause the slate to break prematurely with age.

PEACH BOTTOM: Black slate, from the Pennsylvania/Maryland border, very hard and long lasting. This is an excellent slate with a lifespan that may approach 400 years, and should be preserved on roofs, or recycled, as this region is no longer quarrying roof slate. On close examination, you may see a slight sparkly luster to the slate, especially on the back (unweathered) side.

Approximate Colors of American Roof Slates When New

Vermont Mottled Purple

New York Red

Vermont Gray

Pennsylvania Black or Spanish Black, similar in tone to Virginia slate

Vermont Green

Vermont Purple

Vermont Gray-Black

Vermont "Sea Green"

Note: Roof slate is a natural material and will vary from sample to sample in color and character. Slate samples are available from the quarries and suppliers listed in the back of this book in the "Quick Reference" section. Acquire a sample to accurately determine the color.

"UNFADING SLATE"

Natural slate comes in many colors, including black, gray, purple, green, "sea green," red and others. The industry term for slate that changes color with age and weather is "fading" or "weathering" slate. Roof slate that does not change color, or only changes minimally, is called "unfading," and is often regarded as a higher quality slate within the industry. Common types of American slate include "unfading green," "unfading purple," "semi-weathering gray-green" or "sea green," and "unfading red." Most commercial Spanish slates are unfading black (some are green). German roofing slate is also unfading black, as is French roofing slate. Canada, like Wales, produces unfading black roofing slate (Quebec) as well as both unfading green and purple slates (Newfoundland). China produces many colors of roofing slates and their unfading black slate is gaining popularity throughout the world. Don't be fooled by the hype, however, as some of the weathering slates — sea green, for example — are among the best slates in the world.

Left: Unfading red slate from eastern New York on a church in Foxburg, Pennsylvania. Photo by author.

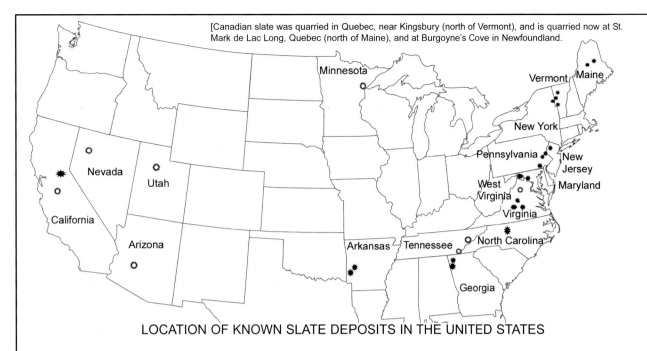

[Canadian slate was quarried in Quebec, near Kingsbury (north of Vermont), and is quarried now at St. Mark de Lac Long, Quebec (north of Maine), and at Burgoyne's Cove in Newfoundland.

LOCATION OF KNOWN SLATE DEPOSITS IN THE UNITED STATES

Solid marks are (or were) productive districts, hollow marks are prospective districts. The five commercially important U.S. roofing slate districts are: 1) Monson District, Maine; 2) New York/Vermont border; 3) eastern Pennsylvania; 4) Peach Bottom (PA/MD border); 5) Buckingham District, Virginia, although only #2, #3 and #5 are still productive.

MAINE: Monson, Brownsville, North Blanchard. Solid black slate with slight luster, (hard, durable).

VERMONT/NEW YORK BORDER: Rutland County, VT (Poultney, Fair Haven, Wells, East Poultney) and Washington
    County, NY (Granville). Green, sea green, gray, red and purple slates (hard, durable).

EASTERN PENNSYLVANIA: Lehigh and Northampton Counties (Bangor, Pen Argyl, Chapman, Slatington, etc.) Black,
    blue-black, dark gray, other shades of black (ranging from soft to moderately hard).

PEACH BOTTOM, PA/MD: York/Lancaster Counties, PA and Harford County, MD. Black slate (hard and long lasting).

BUCKINGHAM, VA: Buckingham County (Arvonia). Black slate (very hard and durable).

NEW JERSEY: Sussex Co. (Lafayette). Black, same slate as easternmost PA

GEORGIA: Polk County (Rockmart), and Bartow County (near Fairmount). Greenish gray hard slate south of
    Fairmount; bluish gray (black) slate at Rockmart (ranges from hard to somewhat soft).

TENNESSEE: Monroe County (Tellico Plains). Purplish, greenish and black slate (hard).

MINNESOTA: Baraga County (Arvon). Black slate (quality uncertain).

UTAH: Slate Canyon. Green and purple slates (quality uncertain).

ARKANSAS: Montgomery County near Norman and Slatington. Red and green slate, and some greenish gray and
    black slate near Mena, Polk County (quality uncertain).

CALIFORNIA: Eldorado County, near Kelsey. Dark gray slate resembling PA slate in color (quality uncertain).

NORTH CAROLINA: Quality and type uncertain.

Note: The original reference materials included no descriptive information for the States shown on the map that were not listed above.

[From a 1914 map by T. Nelson Dale for the U. S. Geological Survey, Bulletin 586, Plate 1; North Carolina data from Moebs, Noel N., and Marshall, Thomas E.,; 1986; Geotechnology in Slate Quarry Operations, USDI, Bureau of Mines RI 9009]

Roofing slate is also presently quarried or mined in many places throughout the globe, most notably Spain (world's largest producer), China, Germany, France, Italy, Argentina, Wales, England, South Africa, India, Norway and elsewhere.

In general, slate can be expected to last from 55 years to hundreds of years on a roof in the United States, depending on the type of slate, roof construction, and proper maintenance. Slate type is determined by where the slate was originally quarried.

BUCKINGHAM: Gray-black or blue-black slate, very hard and durable, can last centuries. Similar to Peach Bottom, shows many tiny sparkling specks in sunlight on close examination, especially on back where air pollution hasn't stained the surface.

Air pollution causes slate to change color with age. Lots of slates are hard to identify simply because of the neighborhood the roofs are in. Smoke from coal stoves or from local factories will cause the roof to develop a peculiar hue which may turn a black slate lighter gray, or a green slate brown, or a red slate black. There's no way to predict this, so in these situations you'll have to resort to breaking open a slate and looking inside to identify the color. Remember, the color inside the slate indicates the color at the time of quarrying. Add to this the fact that some slate turns color on the surface with age, and then may undergo another color change due to air pollution, and you can see why visual identification can be tricky at times. Furthermore, some black slates from eastern PA will exude a chalk that turns them somewhat white (usually around the edges), while other black slates from the same region will weather to a brownish hue.

To add to the confusion, overhanging trees change the color of a slate roof. Leaf drop from an overhanging tree will stain slates, making them a darker color. Also, slate is readily stained to a very dark brown by rusting metal, such as rusting ridge iron or flashing.

Why is it so important that we identify the slate on old roofs? Well, if the roof has Monson, Vermont/New York, Peach Bottom or Buckingham slate on it, then it's highly restorable and the slate could very likely last the life of the building it's on — such roofs should practically never have to be replaced. On the other hand, if the roof is made of the softer, eastern Pennsylvania slate, it may have reached the end of its life (after a century) and no amount of work, money or prayer will save it. The roof will have to be replaced whether you like it or not (although many older Pennsylvania slate roofs may have *decades* of life left in them). When soft slate roofs are replaced, by the way, they should always be replaced with *slate* roofs (not asphalt shingles), and Pennsylvania black slate is an excellent choice if someone wants a roof that will last a long time.

In any case, it's important to know how to identify the slate on a roof because when the roof is restored or repaired, matching slates should be used to do the restoration work. In cases where the original color of the slate is totally obscured by environmental conditions and therefore cannot be matched with any replacement slates, then a relatively invisible section of the original roof must be "cannibalized" in order to provide suitable repair slates. The cannibalized section is then reslated with a slightly different roofing slate. In subsequent chapters we'll look at each of the slate regions in greater detail, as well as at restoration techniques, so if you're confused about the different types of slate roofs, there's still hope!

But first, we should look at how and why slate was discovered and quarried in the United States, and in order to do that we must take a trip to Wales.

▲ Vermont "unfading green" slate roof on the Zimmerman House at the Delaware Water Gap National Recreation Area is 120 years old at the time of the photograph, and the slates themselves are in phenomenally good condition. The flashings need some work, however.

▼ Chapman, Pennsylvania, black slate showing its characteristic diagonal ribbons across the face of the slate. This slate can last about 100-120 years if properly maintained. Read more about this slate in the Chapman section of this book.

Photos by author.

▲ A mix of New York red slate with Vermont purple and Vermont sea green random-width slates gives this roof a beautiful and unique character. ▼ An ornate Vermont sea green slate pattern in a background of either black Pennsylvania or Virginia slate creates a monumental roof that will last for generations if properly maintained.

Peach Bottom slate is shown on opposite page, top, on the Stevenson United Methodist Church in Maryland, eighty-eight years old. Roof on opposite page, bottom, is Ford's Theater in Washington, DC. The slates that are visible in the photo are the 135-year-old original Buckingham, Virginia slates. Much of the remaining roof had been replaced with new Buckingham slate.

Photos by author.

Chapter Four

WALES

"In 1980, a small boat, loaded to the gunwales with slate, was discovered at the bottom of the Menai Straits. It has been dated at 1270, and, although no one can be certain, the slates appear to have come from the Penrhyn Quarries at Bethesda, North Wales."
John Brigden

A study of slate quarries and the phenomenon of the slate industry in the United States must start in Wales. In order to gain a deeper and more thorough understanding of slate roofing, we should take a look at the men and women, the times and the places that started it all for us Americans. In doing so, we shall understand that slate quarrying in America and the use of slate for roofing were not new endeavors brought to fruition by pioneering, resourceful people in a new country. They were instead ancient skills wrought from the knowledge of a timeless people — a people who endured lengthy sea voyages to carry their mastery of stone work across the vast Atlantic Ocean to a new land of hope and promise. And with their proud trades they brought proud traditions: the Welsh language, church and customs. In essence, they brought Wales itself, carving out small enclaves in America that seemed very much like their former home.

The mountainous country of Wales is located along the western edge of England and measures 136 miles in length, varying from 37 to 92 miles in width. North Wales is home to Mt. Snowden, the highest mountain in England and Wales, a rugged, snow-capped mountain which practically casts its shadow over the largest slate quarries ever worked. These quarries are laden with immense slate deposits yielding some of the best purple, green and black slate in the world.

Pangaea broke apart hundreds of millions of years ago, shifting the earth's crust eons before the earliest evolutionary beginnings of the human species. That separation of the planetary surface left a link between Europe and the United States in the form of underground slate deposits that would eventually draw the Welsh people to the new land of America, and in so doing, influence the ethnic and cultural development of a country we now call home.

"Keeping up with the Jones's" is a common American saying, and Jones is a Welsh name. So is Thomas, Roberts, Evans, Hughes, Lewis, Williams, Phillips, Powell, Jenkins, Griffith, Richards, Morgan, Humphries, Edwards, Davies, Hopkins and others. Granted, slate is not the only reason the Welsh emigrated to the United States. They were also master tin platers, coal miners and iron workers. However, their expertise in slate dates back centuries.

The most important slate regions in Wales are in the ancient kingdom of Gwynedd (pronounced "Gwineth" — the Welsh "dd" is pronounced like the English "th"), including Snowdonia in North Wales, the home of Mt. Snowdon which rises 3,560 feet above the Irish Sea to the northwest. There are also fourteen other peaks rising over 3,000 feet in this area, making it a rugged territory where the removal of tons of stone was not just a challenge, but a feat. This harsh terrain in North Wales is particularly notable when one considers the primitive transportation systems available until recent times, and the poor conditions of the roads (or lack of roads) in the nineteenth century, when most of the slate was removed.

Opposite page: Entrance to the now abandoned Vivian Slate Quarry near Llanberis, Wales. Photo by author.

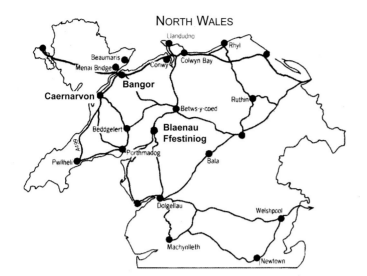

NORTH WALES

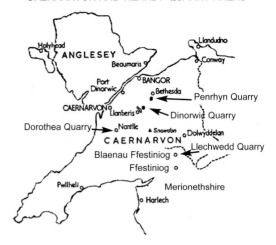

CAERNARVON AND NEARBY QUARRY AREAS

Up until the end of the 1700s, sledges without wheels were used to transport goods throughout the area, and in 1798, Reverend Richard Warner described them as having *"the shape similar to the body of a waggon, capable of containing two or three hundred weight of peat."* These sledges were drawn by a Welsh pony, a small beast of burden owned by almost every cottager.

North Wales was also geographically isolated from both England and South Wales by marshland, moorland, estuaries and the Severn River. Or, as one quarry manager put it, *"[South Wales] is a foreign country as far as the North Walian is concerned."* The uplands of Caernarvonshire were suitable only for grazing sheep, or for the scenery, and visitors have been drawn there for centuries in order to get a glimpse of the magnificent countryside, which boasts seaside and mountain range, green valleys and quaint villages.

What all this all adds up to is a slate industry that, although somewhat active for centuries, did not boom until the 1830s. Once the roads and rails were in place the industry expanded rapidly, only to decline just as suddenly in the early 1900s. The annual output of slates in North Wales in 1832 was 100,000 tons, which rose to about 450,000 tons in 1882. By 1972 the output had fallen to about 22,000 tons, due to market competition from ceramic tiles and concrete roofing materials (not from asphalt shingles, by the way, which are a cheap, temporary, petro-chemical roofing material extremely popular in the United States but almost unheard of in Europe).

One early description of the resources of Wales dates from 1387, in the words of John Trevisa as translated from Ranulf Higden:

"Valeys bryngeth forth food, And hills metal right good, Col groweth under lond, And grass above at the hond, There lyme is copious, And sclattes also for hous."

The word "slate" comes from the Middle English "slat" or "sclate." After around 1630, usage of the latter version became exclusively northern or Scottish. The word is related to the French word "esclater" which means to break into pieces, and refers to the cleaving characteristic of the rock. Therefore, in the verse we just read, the last line refers to the abundance of slate used for roofing as well as other house construction purposes, such as walls, in Wales.

In Welsh, cleavable rocks such as slate can be called "llech" (the "ll" is pronounced like a guttural "cl"), and the name is reflected in the name of at least one slate quarry, the "Llechwedd Quarry" near Blaenau Ffestiniog. The older Welsh name specifically referring to slate is "ysglatus," "ysglats," or "sglatys" and dates from the fifteenth century. One poet of the time (late 1400s) referred to Welsh slates as *"jewels from the hillside,"* and *"warm slabs, as a crust on the timber of my house."* The cleavage of the slate is called "hollt" in Welsh.

Slate is located in Wales in the counties of Caernarvon, Denbigh, Merioneth, Montgomery and Pembroke, although the largest quarries are located in Caernarvonshire (Caernarvon County), which also is home to the slate exporting cities of Caernarvon and Bangor. The two largest slate quarries in the world are located in Caernarvonshire, and are known as the Penrhyn Quarry and the Dinorwic Quarry, located on opposite sides of the Elidir Fawr Mountain. The Penrhyn Quarry, which is still in

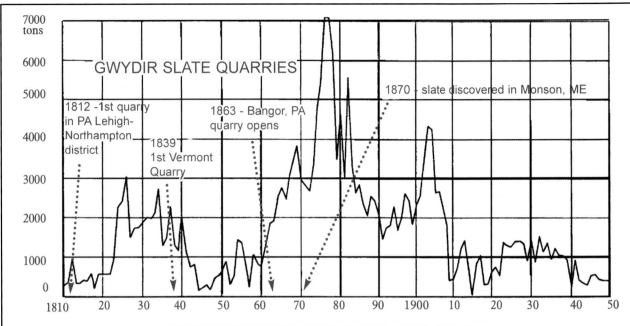

SLATE PRODUCTION IN GWYDIR QUARRIES, WALES, 1810 - 1950

Although this graph only shows the production of one *small* Welsh quarry (Gwydir), it follows a *pattern* much in line with other areas of Wales. Note the slump in production between 1840 and 1860. It is perhaps no coincidence that much of the early immigration of quarry workers from North Wales to the United States took place during the mid-1800s. The Gwydir slate quarry was a small quarry which produced about 7,000 tons of slate at its peak. In comparison, Wales' largest quarry, the Penrhyn Quarry, which was responsible for the production of one quarter of all Welsh slate, produced 130,000 tons in 1862. The Ffestiniog Quarries produced 145,000 tons in 1882.

AND IN AMERICA

Slate in America was discovered at Peach Bottom by the Welsh in 1734, and the first commercial quarry opened there in 1785. The first slate quarry in the Lehigh-Northampton district of eastern PA opened in 1812. Owen Jones, a Welshman, discovered slate along the Lehigh River in 1846, and slate was subsequently discovered near what is now Bangor, PA, in 1853. The Bangor Quarry was opened by a Welshman named Robert Jones in 1863. Slate was discovered in Vermont by a Welshman in 1835, and in 1870 slate was discovered near Monson, Maine, once again, by a Welshman.

[Graph: Williams, M. C., and Lewis, M. J. T., 1989, Gwydir Slate Quarries, Snowdonia Nat. Park Cr., Plas Tan y Bwlch, Maentwrog, Blaenau Ffestiniog, Gwynedd LL41 3YU, Wales]

production today, eventually grew into a hole a mile long, a third of a mile wide, and 1,300 feet deep, and out of this hole 300 million tons of slate and rock have been removed, mostly by hand. The Dinorwic quarry was so large, it had as much as fifty miles of rail track within it, covering a total area of about 700 acres, eventually leading to the removal of an entire mountaintop.

The slate in central Caernarvonshire dates from the Cambrian age (600 million years old), making it a very hard and durable material. The slate towns in this area include Bethesda (the Penrhyn Quarry), Llanberis (the Dinorwic Quarry) and Nantlle (the Dorothea Quarry).

A second slate producing area exists near Blaenau Ffestiniog in Merionethshire and in the adjacent parts of Caernarvonshire. Here the slate is of Ordovician age, which, at roughly 500 million years old, is younger than Cambrian, although still very durable. A slate worker at the Llechwedd Quarry once told the author that Blaenau Ffestiniog slate would last 400 years. These are the most impor-

tant Ordovician slates in the Snowdonia region, and the deposits dip into the earth under the mountains, so most of the slate has been deep mined. The Llechwedd Quarry dates from 1846 and boasts of a valuable "old vein" which was purported by a worker there to have once produced thin sheets of slate long and pliable enough to bend into a circle, although this claim is found unbelievable by American slate quarriers. The underground tunnels in this mine totaled 25 miles in length, and descended to 900 feet. Some of the original tunnels have been restored for historical purposes and can now be toured on an electrically powered underground tram for a distance of about a half mile.

There are also three other commercially important slate producing regions of Wales: the area between Towyn and Dinas Mawddwy, the area between Llangollen and Corwen, and the Presely district of Pembrokeshire. Incidentally, slate is also located in England in Cornwall, Devon, the Lake District, Aberdeenshire and Argyll (see Chapter 10 for additional information about UK slate).

Welsh slate can be of a variety of colors in addition to purple, green and black, going by such quaint names as silky red, hard spotted blue, royal blue, curly red, old blue, red hard, hard red bronze, purple green wrinkled, sage, willow, barred red, old quarry blue and blue grey mottled, in the Llanberis area alone. The Nantlle area has a green vein, silky vein, blue mottled vein, red and blue striped vein, red spotted vein and red vein. The Penrhyn area has green, purplish blue, grey, grey mottled, mottled and striped, and blue. The Ffestiniog area has mainly blue-grey. In Corris, Abergynolwyn and Aberllefenni, the slate is pale blue, and in Glyn Ceiriog and near Llangollen, the slate is blue. Mind you that these descriptions are industry terms which tend to describe black as "blue," and the slates they describe as red and blue striped slates do not look like the American flag. In fact, the slates from Wales can be generally categorized as dark purple, light green or black, although some of the purple is reddish and some bluish, and variations exist in all the colors. In short, Welsh slate is quite similar to some of the slates of the American continent, especially Vermont slate, and it is practically identical to the slate of eastern Canada, at Burgoyne's Cove in Newfoundland.

HISTORY

One of the oldest slate quarries in Wales is the Cilgwyn Quarry in Nantlle, dating from the 12th century. Other records mention a slate roof installed in 1317 on the huge Caernarvon Castle by "Henry le Sclatiere." In 1399, a fellow named Creton, in reference to a trip to a Welsh town just north of Bangor, (Conway), wrote:

*"So rode the King, without making noise,
That at Conway, where there is much slate on the houses, He arrived with scarce a pause,
At break of day."*

Records show a man named Sion Tudor ordering 3,000 slates in 1580 from Bangor, Wales, to replace the thatch roof on his house. In 1682, parts of the St. Asaph Cathedral were slated with Penrhyn slates, and in 1930, nearly 250 years later, the same slates were still good enough to be replaced on new wood on the same roof when the roof was redone, or so the story goes. A record of a lease in Wales during the period 1568-74 reads:

"I did demise unto Gruffith one tenement called

Lloyn y bettws . . .20 shillings to me towards the slating of the dwelling house, and he to send the carriage of the slates and the meat of the slaters and carpenters."

It's anyone's guess what "meat of the slaters" is supposed to mean (probably their dinner) — the point, of course, being that slate roofs have been common in Wales for centuries. Other records speak of slate roofs in 1536-9 when John Leland stated, *"The houses within the town of Oswestre be of tymbre and slated,"* and *"they dig oute slate stones to kyver houses,"* and in 1597, when John Wynn's memoranda included a note to *"slate the cattle houses."* Such records continue through the 1600s.

In the late 1500s slate was being exported to Ireland from Wales — about 100,000 were exported in 1587. A hundred years later, ten times that amount would be sent over. Other shipping records show slates being exported from North Wales throughout the 1700s, and between 1729 and 1730, over two and a half million slates were sent from Welsh ports, a million of these to Ireland.

The majority of the medieval roofing slates that have been examined have square corners, with sides that are more or less parallel, and are quite small and thick. Their usual size is about seven inches by three and a half to four inches. Today, such tiny slates are not available and even a six inch by twelve inch slate is considered quite small. The early Welsh slates remained small until the skill and the expertise of the slate workers increased. In 1740 the slates doubled in size, and later doubled again. These slates were called "doubles" and "double-doubles." In the 1500s, Welsh shipping records showed slates ranging in size from 5" X 10" to 6" X 12" and ½ to ¾ inch thick. In contrast, most slate roofs installed in America during the late 1800s and early 1900s have uniform slates (all the slates on the roof are the same size) ranging from 9" X 18" to 14" X 24", and averaging 3/16" thick.

Slate had other uses in Wales besides roofing. Many houses and buildings had walls built of slate stones still standing today in perfect condition. A slate spindlewhorl used for the hand spinning of wool was found in Caernarvonshire in 1944 and is now displayed at the National Museum of Wales at Llanberis. Slates were also used as bake stones, probably from medieval times. They were similar to our modern pizza pans, about an inch thick, and were used to bake bread known as "bara llech" or "slate bread."

Ironically, the predominant type of roofing in Wales was not slate until the nineteenth century.

▲ A sea of slate roofs best describes the town of Blaenau Ffestiniog, which rests stoically amidst the barren, misty mountains of North Wales.

▼ A tunnel in a Blaenau Ffestiniog slate mine. Some descend to 900 feet below the surface. Many have rail tracks for slate cars and trams. The tunnels open into cavernous rooms where the slate was removed by hand, in candlelight, then winched to the surface and split into shingles.

Photos by author.

Prior to that time it was thatch, which is a thick and beautiful covering of reed that can absorb enough water to become quite heavy, heavier even than slate. It's also flammable when dry, and it contributed to some of the great fires of the time, such as the Great Fire of London. It doesn't take too many Great Fires before people start looking for a fireproof roofing material such as slate. At one point in time it was declared in London that, *"Every person who should build a house should take care that he did not cover it with reeds, rushes, stubble or straw, but only with tiles, shingles, board or lead."* Wooden shingles were also used for roofing, and around 1810, Thomas Pennant, while touring Wales, reported that in many regions *"shingles, heart of oak split and cut into form of slates,"* was *"the ancient covering of the country."*

The sale of roofing slate rises proportionally as the construction of new buildings occurs, and when new construction grinds to a halt, as during war-time or economic depression, the production of slate is adversely affected. Slate production in Wales boomed in the 1800s, and the number of slate workers nearly doubled from 1861 to 1881. But in the 1880s, a depression in the building trades in Great Britain reduced the demand for roof slate. Although the building industry stagnated in the 1880s, it picked up again in the 1890s, when new houses averaged about 130,000 a year. However, by 1909, new houses had dropped to 90,000 a year, and by 1913 they were down to 62,000.

While the slate industry was subject to the vagaries of the building trends, it was also affected by foreign developments. Exports of Welsh slate dropped off between 1889 and 1918 from nearly 80,000 tons to 1,500 tons. This was largely due to World War I and the loss of Germany as a customer, as Germany accounted for 72% of Welsh slate exports in 1876, and many buildings in Germany were covered with such slate. The Germans were partial to the black slate of the Ffestiniog area, so their loss as a customer hurt this area the most.

American slates began cutting into the Welsh slate market by the end of the 1800s. In 1898, American slates were being sold in Dublin, Ireland, at 25% less than Welsh slate, doing serious injury to the Welsh slate market there. In 1897, Pennsylvania was the leading slate exporter, followed, in order, by Vermont, Maine, Virginia, New York, Maryland, and Georgia. Slates were also being imported into Great Britain from France, Belgium, Norway, Portugal, Italy, the Netherlands and Germany. The Welsh slate industry fought back against the import of foreign slates by pointing out their disadvantages. They

claimed, with considerable exaggeration, that France's blue slates soon turned "dirty grey," while Germany's red slates were "soft as clay," and America's slates were so rough as to be unsuitable for roofing houses. Admittedly, Welsh slate was quite durable and lasted many years even in smoky, acidic city environments, while some American slates "only lasted twenty years," *according to Welsh slate workers*, and some German slates were "even worse." The best French slates were an exception, being much more durable.

In 1901, a depression began that was in part attributed to the Boer War, leading to a scarcity of money and increased unemployment. In 1906, the arrival of asbestos tile on the market further encouraged the downward trend in slate production, which ultimately carried on into the first World War, never fully to recover. An article in a 1910 *Beautiful Homes* magazine described these new asbestos shingles as being made of asbestos fiber and portland cement and coming in shades of either gray or bright red, while costing no more than a slate roof. Incidentally, many asbestos roofs were also installed in the United States during the early 1900s, and many remain on American roofs today. People tend to mistake them for slate roofs, and some even insist on calling them "asbestos slates." However, asbestos roofs are not slate roofs and only vaguely resemble slate — although asbestos roofs will usually last at least 75 years.

The number of Welsh slate mines dropped from 29 to 12 between 1914 and 1918. Building came almost to a standstill during the war, and in 1917, as in the United States, slate quarrying was officially declared a "non-essential industry." As a result, skilled quarry workers were drafted into the military, and many quarries closed for the duration of the war.

Although slate is still quarried in Wales today, in many areas, as in the United States, only abandoned quarries remain.

LIVING AND WORKING CONDITIONS

Much of the slate mining in Wales was done in three ways: the slate could be worked in terraces or shelves cut into the sides of a mountain as at Penrhyn or Dinorwic or large holes or pits could be sunk into the ground as at Dorothea or the slate could be brought out of the earth in tunnels or mines.

When the slate beds plunged deep underground, they had to be followed down via tunnels by the quarrymen in order for the slate to be extracted. The Llechwedd quarry near Blaenau Ffestiniog was one of the deep mines, and the interior of the mine was damp and *totally* dark. Its 25 miles of underground tunnels connected 60' chambers from which the slate was removed. The workers adhered candles to the rock with clay in order to see what they were doing. The candles were expensive, and often only one candle was shared by two people, especially if the second person was a young apprentice.

The men drove holes into the rock face using six foot long iron bars called "Jwmpars" or "jumpers" in English. The bar was repeatedly pounded against the rock to produce a hole, progressing six to twelve inches per hour. Blasting powder was tamped into the holes, but not with an iron bar which could spark a disastrous explosion. A fuse was then laid, which burned at a rate of an inch a minute, giving the men time to get out of the cavern

NAME	SIZE
Kings	.20" x 36"
Queens	.20" x 34"
Princess	.14" x 24"
Duchess	.12 "x 24"
Marchioness	.11" x 22"
Countess	.10" x 20"
Viscountes	.9" x 18"
Lady	.8" x 16"
Doubles	.7" x 12"

Individual names were once assigned to each standard size of roof slate. In 1933, however, the British Standards Institution decided to dispense with such names and to define sizes instead by length and width. Discrepancies existed when these names were used. Some sources listed ladies as 10x16, some as 12x14, small ladies as 8x14, wide ladies as 12x16, and similar variations existed with the other names, which is probably why this nomenclature was abandoned.

Source: Brigden, "Turning Stone into Bread"; and Lindsay, Jean, (1974), A History of the North Wales Slate Industry, p. 295; and Ffestiniog Slate Quarry, Blaenau Ffestiniog, Wales

The author at the Blaenau Ffestiniog museum where the various sizes of roof slate made in Wales over the years are displayed. Sizes range from 10"x 4" to 36"x 20".

Photo by author.

▲ The Welsh Slate Museum at Llanberis, built and roofed of slate, was once the mill where slate from the huge Dinorwic quarry was processed. Note the mountain of slate rubble in the background. The museum now displays the old tools and equipment, including the 50' diameter water wheel which powered the mill's 1/8 mile of line shaft.

Dafyyd Davies (right) a museum worker, demonstrates the rhythmical motion needed to trim slates with a "stool and traverse." The "slate knife" he holds can still be bought in hardware stores in Wales today.

and find shelter in the tunnel during the explosion.

Large blocks, two to three feet thick and up to ten feet long, were brought down by the blast and were broken up by the rockmen with a hammer and chisel, then loaded onto trolleys by a chain hoist and tripod, and winched to the surface. "Badrockmen" were paid by the ton to remove non-productive rock, and "rubblers" cleared the waste.

At Llechwedd, the men worked each chamber in teams of four, two in the chamber and two on the surface in the mill. The two rockmen below the surface extracted the slate and sent it up to their partners above, who broke the slabs down further, finally splitting and trimming out the roof slate using only hand tools. Other quarries used four, six or eight men to a team.

In 1876, a quarry worker wrote, *"When I was a child, I had to walk five miles before six in the morning, and the same distance home after six in the evening; to work from six to six; to dine on cold coffee, or a cup of buttermilk and a slice of bread and butter. Some of the men had to support a family of perhaps five, eight or ten children on wages averaging 60 pence to 80 pence a week."*

Apprentices received no pay for the first six months, and had to wait nine years before they received full pay. Some quarries had their own wood shops, pattern shops and foundries, where they made their own tools and parts for their equipment.

Men living too far from the quarry to travel home stayed in barracks and only traveled home on weekends. The men worked outside during all types of weather, and some quarries provided eating houses and places with a fire where the men could dry their clothes. Typical meals consisted of tea, egg, and bread and butter for breakfast; tea and bread and butter for lunch; bread and butter for teatime; and potatoes, sometimes with beef, bacon or buttermilk for supper.

Quarry work was hard and dangerous; over four hundred men and boys have been killed in the Penrhyn Quarry since 1782, and one quarryman, Robert Williams, in 1792, had as his gravestone the piece of rock that fell on him. Fatal accidents at the Bryneglwys Quarry also occurred year-round, and varied in cause, although records reveal typical causes of death: a worker named Owen fell 18 yards down a shaft and died. A few years later, William Owen was knocked over a ledge by a falling rock while boring a hole. He injured his spine and died the next day. A few months later, a piece of rock weighing 7 tons fell on David Evans and Hugh Jones and crushed both of them to death. Ten days later,

Edward Davies fell 20 yards while trying to unhook chains from a wagon, and died. This was about the same time another man was killed at the quarry by a runaway truck. Later, John Watkin was killed when the earth above him collapsed as he excavated a reservoir, and a few months later, Tom Rogers got his coat caught in a rotating turbine shaft and was killed. John Lewis was later trapped by a falling rock and killed; David Roberts was hit and killed by a falling stone weighing 30 pounds; Richard Davies was struck and killed by a rock falling from a chamber roof; David Evans was killed by a rock weighing 25 tons; Owen Ellis fell about 50 feet and was killed when climbing down a chain being used to lower a block of slate; Tom Ellis was killed a few months later when hundreds of tons of rock from a chamber roof fell on him; David Owen fell down a shaft he was exploring out of curiosity, and died; David Davies fell backwards over a ledge when his crowbar slipped as he was working a stone; Edward Lewis died when the ledge he stood upon gave way, crushing him; and numerous others died from similar causes.

In the deep mines, slate had to be left in place as supporting walls in order to prevent undermining and consequent disaster. However, some profit-hungry quarry operators robbed these support walls, thereby thinning them or removing them altogether. In 1883, no less than six and a quarter millions tons of overburden and slate collapsed on the workings of the Welsh Slate Company due to irresponsible undermining. The owner of the land, William Edward Oakely, sued the mining company for compensation for the damage done to his assets, and won his claim in one of the longest arbitration disputes in legal history.

Men of North Wales who didn't work in the quarries lived to an average age of 67 years, while quarry workers were likely to die by the age of 38. Many of the quarrymen suffered from respiratory diseases because of the dust in the quarries. Ironically, quarry medical personnel disagreed. At the Penrhyn Quarry Hospital in Bethesda, medical officer J. Bradley Hughes issued a written statement on June 1, 1922, stating, *"We have no case of silicosis in this quarry of which I am aware, and I became convinced after four years' experience here that slate dust is not merely harmless, but beneficial."* Dr. E. Shelton Roberts, who had a very long acquaintance with the conditions both at Penrhyn quarry and at other quarries in the district, further maintained that he had not met a single case of silicosis or lung trouble among the quarrymen attributable to their work at

A "drumhouse" (right) housed a large drum (below), with heavy cables wound against each other. Slate cars attached to the cables were let down from the mountain top on iron rails. The weight of the cars loaded with slate pulled the empty cars back up the mountain. A brake lever controlled the operation. These long abandoned relics remain intact at the Dinorwic Quarry, adjacent to the Welsh Slate Museum.

Photos by author.

(Bottom left) Looking up the first incline from the bottom of the mountain at Dinorwic. Old rails are clearly visible. The lowest drumhouse, pictured on the previous page, sits at the top of the photo.

(Top left) Looking down toward the lowest drumhouse, which now sits around the corner to the right.

(Top right) Looking up another incline toward the top of the mountain. Abandoned slate mill sits at right. This quarry (Dinorwic), covering 700 acres, had 50 miles of iron track.

Photos by author.

ANALYSIS OF TYPICAL SLATE OF NORTH WALES (%)

Silica 55.30
Oxide of Iron 10.00
Alumina 24.84
Lime 0.36
Magnesia 2.46
Carbonic Acid . . . . .nd
Sulphuric Acid 0.21
Potash 1.47
Soda 0.53
Water of Hydration .4.70

[Source: Lindsay, Jean, (1974), A History of the North Wales Slate Industry, p. 292]

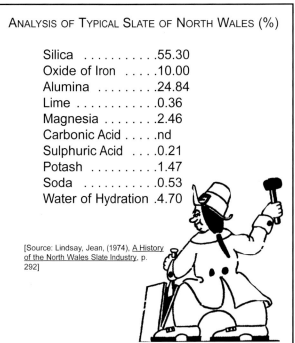

The card of prices (left) is an authentic document from the North Wales slate industry, as are the documents on the following pages related to the attempts to unionize the quarries in the 1800s.

Three jobs involved in producing roof slates: the fellow on the right is splitting blocks into smaller pieces, while the man in the center splits them down further into roof shingles. The man on the left is trimming the slate with a slate knife and a "stool and traverse." The Welsh don't punch ("hole") their slates at the quarry as Americans do. They let the roofers do it on the job site. (Courtesy of Old Line Museum, Delta, PA)

the slate quarries. Dr. John Roberts, general practitioner with 39 years experience in a quarry district wrote, *"During my experience it is remarkable to say that no quarryman has ever complained of cough or choking sensation due to inhalation of slate dust. . . I have not examined slate dust microscopically, but am of the opinion that the particles are soft and none irritant, also it contains very little silicate and a good deal of iron and sulphur, the latter is regarded as a germicide. . . The facts already stated and my experience convinces me, that slate dust is neither an irritant exciting or a predisposing cause of Tuberculosis. There is no doubt that the quarryman's mode of living and poor feeding consisting mainly of tea, bread and butter reduces their resisting power against Tuberculosis."*

Of course, the doctors who issued these statements were on the payroll of the slate quarries, and their reports were used by the North Wales Quarry Proprietor's Association to downplay any detrimental effect on the health of quarry workers caused by the quarry environment. The North Wales Quarrymen's Union responded to the above statements by issuing a statement of their own, declaring, *"That we cannot agree with the statement of the case as submitted by the North Wales Quarry Proprietor's Association. It is perfectly obvious to us that no reliance can be placed on the statements submitted by them . . . In our view there is not sufficient data available to form a definite opinion one way or the other as to the effect of slate dust on the internal organs, although our members are convinced that it is chiefly responsible for bringing on some forms of Phthisis [Tuberculosis]."*

By the 1950s, however, according to J. G. Isherwood in <u>Slate From Blaenau Ffestiniog</u> (1988), "the quarries had been decimated. . . All were pale shadows of their former selves, their workforces ravaged by the effects of silicosis, a disease ignored for too long by many quarry owners until the evidence was beyond doubt; the most experienced men had received the greatest exposure to the killing dust and so the dearth of experienced miners and rockmen increased."

The struggle between the Welsh quarry workers and the quarry owners was bitter at times. The men who dared to complain about the working conditions could simply be fired from their jobs and possibly evicted from their homes, as quarry owners were often both employers and landlords.

Despite the tough working conditions, a principal complaint among the workers concerned the corruption, bribery, favoritism and discrimination which seemed to prevail among the quarry officials, who would grant a good rock face (a "bargain") to a worker of similar religious or political bent, while poor rock faces went to others. This forced the men to form unions and to take their complaints to the management by union committee in order to protect the individuals with grievances from losing their jobs.

The first union attempt occurred at Penrhyn Quarry in 1865, where 1,800 men joined, and soon lost their employment as a result, until the union was broken. The union organizers were subsequently fired. This attempt at unionizing caused quarry owner Lord Penrhyn to issue a statement on December 2, 1865, advising the men *"to consider well before listening to agitators,"* and offering *"a word of caution to avoid having anything to do with such a movement as a trade union in future, as on the very first rumor of such a state of feeling, he will immediately close the quarry, and only reopen it and his cottages to those men who declare themselves averse to any such scheme as a trade union."* The closing of the quarry and the men's cottages would subject the men to extreme hardship, especially in the winter months, while Lord Penrhyn could simply wait it out in the Penrhyn Castle, relying on his wealth and servants to continue an opulent lifestyle. A saying eventually arose, attributable to Lord Penrhyn's acquisition of land holdings, *"Steal a sheep, they hang you, steal a mountain, they make you a lord."*

In 1870, over 80 Penrhyn quarrymen, among the best in the quarry in character and skill, were fired without reason, although it was widely believed that the men were fired for political reasons. A letter

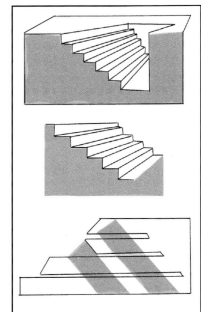

THREE TYPES OF QUARRY OPERATIONS, SIMPLIFIED

Top: Pit Quarry (Dorothea and Penrhyn).
Middle: Open quarry showing galleries or shelves (Dinorwic).
Bottom: Deep mine showing horizontal shafts entering into the side of a mountain (Blaenau Ffestiniog). Shaded areas show slate deposits.

Source: Williams, M. (1991)

▲ Abandoned buildings at the Dinorwic Quarry. Note entire mountainside removed in background.

Opposite: Slate walls, some with slate shelters built into them, line most pathways.

▼ Workers barracks at Dinorwic, long abandoned — no trace of the roofs remain.

Photos by author.

written by an apparent mediator in 1874, during another strike, perhaps sums up the situation succinctly: *"Over 2,200 quarrymen are out on strike and about 300 old men have been 'turned out' by Lord Penrhyn for whom some provision must be made to keep them from the Union House. Some of these old men have been working at Lord Penrhyn's quarries for 40, 50 and 60 years, and some of them even more than this; and simply because the bulk of the quarrymen resolved to ask for an increase of wages, these poor, innocent old men were being cruelly punished by being thrown upon the Parish for any provision Lord Penrhyn was ready to make for them. This certainly is not charity!"*

A more successful, broader union formed in 1874 at the Dinorwic Quarry, attempting to unionize all the slate quarries in North Wales, and was met by the uniform hostility of the quarry owners. The owner of the Dinorwic Quarry attempted to exclude all unionists from the quarry, leading to a five week walk-out until the union was recognized. Another strike occurred at Dinorwic in 1885-6, as a result of complaints by the workers of favoritism. *"We respectfully invite you to examine the grounds for the following complaints; and if you approve of it we shall be ready to appear before you to discuss this, and to enter into particulars if required. We complain that at present, and for some time past, we are not treated as workmen, but that the question of our politics and religion are allowed to affect the question of our work and our wage. In this matter we maintain that clear and offensive favoritism is shewn. . . Let honest work be paid for and not the Political or Religious creed of any man . . ."* Shortly thereafter, it was agreed that no worker would be discriminated against because of religion or politics, after which Dinorwic experienced a considerable period of peace.

Penrhyn Quarry was not so lucky. Lord Penrhyn refused to recognize the right of union committees to take up grievances on behalf of the workers. A lengthy strike followed in 1896, but work eventually resumed without any grievances being settled. Worker discontent culminated in a three-year strike at Penrhyn from 1900-1903, having a disastrous effect on the local town, Bethesda. The military was called in to keep order as thousands of quarry workers and their families were threatened with starvation. The specter of "Socialist experiments" was raised in the Tory press, depicting the work of the unions to be a creeping menace. Lord Penrhyn adamantly refused to give in to the demand that he recognize union grievance committees, and the strike eventually failed. The union carried on, however, and Lord Penrhyn eventually became *"The most hated man in the world,"* according to a worker at the Welsh Slate Museum, who added, *"We Welsh don't like to talk about him."* And that was the sentiment still in 1996, nearly a century later!

The British government did pass the Quarry

To the late Employés at the Penrhyn Quarries.

I am again instructed by Lord Penrhyn to make you the accompanying offer of work on the same condition as before, viz.:

That there shall in future be no attempt on the part of any Committee to interfere with the management of the Quarry, or to prevent Employés from obeying the orders of the Managers.

Upon the above distinct understanding, all applications (including those of the 71 men who were suspended on 28th Sept., 1896) will be impartially considered, and as many of the late employés as there can be found room for, will be re-engaged, without reference to the events connected with the strike.

As misapprehension appears to exist with regard to the 71 men to whom notice of suspension was given (Sept. 28), I must point out that the suspension of those men (*ipso facto*) ceased immediately the present strike was declared, whereby the men severed their connection with their employer, thus making it obviously necessary for you all to apply individually for re-engagement if you are desirous of obtaining work at the Penrhyn Quarry.

E. A. YOUNG.

NOTICE.

WANTED AT

THE PENRHYN QUARRIES

Blacksmiths, Badrockmen, Boys, Brakesmen, Engine-drivers, Fitters, Foundry-men, Joiners, Journeymen, Loaders, Labourers, Masons, Machiners, Miners, Quarrymen, Rybelwyr, Sawyers, Stokers, Tippers, &c.

Applicants for work can apply on Monday or Tuesday next (February 15th and 16th), between the hours of 10 a.m. and 3 p.m., at the following Offices :

To apply at the Yard Office - - -	Blacksmiths, Fitters, Foundrymen, Joiners, &c.
To apply at the Slab Mill Office - -	Sawyers, &c.
To apply at Tross-y-fordd Office. - -	Badrockmen and Journeymen.
To apply at the Pay Office - - -	Quarrymen and Journeymen.
To apply at the Pay Office - - -	Boys.
To apply at the Marker's Office - -	Loaders, Masons, Miners, Machiners, Engine-drivers, Stokers, Brakesmen, Tippers.
To apply to William Parry, Overlooker's Office - - - - - - -	Rybelwyr.

PORT PENRHYN, BANGOR,
   8th February, 1897.

E. A. YOUNG.

Fencing Act in 1887, followed by the Quarries Regulation Act in 1889, and the 1894 Quarries Act, initiating government inspection of quarries and quarry machinery. In addition, the Workman's Compensation Acts of 1880 and 1911 provided greater protection for the workers.

Otherwise, workers turned to their church and/or pub for consolation during hard times. In one small mining village there were more than 29 places of worship within three miles of the center. The chapels were the centers of the community, and were used for important meetings as well as for song and worship. On the other hand, pubs provided some competition as meeting places, and in some towns the pubs matched the places of worship, pub for chapel. By 1880, there were 35 pubs in the tiny town of Bethesda alone (near the Penrhyn quarry).

It's no wonder that the lure of America became so strong to many of the Welsh people. In America, they could own their own land and their own home with a garden and livestock, and without the meddling of landlords. Letters of Welsh immigrants spoke of the opportunity they found in America. On January 29, 1869, one wrote, *"I am amazed by the efforts made by the Welsh in Wales to get a farm. When one comes vacant, there are hundreds trying to get it. But here you can be your own master without fear of being turned out and you can do what you like with your own land."* On January 2, 1871, another wrote, *"My old friends in Lleyn [North Wales] can have a small holding for themselves [in America] for the money they pay in rent for one year in Wales. One need not fear any notice to leave from any landlord or steward for voting according to conscience. I never met anyone yet who regretted coming to this country, but only many who were sorry they had not come before."*

And so, the Welsh immigrated to America, discovering slate in York County, Pennsylvania, in the 1700s, and establishing the town of Bangor, which became Bangor West, and is today known as Delta, PA, where a large Welsh church graces the main street. In the Lehigh-Northampton district of Pennsylvania, the Welsh discovered slate in the 1800s, and once again a town named Bangor sprung up where it still exists today, just a few miles from another town — East Bangor. Both Bangor and Delta, PA, remain as strong centers of Welsh influence in the US. Another Welsh settlement grew at Wind Gap, PA, in Northampton County, not far from Bangor.

Vermont fared much the same way. The border between Rutland County, VT, and Washington County, NY, became a Mecca in the 1850s for Welsh immigrants drawn to the slate quarries. Many small Welsh settlements sprang up, including Fair Haven, Blissville, Poultney, South Poultney, Pawlet and

▼ A sign that speaks for itself.

Photo by author.

West Pawlet, all in Vermont, and in Granville and West Granville, New York. One strictly Welsh organization from the region, the Poultney Welsh Male Chorus, was still being organized as late as 1939.

Other Welsh immigrants were drawn to the Virginia slate regions, concentrating in Arvonia, a town named after Caernarvon, Wales. Others were drawn to the Rockmart and Fairmount areas of Georgia. Welsh emigration to America reached its peak by 1900, with nearly 94,000 total immigrants and over 170,000 children of Welsh immigrants present by that time. The significance of this emigration is drawn into perspective when we realize that the entire population of Caernarvonshire in 1871 was only 106,000, and in Merioneth only 46,000, while the population of Wales in its entirety in 1871 was only 1.2 million. This means that nearly 7% of the entire population of Wales had emigrated to America by 1870, or in today's terms, that would be like 18 million Americans emigrating to (for example) China.

Pennsylvania, Ohio, and New York attracted over one third of the Welsh immigrants and their children, while Illinois, Wisconsin, Iowa, Utah, Kansas, Colorado, Indiana and California attracted lesser quantities, in that order. Welsh were most numerous in Pennsylvania in the coal mining and steel areas such as Pittsburgh, as well as in the slate areas already mentioned. The census of 1900 shows the Welsh most concentrated in Pennsylvania in the Wilkes Barre-Scranton area, Pittsburgh and other coal mining areas as well as slate regions.

The Welsh quarry workers who stayed in Wales continued to put in a good day's work, although the quarry managers could never squeeze enough out of them. In 1913, for example, each Welsh mine worker averaged 33 tons of finished slate per year, and each quarry worker averaged 32 tons. Despite the workers' long days, the manager of the Penrhyn Quarry insisted that they only worked seven and a half hours a day *"after deducting the meal hour. And then there is the time when the men are not working, such as blasting time, and if you add to this the loss of time such as holidays, attending funerals, hay harvest, and rough weather, the actual working time becomes small."* Note that he even complained about the men attending funerals! With management like that, it's easy to see why the workers wanted to leave the country.

▲ An old section of the Penrhyn Quarry as it looks today.
The pit has filled in with water, but the terraces cut into the mountainside are clearly visible.
▼ Meredith Tanyard Cottage in Dollgellau, Wales, an old cottage with a restored roof.

Photos by author.

▲ The author on a self-guided tour of a hotel in Wales.
Opposite: Welsh slate walls and roof of the Llanberis area near the Dinorwic Quarry.
Welsh slate is very similar to Vermont and Canadian slate in color and durability.

Top photo by Jeanine Jenkins, opposite by author.

⊹ EUREKA ⊹

SLATE COMPANY.

MANUFACTURERS OF ALL COLORS OF SLATE,

THE ONLY

EUREKA!

UNFADING

HUGH G. HUGHES,

R. WYNNE ROBERTS.

POULTNEY,

18 Little Tower,

VERMONT.

LONDON, En

ESTABLISHED 1852.

EUREKA SLATE QUARRIES

Are now the Oldest existing in the State. These Quarries are now produc
the best, and, in fact, the only unfading Green Slate in the world,
and have gained this reputation throughout the States
and Foreign Countries as well.

POULTNEY, - - VERMONT.

## Chapter Five

VERMONT – NEW YORK

The "Slate Valley" of Vermont lies mainly in Rutland County, but also straddles the state border to include some of Washington County, New York. The 24-mile-long, six-mile-wide valley extends from Granville, NY, and West Pawlet, VT, north to Fair Haven, VT. The roof slates from this area are well known both for their durability and for their variety of colors, ranging from deep red to solid purple, purple with green flecks or streaks, gray-green, solid green, "sea green," gray with black streaks, gray-black and black.

This differs greatly from the slates of Monson, Maine, eastern Pennsylvania, Peach Bottom, PA, or Buckingham, VA, which are all a shade of black or gray. Although some green, purple and blue slate may be found in eastern Canada, at Burgoyne's Cove in Newfoundland, as well as in small deposits in Arkansas, Tennessee, Georgia and Utah, the only commercial source of colored slate in the United States, of any significance, lies in this area straddling the border between New York and Vermont.

The bulk of the green and the purple slate is found on the Vermont side of the valley, and all of the red slate is found on the New York side. It can be fairly said that all of the true red slate in America, and perhaps the world, comes from Washington County, NY. First discovered in the 1850s near Green Pond in Hebron, in an area now known as Slateville, red slate is now one of the most durable (and most expensive) slates available anywhere.

The slate quarried from the Vermont side of the valley is primarily green, "sea green" (green that develops a mottled reddish cast over the years) and purple, but also includes some Vermont black and gray.

The first quarrying in the Slate Valley took place in 1839 near Fair Haven, Vermont. A fellow by the name of Colonel Alonson Allen started the quarry at a place called Scotch Hill, just north of Fair Haven. Allen was born in Bristol, Vermont, on August 22, 1800, settling in Fair Haven at the age of 36 as a proprietor of a small general store. He started his slate quarry with Caleb B. Ranney.

For a few years, Colonel Allen dabbled experimentally with slate; then in 1845, he went into the business of making school slates. In the mid-1800s, hand-held pieces of slate were used by school children instead of paper writing tablets, and much of the slate quarried at that time was used to produce these school slates. In fact, slate was also used to make slate *pencils*, which were made entirely of solid slate (no wood or graphite). A slate pencil was about three or four inches long and a quarter inch in diameter, pointed at one end and used for marking school slates, much the same as a pencil is used to mark a piece of paper today. When a sharp piece of slate is dragged across a flat piece of slate, a white mark remains on the flat piece, which is easily rubbed off by a piece of cloth or even a child's hand. Chalk is not needed when one has a slate pencil. And so, thanks to the quarrying of slate, many schools were provided with both "tablets" and writing utensils for the children. In addition, slate "blackboards" hung on walls and were used by teachers and students. Today we have sayings such

Photo by author.

FIRST SLATE QUARRY IN WESTERN VERMONT
This area of Vermont is known for its high quality slate; the first quarry was opened on Scotch Hill in 1839 by Alonson Allen & Caleb Ranney. Allen began the first manufacture of roofing slate in Vermont in 1848. By 1869 there were seventeen quarries in Fair Haven of which eleven were on Scotch Hill. Quarrying of slate was important to the economy of the area and brought in many skilled Welsh immigrants who were familiar with the quarrying of slate in their native Wales.
VERMONT DIVISION FOR HISTORIC PRESERVATION - 1997

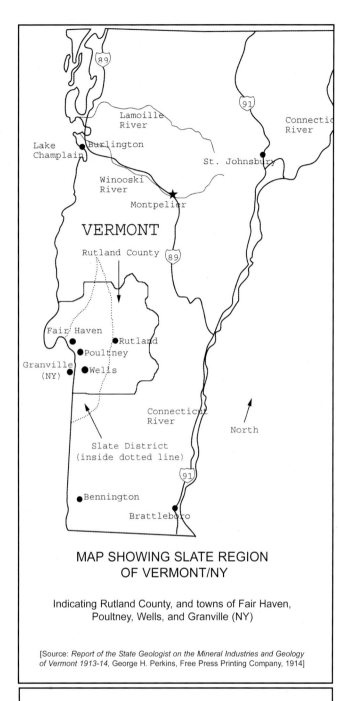

MAP SHOWING SLATE REGION
OF VERMONT/NY

Indicating Rutland County, and towns of Fair Haven,
Poultney, Wells, and Granville (NY)

[Source: *Report of the State Geologist on the Mineral Industries and Geology of Vermont 1913-14*, George H. Perkins, Free Press Printing Company, 1914]

as "wipe your slate clean," or have a "clean slate," or events are "slated," or a "slate" of candidates. These are all carry-overs from the days when slates were used as writing tablets.

Colonel Allen didn't stay in the school slate business for long, however, because he found he could make more money making roofing slates, which were about the same size and thickness as school slates. He began roofing slate production in 1847-48, just two years after starting the manufacture of school slate; then he abandoned his school slate production in 1848.

The first slate roof covered by Vermont slate was done by Colonel Allen in 1848, but the farmer who owned the roof was a skeptic. He insisted that Colonel Allen wait a full year for payment in order to determine if the roof would collapse under the weight of the slate. If the roof did collapse, Colonel Allen was to receive no pay, and he was instead to pay for all damages. It turns out that the roof passed the test and lasted quite a bit longer than the skeptic's one-year trial period. In fact, it reportedly still exists in good condition today, over one hundred and fifty years later.

Allen also served as state senator in 1842-3, and as an assistant judge of the county court in the early 1860s. In 1886, eight years after Allen's death, a Rutland County historical account stated, *"but for his boldness and courage, to this hour not one slate would have been shaped from Fair Haven to Salem."*

Colonel Allen opened a floodgate in 1848, after which slate quarries, following his example, began popping up everywhere. The "Eagle Quarry" was opened just outside of Poultney, Vermont, in 1848; then in 1851, another quarry opened on a farm owned by Daniel Hooker, three miles north of Poultney village. Hooker and Son quarried a few slates in 1851, then a few more in 1852, as they were able to sell them to their neighbors. By 1854, they had a full-fledged operation going, manufacturing mainly roofing slate and eventually employing as many as 60 men. Some of their slates were of such high quality they were even exported to England. Then Welshmen began opening quarries right and left. By 1885, forty-six years after Allen started the first quarry in Vermont, 67 slate quarries were operating in Rutland County.

By 1871, slates were also being manufactured for mantels, billiards tables, hearths, table tops, blackboards, floor tiles, door steps and other articles. Some slate was "marbleized," which involved a process of painting the slate black, heating it to 175° F, then dipping it in water that had paint floating on the surface. The slate was baked again, varnished, baked again, polished and baked a fourth time until the final product looked incredibly similar to marble. An 1875 description of the process states, *"The ingenious process of marbleizing is one of the recent inventions. By a certain chemical process the surface of the slate, after it receives a polish, is converted into the exact semblance of the most beautiful of the foreign and domestic marbles, and also made to imitate rosewood, mahogany and ash, and so exactly that the most experienced would be puzzled to detect the difference from sight."*

One typical quarry by this time was being worked to a depth of 150 feet from the surface, all dug without today's modern machinery. From this pit opening, six tunnels were dug into the hillside to a distance of 600 feet, the slate being removed on railways operated by steam. The slate deposits in this area were considered *"inexhaustible,"* and still exist in huge quantities today.

By 1872, over 11,000 tons of roofing slate were being shipped annually from Poultney, Vermont, amounting to about 35,000 squares of roofing — enough to cover roughly 3,000 average houses. The cost per square was then about $4.50. The cost today (2003) is about $350.00.

Visitors to the Slate Valley today can camp at Vermont's beautiful Half Moon Pond State Park, and see the old Scotch Hill quarries along Scotch Hill Road, which connects the park to the lovely town of Fair Haven. While in the area be sure to visit the Slate Valley Museum, 17 Water Street, Granville, New York 12382 (Ph: 518-642-1417); or visit the web site at http://www.slatevalleymuseum.org.

Eleanor Evans McMorrow was born in Wales near Caernarvon, and came to Vermont with her parents in 1924, at the age of seven. She reminds us that "Caernarvon" Street in Fair Haven is an example of the Welsh influence in the development of this area of the United States. Almost all of the Welsh in the Vermont/New York slate region were from North Wales where the Welsh slate quarries are located, she says, while those Welshmen from South Wales, who mostly worked coal, immigrated to Pennsylvania to work the coal mines.

Eleanor's Welsh maternal grandparents had lived in Vermont in the mid-1800s but they eventually returned to Wales. Her maternal uncles had also moved from Wales to Poultney, and in doing so influenced Eleanor's parents to make the great Atlantic journey themselves. Her father started working in slate in Wales when he was eleven, not an uncommon age for Welsh lads to start in the "pit," as

they called it. He became quite an expert at slate, and helped many a person who wanted to start a slate quarry. *"He could look at the rock on the top and tell them how to start whatever they did to the slate. They have to get it in a certain vein or something and he could do that,"* she said. He worked in the pit in Vermont until about a year before he died.

Once in Vermont, Eleanor found that her church was Welsh and all the services were in Welsh. She went to church seven days a week, as it was the center of her social life. *"The Welsh sang all the time,"* according to Eleanor, whose dad, uncle and brother sang in choirs and choruses.

The slate they quarried was for roofing material. *"This whole area is slate. There are many, many, many colors, beautiful colors. Middle Granville [NY] has a beautiful red. And then there's purple and black, gray. Then there's one that's sort of mottled green and purple. It's awfully pretty."*

Life in the quarries could be very hard, according to Eleanor. *"I'm very happy I don't have anyone in the slate quarries anymore. We were brought up to get your bed made, your dishes done and your floor swept, in case somebody gets hurt and they bring them home from the quarry. That was the way you were brought up. My father was hurt badly. He was unconscious for 21 days. They operated on his brain. He was*

Since there are so many century-old sea green slate roofs in the United States, it's important to know that 100-year-old sea green slates are usually still very durable, and may yet last another 100 years. Although most sea green slate is classified as hard, some sea green quarries produced a softer slate which may appear brownish and flaking with age. These may be relatively soft and may only last 150 years, or less.

ANALYSIS OF VERMONT/NY SLATE (%)

	Sea Green	Unfading Green	Purple	Red
Silica	65.02	64.71	62.37	73.93
Protoxide of iron	5.44	5.44	4.21	1.74
Peroxide of iron	2.99	7.23	7.66	10.17
Alumina	16.02	7.84	13.40	5.16
Manganese Oxide	0.31	0.30	0.20	0.10
CA Carbonate	1.38	3.00	2.50	1.25
CA Sulphate	1.31	1.55	0.16	1.06
Phosphoric acid	trace	trace	trace	trace
Alkalies (NA)	4.16	6.92	7.20	3.92
Water	1.37	1.38	1.50	1.24
Magnesia	2.00	1.63	0.90	1.43
Carbon:	None			
Ferrous Oxide:	4.71%			
Strength:	11.040 lbs/sq. in. (modulus of rupture)			
Porosity:	.220% water absorbed in 24 hours			
Corrodibility:	.722% of wt lost in acid sol. for 63 hrs			

(Source: <u>History of Rutland County, Vermont</u>, edited by H. P. Smith and W. S. Rann, D. Mason and Co., Syracuse, NY, 1886, p. 198. Data attributed to Professor J. Francis Williams, Rensselaer Polytechnic Institute, Troy, NY.; and U. S. Geological Survey, date unknown)

▲ Mountains of waste slate like the one above near Poultney, Vermont are a common sight in quarry regions.

Photo by Jeanine Jenkins.

Slate splitter — Hilltop Slate Co., Middle Granville, New York

All photos by author unless otherwise specified.

Much of the history of the Vermont/New York slate valley can be studied at the Slate Valley Museum, 17 Water Street, Granville, New York 12832 (phone: 518-642-1417). The engraved slate plaque, above, is on display at the museum.

▲ Modern block cutter at U.S. Quarried Slate Co., Fair Haven, VT, splitting an incoming block using a hammer and a chisel.

▲ James Kelly of Wells, VT, is one of the few remaining blacksmiths still making tools for the slate industry.

home for nine years after it. He had to learn to write and everything over again, learn to talk."

"They have these big blocks they send up. It was a bad day, it was sleeting. A block came out of the chain and hit him in the head. My father got $9.00 a week in compensation. I don't know how they made ends meet, if they did."

There were always injuries, says Eleanor, who tells of broken arms and broken legs. *"Somebody was always getting killed, it was terrible. More so here than in Wales, I think, from what I've heard them tell. Over there the quarries were mined in shelves. And here, right down."*

The huge dumps of waste slate around the quarries were also hazardous, according to Eleanor. *"The man that was boarding with us, ten days after my father was hurt, this guy made him go and work in the same spot my father was in. These dumps get icy, they slide. And it slid and killed him. He was knocked right off the ledge and into the quarry. He was killed ten days after my father was hurt. Today they wouldn't be able to allow that because of your safety codes. But they didn't have safety codes back then, in 1938."*

SEA GREEN SLATE

Vermont is a beautiful state, and it has produced perhaps more hard slate than any other. "Sea green" (now known as semi-weathering gray or gray-green) slate roofs can be found at almost any location in New England served by railroads at the time the slate was quarried, although many areas not served by railroads at that time do not have slate roofs today, and never did. For example, just a few miles east of the slate quarrying region of Vermont, over the Green Mountains, slate roofs are hard to find, as there was no railroad access to that area during the heyday of slate quarrying. On the other hand, Vermont's sea green slate can be found hundreds of miles away to the south and west where the rail lines routinely hauled the slate to Pittsburgh, and farther west into Ohio and beyond.

Sea green slate is easily identifiable because the roof will appear light gray with reddish slates scattered throughout — it only appears "sea green" when freshly quarried. The reddish color is also called buff, tan, pink, orange, rust or brown, but since it is caused by the oxidation of iron layered into the slate itself, we may as well call it reddish. People in the slate industry do not like to use the term "red" when describing the mottled color of sea green slate because true "red" slate, quarried in New

York near Granville, is a deep, solid (not mottled) red similar in color to red brick, and does not resemble sea green slate at all.

Thanks to Margot McKinney, a reference librarian at the Green Mountain College in Vermont in 1976, a direct history of the sea green quarries has been put on record. We are even more grateful to Owen L. Williams, the author of the history, who, at the age of 89, in 1969, penned his fifth revision of *"The History of the Sea Green Quarries,"* on a six inch by nine inch dime store tablet. Margot McKinney found and typed the contents of the tablet after Mr. Williams died in 1975 in Wells, Vermont, and recorded it in the school records of 1976, where it found its final resting place in a manila envelope in a drawer in an upstairs room in the school library. Here it is now excerpted and paraphrased to provide a first-hand account of the sea green quarries *"by an old slate maker of over sixty years experience. . ."*

Owen Williams was born in 1880 in Caernarvonshire, Wales, and emigrated to the United States in 1901. Twenty-six years later he sat in the Vermont Legislature, despite the fact that he never had any formal education beyond the seventh grade.

He wrote *"The History of the Sea Green Quarries"* five times, each time a revision of the previous version, all handwritten on 6 x 9-inch tablets. The final version, written in 1969, bore the words, *"This is the last I wrote."*

The sea green quarries were located in the southwest part of Rutland County, Vermont, specifically near Fair Haven, Pawlet, Poultney and Wells. The slate deposit covered about 10 miles or more, and varied in width from a few hundred feet to half a mile or so.

According to Mr. Williams, various records show the opening of the first sea green quarry to have taken place in the late 1860s or early 1870s. The sea green slate deposits were, of course, below the surface of the earth, and an excavation had to be sunk into the ground in order to access the slate and begin to open a quarry. It was necessary for the quarrymen to put one or several holes down and blast out a section of overlaying rock, in order to reach the slate and determine whether the slate was of adequate quality to continue. If the slate looked promising, then more holes were drilled at an angle corresponding to the layers of the slate, and blasting was continued until the quarrymen had cut a *"butt and a fireside."* *"These terms,"* states Mr. Williams, *"are used by quarrymen. Please bear in mind that these terms that*

I use in this history prevailed in the early days with nothing but hand labor."

When the overlying rock was covered by soil, sand or clay, a team of horses with a scoop shovel would drag a trench across the vein to expose the rock. Then, using the same technique, the trench was widened until the rock could be reached and blasted. The overlying rocky slate was often not solid enough *"as quarrymen say to make roofing slate of good quality and durability"* to a depth of ten to fifteen feet, although in some years the top rock was in demand to make antique-looking ("freak") slate.

In those early days of slate quarrying, all labor was done by hand. Hand shovels were used to take care of the rubbish, and elm or oak plank boxes, braced with strap iron and measuring four feet wide and five feet long, were used for moving slate, and for dumping scrap. At the turn of the century (1900), self-dumping boxes came into use, made of iron. Other tools included both double and single-hand hammers (requiring one or two hands), crow bars, drills, wedges, chisels, gouges, rope, black powder and fuse. Dynamite was used on poor rock, but never on good slate rock *"on account of the shattering effect."*

Drilling was all done by hand. One man held a metal rod fitted with a drill bit on its end, while another man (or two) struck the rod with a double-handed hammer. The man holding the rod turned it a little after each strike, boring a hole into the rock suitable for holding explosive black powder. This drilling process was very dangerous and required the strict attention of the three men; otherwise, the man holding the drill could be seriously, even permanently, injured by a blow from one of the heavy hammers.

The fuse was then placed into the hole, followed by the black powder. After the blast, the slate had to be *"cleaved"* or split apart. The cleaves were ten or twelve inches in thickness, and if a piece of stone was too large to handle, it was *plugged* — split on a plane perpendicular to the layers and broken in two. Plugging again involved drilling a hole by hand, although in this case only one man drilled the hole using a drill about two feet long with a single-hand hammer. This hole was not to be used for explosive powder, however, but instead a *plug and feathers* were inserted into the hole to force the block of slate to split in two.

A "plug and feathers" is a set of tools consisting of two thin steel sleeves (the feathers) that slide down either side of the hole, after which a tapered wedge (the plug) is forced between the feathers with a hammer. The tapered wedge caused the block of slate to split in two, while the feathers provide a smooth surface allowing the wedge to be driven into the hole. This way, slate can be split along a plane perpendicular to the layering of the stone.

Instead of a drill, a *"jumper"* could be used to plug a piece of slate. A jumper is a steel rod about six feet long with a small, but heavy ball of iron welded into it about fifteen inches from the end to provide additional weight in the drilling process. This tool required a good deal of experience in order to drill a hole suitable for a plug.

Another tool frequently used by quarrymen was the *gouge*. According to Mr. Williams, the gouge was made of 5/8-inch steel, about 10 inches long (when new) with a point suitable to cut across the endgrain of a stone, in order to *sculp* it — make a cut the length of the stone *perpendicular* to the obvious cleavage plane. If the resulting *ditch* on the end of the stone was cut smooth and straight, the chiseled split would be very good and made in the direction desired, but if the ditch was not made right and straight, the result would be very unsatisfactory, and if so would require more labor and a waste of valuable stock.

Once the stone had been split to a size that was workable, it had to be hoisted out of the pit. This was done using a crow bar to manipulate the block while a chain was securely wrapped around it. Here was another hazardous task, as a stone could slip from the chain and injure or kill someone as it was being hoisted out of the quarry. The chain was about 12 feet long and made of half inch wire, with links two and a half inches across. *"Sometimes that rockman had to use a rope to hang on to in order to put the chain around the stone secure."*

Slate-making itself was divided into three operations: block cutting, splitting and trimming. As a rule, says Mr. Williams, there were three men in each group. The function of the block cutter was very important in the production of roofing slate, requiring a great deal of experience. It was a job that could wear a man down very quickly unless he knew what he was doing. Blocks came out of the pit in irregular shapes and the block cutter had to make the blocks suitable for splitters to work. If the block needed to be split using a plug and feathers, the block cutter needed to measure for the proper place to put the hole.

If the stone needed to be sculpted, then the block cutter had to use his gouge to cut a groove or ditch in the end or side of the stone. From this groove, the block cutter cut into the block with a chisel in the direction he wanted. Then he had to cleave the block (split it along the obvious splitting plane) to make the pieces light enough to carry into the splitting shanty. This was all done outside.

A person referred to as the *"splitter"* sat in the splitting shanty. His tools consisted of a low stool to sit on, a wooden mallet with iron rings at both ends, and two or three thin, flat iron chisels about three inches wide at the wide end, which were used to split out the individual pieces of roofing slate. The splitter (or "slitter" in Wales), sat on the stool and leaned the slate block against his leg with the smoothest edge upward. The block had now been split to a thickness of about one and a half inches by the block cutter. The splitter then split the block along the cleavage plane into two equal halves with the hammer and chisel. A good piece of slate will split right into two halves with a good tap or two of the hammer. Each of the halves was again split in half, and so on, until the desired thickness of about 3/16 inch was reached. Naturally, some of the finished slate would be thicker or thinner, but for the most part, the finished slates were amazingly uniform in thickness. Again, this craft required a lot of experience — some consider it an art. The splitter kept the unsplit blocks piled on his left side, and the *"chips"* (untrimmed slates after splitting) piled on his right on a small bench within easy reach of the trimmer.

A person known as a *trimmer* then took the chips one at a time and trimmed them to make the largest possible size roof slate. The trimming was done by a trimming machine run by a foot treadle, so that the trimmer had to hold and position the slate with his hands while he operated the foot treadle at the same time. The maximum *standard* size of a roof slate in the USA is 14" wide and 24" long, and the standard sizes range down to about 6" wide by 10" long. In addition to these widths are 12", 11", 10", 9" and 8" wide slates of different lengths.

Although it was a very arduous job, some men could trim 10 to 15 squares of roofing slate in a ten-hour day, amounting to about four to five tons of rock. The trimmers also had to shovel their rubbish as well as the splitter's, and sometimes help the block cutter with his work.

After trimming, nail holes were punched in the slate. This was done by placing each slate, one at a time, on a slate punch. A foot pedal operated the machine which caused two sharp, metal punches to poke nail holes in the slate about one fourth to one third of the way down from the top and about an inch and a quarter to two inches from the sides. The punch breaks out the back of the slate when it pokes the holes through, and this breakage was advantageous as it allowed for the head of the nail to be countersunk into the slate. This kept the nailhead from rubbing on the overlapping slate after being nailed to a roof, thus prolonging the life of the roof.

A foreman kept a record of the production of each slate shanty, resulting inevitably in an atmosphere of competition and rivalry between the shanty crews when their individual rates of production were compared. The bosses encouraged these rivalries in order to gain a greater level of production. In some companies, a group of about six men, known as a *slate gang*, had the job of counting and piling the slate and loading up the teams of horses.

The mornings began with the whole gang in the quarry carrying the previous day's production out into the slate yard, before the pitman went into the pit. Because of the great weight of slate, this was heavy work, and the custom prevailed that there was a ten-minute break after the slate was carried out. The slate was then loaded onto wagons drawn by teams of horses and taken from the slate yard to railroad cars. Here the slate had to be packed into the cars properly — they were stacked three high on the long edge with wooden lath placed between the rows. An article in the Granville Sentinel, dated

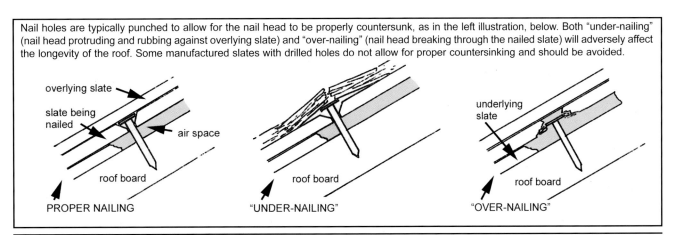

Nail holes are typically punched to allow for the nail head to be properly countersunk, as in the left illustration, below. Both "under-nailing" (nail head protruding and rubbing against overlying slate) and "over-nailing" (nail head breaking through the nailed slate) will adversely affect the longevity of the roof. Some manufactured slates with drilled holes do not allow for proper countersinking and should be avoided.

overlying slate

slate being nailed

air space

roof board

PROPER NAILING

roof board

"UNDER-NAILING"

underlying slate

roof board

"OVER-NAILING"

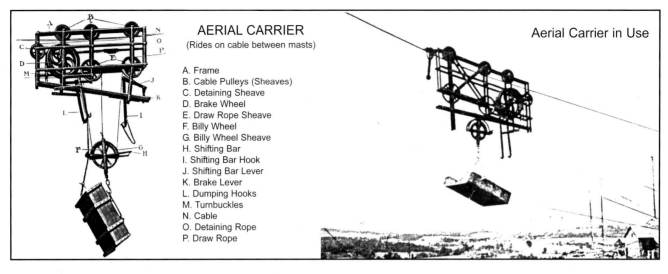

AERIAL CARRIER
(Rides on cable between masts)

A. Frame
B. Cable Pulleys (Sheaves)
C. Detaining Sheave
D. Brake Wheel
E. Draw Rope Sheave
F. Billy Wheel
G. Billy Wheel Sheave
H. Shifting Bar
I. Shifting Bar Hook
J. Shifting Bar Lever
K. Brake Lever
L. Dumping Hooks
M. Turnbuckles
N. Cable
O. Detaining Rope
P. Draw Rope

Aerial Carrier in Use

▼ Masts and Aerial Carriers with Scrap Piles Below

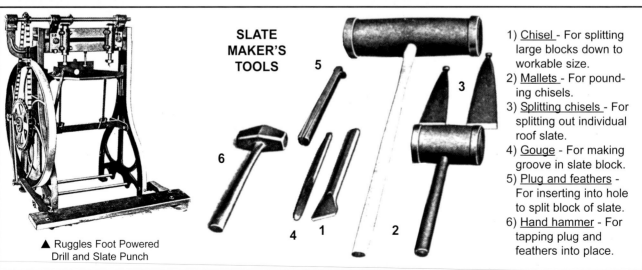

SLATE MAKER'S TOOLS

1) Chisel - For splitting large blocks down to workable size.
2) Mallets - For pounding chisels.
3) Splitting chisels - For splitting out individual roof slate.
4) Gouge - For making groove in slate block.
5) Plug and feathers - For inserting into hole to split block of slate.
6) Hand hammer - For tapping plug and feathers into place.

▲ Ruggles Foot Powered Drill and Slate Punch

Wooden Dump Car

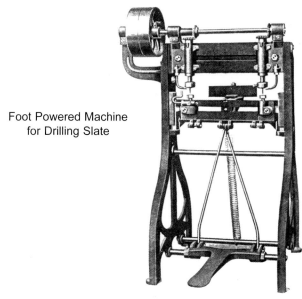

Foot Powered Machine
for Drilling Slate

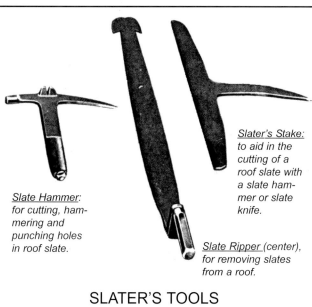

SLATER'S TOOLS

Slate Hammer:
for cutting, hammering and
punching holes in roof slate.

Slater's Stake:
to aid in the cutting of a roof slate with a slate hammer or slate knife.

Slate Ripper (center), for removing slates from a roof.

The "Lightning" Slate Dresser punches holes in slate.

Ruggles Original
Slate Trimmer

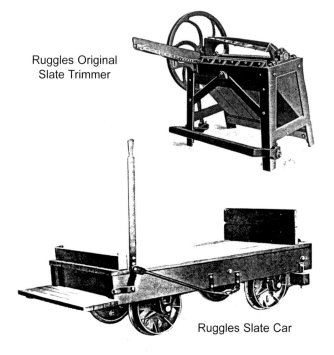

Wooden Dump Car With Skip

Ruggles Slate Car

March, 1890, stated, *"Charles Peck drew 24 and a half squares of 18 by 9 slate on one load from a Rising and Nelson quarry to the station."* This load, weighing nine tons, was probably drawn on a sled, and certainly pulled by horses. During the first quarter of the 20th century, so much slate was being shipped out of Vermont's slate valley that the Delaware and Hudson Railroad ran a freight train called *"The Slate Picker"* primarily to pick up loads of slate. Eventually, all of these slates ended up on someone's home, barn, garage, church, grange hall, carriage house or other building.

In 1889, the going price for a square of slate was around $3, while the average pay of a quarry worker was about $25 per month. In 1890, one slate company sold 170,000 squares of slate, enough to cover 14,000 roofs, weighing more than 10 million pounds and requiring 3,400 freight cars to haul to market. The peak of slate production in the slate valley occurred in 1927-28, when a square of roof slate went for around $15, compared to about $350 today.

Often a horse provided the power to hoist stone out of the quarry, but as the pit got deeper, a steam boiler and engine were used. Eventually, an A-frame supported a cable that extended across the pit, and a *carriage* was hung on the cable. The carriage was attached to the engine with a wire rope, enabling the power of the engine to pull the carriage along the cable. The carriage could be brought to a stop by bolting a wooden block on the cable, and this stop was located near where the men needed to unload stone or rubbish boxes. The slate blocks and rubbish boxes were put on a small railcar; the blocks were rolled to a shanty while the rubbish was rolled to the dump. Eventually, a self-dumping carriage was invented by a blacksmith in Poultney, Vermont, and it came into universal usage in slate quarries in a very short time, bringing with it a brand new job. The self-dumper saved a great deal of human labor as the engineer could stop the carriage anywhere along the cable and set the block down in a convenient location for it to be worked on.

As the pits got deeper, they began to fill with water creating a nuisance that had to be pumped out. The water was not wasted, however — it was pumped to the boiler house to be used to make steam to power the steam engine.

A very high *mast*, or *stick*, was needed in order to use these carriages. The masts varied in length from less than a hundred feet to over two hundred feet tall, depending on the topography of the land. They had to be located four or five hundred feet from the edge of the pit to allow for dumping of the rubbish stone. There was a demand for qualified men who could raise the mast, which was quite a feat of engineering in those days, considering that the stick may have been 200 feet long. The *"tallest quarry pole in the world"* was 235 feet high, erected at Rising and Nelson's Quarry #4 in 1908. A fellow by the name of Henry Vogel of Truthville was a pioneer in this business, and he was so highly regarded that his services were in demand — and he accomplished them even after he became handicapped in one leg.

The sticks as a rule were native green pine made of two or three poles spliced together to form one piece. The splices were made of eight-inch square oak timbers about sixteen or twenty feet long, bolted to the stick. A cable, two guy wires, *plus "saddles"* and *"sheaves"* (pulleys) had to be attached to the top of the mast before it went up. All of this together made for quite a heavy burden, and raising it was extremely difficult. One mistake and it could come down in pieces, and the work of a team of men who had toiled for days would be lost. One can still see the old masts standing tall among the scrap slate piles in Vermont today.

The cables were strung high in the air from mast to mast, and the aerial carriers rode on the cables. Despite their height, these carriers could reach into the deepest pit by dropping a box attached to another cable. The slate block or rubbish could then be loaded onto this box and raised up out of the pit, moved along the cable, then let down near the splitting shanties (if a slate block) or dropped into the dump (if rubbish). All this was done by the engineer, who operated the aerial carriers, perhaps three at one time, powered by a steam engine.

Electric power came into the slate valley in 1913. Eventually, air-powered jack-hammers came into use for cutting slate, and these could drill a hole in a slate three times as fast as a man could by hand. Around 1919, trucks began to be used to haul slate to the railroads; the first ones had hard, solid tires.

The early work forces involved in the sea green quarries were of Irish and Welsh nationalities. These people learned the trade of slate quarrying in Wales and emigrated to the United States, both as skilled workers and common laborers. Toward the end of the nineteenth century, Eastern Europeans began to emigrate to Vermont to work the quarries, and by the turn of the century, the Eastern Europeans came in large numbers. Italians soon joined them in the quarries and also became very good at rock work. There was only a sprinkling of other nationalities working slate in Vermont.

These four ethnic groups did not mix social-

▲ Clark Hicks, president of the Evergreen Slate Company (2003) in Granville, NY, gives the author's wife a tour of one of Evergreen's roof-slate stockyards. Roof slates are trimmed down to size in an Evergreen Slate Co. trim room (right).

Photos by author.

ly, except perhaps to share a drink at a local bar. Each had different religions, which created the main social barrier, as each group's church provided their social center. The Welsh were Protestants and the Irish and Italians were Catholic. The Eastern Europeans had their own church and fraternity.

Throughout the history of the sea green quarries and up until the 1930s, there were no health or safety regulations. There were also no public conveniences available to the workers. The men had to provide their own drinking water *any way they could get it, and sometimes under very unsanitary conditions, in some cases melted ice water from some hole under an old dump which would dry up when the hot weather came.*

Slate quarrying was a hazardous occupation due to the cleavage of the stone and the different natural cuts and joints. Undermining was a common cause of accidents, as was carelessness and faulty equipment. The use of black powder and dynamite added to the perilous nature of this trade. There was no workers' compensation, nor were there safety inspections in the early years. The responsibility for safety rested upon the shoulders of the operators and supervisors.

As a result, accidents were frequent, and it was not uncommon for a man to be killed or maimed for life. In 1903, for example, eleven men were killed in a cave-in at the Vermont Slate Co. Quarry, located near the Warren Switch section of the sea green quarries. A section of ground where slate was stockpiled and where the part-owner of the quarry, Mr. Williams, happened to be standing, was undermined and collapsed, taking Mr. Williams down and piling the slate on top of him. In another accident in 1910, five men were killed in a cave-in in the Owens Quarry on Briar Hill, again the result of undermin-

ing. One man was found alive the next day after being buried for 24 hours, but died the following night from exposure.

Another incident occurred in the spring of 1916. The quarry was situated such that the engineer who operated the steam engine and dumped the rubbish from the aerial carriers onto the dump pile, could not see the top of the dump. The quarry boss saw a slab of slate on the dump which he thought should have been worked into slate shingles, but the block cutter disagreed. The two of them proceeded to argue over the block while standing in the dump, not hearing the familiar sound of the rubbish box approaching overhead. The engineer dumped the rubbish right on top of them, not aware that the men were standing on the rubbish pile; both were killed. From then on, the engineer gave two blasts of the quarry whistle as a warning when the rubbish was about to be dumped.

The lack of worker's compensation during those days made for some tough times for families that lost a family member to death or injury. A custom prevailed in which men could pledge money to be taken directly from their pay to benefit those unfortunate families. This money went directly to the affected families, with no deductions taken from it.

Today's slate quarries are a world apart from the old mines that produced the slates for most of the older roofs in America. Modern quarries are much safer, and the workers are protected by insurance and other benefits. They now use diamond saws, forklifts, trucks, bulldozers, highlifts and other modern equipment. The one operation still done by hand, however, is the splitting of the slate, still achieved by hammer and chisel.

Vermont's famous colorful, durable roofing slate is abundantly available new from a variety of slate companies. Above left, Craig Markcrow of Vermont Structural Slate Co., in Fair Haven, VT, points out the cleavage line on a slate block. Above right, David Thomas of Hilltop Slate, Inc. in Middle Granville, NY, inspects their stockyard. Below left, Jonathan Hill of Greenstone Slate Co. in Poultney, VT, examines slate blocks that are awaiting final splitting, and below right, Chuck Smid of the New England Slate Co., Pittsford, VT, demonstrates the impeccable production characteristics of their North Country Black from Quebec. For a full listing of Vermont slate suppliers see page 295.

NEW ROOFING SLATE IS READILY AVAILABLE TODAY FROM VERMONT SUPPLIERS

Top left: Shawn Camara of Camara and Sons Slate Co., in Fair Haven, Vermont, inspects a pallet of new, extra-thick roof slate. Top right: Steve Taran of Taran Brothers Slate Co. displays an unusually large mottled purple roof slate. Large sizes and thick slates are readily available from Vermont producers. Bottom left: Stephen Williams of Rising and Nelson Slate Co. in Middle Granville, New York, inspects a pallet of unfading green roof slate. Bottom right: Wm. Drew Turner, president of U.S. Quarried Slate in Fair Haven, Vermont, observes his quarry pit where some of the famous Vermont "unfading" slate originates. For a complete listing of Vermont/New York slate suppliers see page 295 or visit jenkinsslate.com on the internet.

Photos both pages by author.

Chapter Six

PENNSYLVANIA SLATE

"When the ribbon turns toward the mountain, that's the best slate."
Randy Rowe

Pennsylvania slate quarries exist in two districts. The largest, most productive and well known of these is the Lehigh-Northampton district, which extends about thirty miles from Slateford in the north on the Delaware River, to beyond Slatington in the south on the Lehigh River, including the quarry towns of Bangor, Pen Argyl, Belfast and Chapman, among others. The slate belt here is about five miles wide, lies on the south side of the Blue Mountain, and may approach 500 million years in age. Approximately 90% of the slate produced in Pennsylvania came from this district, which is still productive today.

It should be noted that this slate deposit extends into New Jersey, and slate quarries have been operated near Lafayette, in Sussex County, New Jersey by the Lafayette Slate Mining Corporation of Port Chester, NY. However, as the New Jersey productions were dwarfed by the Pennsylvania productions and are not in operation today, the New Jersey slate deposit will be considered in this book as equivalent to the Pennsylvania slate of the Lehigh-Northampton district.

The second slate district in Pennsylvania is the Peach Bottom district in York and Lancaster counties, which straddles the Pennsylvania and Maryland border along the Susquehanna River near Delta, PA (see map, next page). This slate deposit is 10 miles long and may be nearly a billion years old. Peach Bottom slate has a higher degree of metamorphism and is higher in silica, making it a harder, more durable slate than the average Lehigh-Northampton slate. In fact, Peach Bottom slate was once considered "the best slate in the world," but is now no longer quarried. Peach Bottom slate will be discussed in greater detail in the next chapter.

Pennsylvania has long been the nation's leading producer of slate. Although at least ten states were producing slate as late as 1944, Pennsylvania's slate production averaged 40-50% of the nations output during the first four decades of the 20th century. Today, however, slate production has markedly declined, not only in Pennsylvania, but also throughout the other slate-producing regions of the U.S. This is due to the market competition of synthetic roofing materials, as well as the tremendous amount of work, and waste, involved in slate quarrying. Some estimates put the amount of

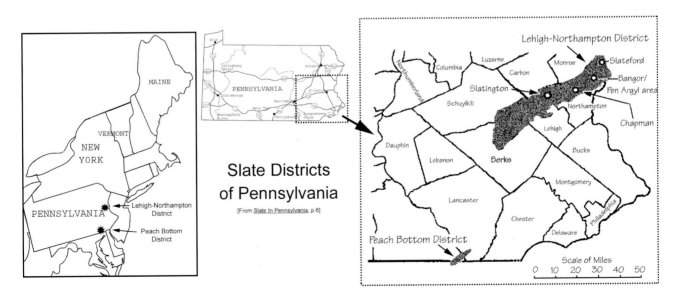

Slate Districts of Pennsylvania

[From Slate In Pennsylvania, p.6]

◀ Slate block being hoisted to the surface, probably in a Pennsylvania mine, illustrating the degree to which quarry workers may be exposed to overhead hazards. [Source: United States Department of the Interior, Bureau of Mines Report of Investigations 9009, 1986; p. 7.]

Pete Papay (right) of the Penn Big Bed Slate Company in Slatington, PA, producer of Pennsylvania's well-known black slate and importer of foreign roofing slates, discusses block-cutting strategies with his diamond saw operator prior to sawing a huge block of PA slate rock.
Photo by author.

waste rock in slate quarries at 70-90% before the invention of the wire saw in 1926, which subsequently enabled slate to be cut more efficiently. In 1985, a full 65% waste was *still* generally considered acceptable in the Lehigh-Northampton district. Due to the decline in business, many quarries have shut down, and entire regions now lay idle.

Some slate regions adapted to the decline in demand for roof slate by producing slate granules for asphalt shingles. Nearly a million tons of roofing granules were produced in 1942, for example, and ground slate was also put to use as an industrial abrasive, as a filler in paints, asphalt mixtures, roofing mastic, oilcloth and other products. Slate has other uses, including billiards tables, blackboards and for structural purposes, and Pennsylvania has been the nation's leading producer of these slate products.

The slate of Pennsylvania is primarily black or dark gray. In the slate industry much of the PA slate is known as "blue-black," as it looks slightly bluish when first quarried. It ranges in durability from soft (S2) slate with as little as a 50-year lifespan, to an extremely hard (S1) slate with a lifespan that may approach four centuries or longer.

The vast majority of the Pennsylvania slate quarried for roofing is soft black slate with a life span on roofs ranging between 50 years and 125 years, originating in what is known as the "soft vein" region of Lehigh and Northampton Counties. This region includes Bangor, East Bangor, Pen Argyl, Danielsville, Slatington and Slatedale. Only a relatively small amount of slate from the Keystone State

originated at the famous Peach Bottom quarries. Other Pennsylvania slate comes from the "hard vein" belt, which occurs near Chapman and Belfast, and which falls somewhere between soft and hard on an overall durability scale. Then there is the "gray-bed" slate, which is not as durable as Peach Bottom, but may be considerably more durable than the Pennsylvania soft slate.

Unlike Maine slate, which is commercially one primary type of black slate, Pennsylvania's vast slate deposits produce many types of slate with many characteristics, although all are black (or close to it) and may not seem very different in appearance to the layperson. As mentioned in Chapter Two, the main slate deposit in eastern PA, which produces primarily soft slate, is 2,800 feet thick, two to five miles wide, and thirty miles long. This is such a massive mineral deposit that considerable variation can occur throughout its range. In the Pen Argyl region, for example, the following slate runs appear in succession: the *Pennsylvania Run, United States Run, Diamond Run, Albion Run, Acme Run and Phoenix Run,* each run bearing a slate of a different quality, although the difference may not be significant. The runs are separated by intervening beds, 75 to 280 feet thick, of unworkable slate-like rock.

Each run is itself then separated into individual *beds*. The Albion Run, for example, consists of 12 beds combining to form a total run thickness of 184 feet. On a visit to the PA slate quarries in July of 1993, the author came upon an old list naming the veins of the Albion run. A quarry worker had evidently scrawled the list in pencil on a piece of cardboard and tacked it to the wall of an old slate shanty, long since abandoned when I stumbled upon it. After blowing the dust off the cardboard, I found a full 43 veins listed, each with a colorful name either pertaining to the qualities of the slate in that vein, or honoring a friend or family member.

If you've read Chapter Two of this book, you understand how slate was formed and realize that the rock was originally clay which settled under water in layers over periods of millions of years. Because of the immense amount of time it took for those layers to form, they were subjected to different environmental factors over time, and today those layers of finished slate will show different characteristics as one digs deeper into them. As a rule, the deeper the slate, the older it is and the harder and more durable it is (providing that the slate deposit hadn't been heaved upside down by geological forces). This is one reason why Peach Bottom slate is more durable than the softer slates farther north in the PA slate belt. It is geologically much older.

In any case, the list of slate veins in the Albion Run near Pen Argyl, PA, provides an excellent illustration of the variation that can be found as one digs deeper into a PA slate deposit. When one understands that black slate from one quarry can have a variety of characteristics depending on the vein in which it was found, then it is easy to understand how difficult it may be to pinpoint the exact origin of a slate currently serving on the roof of a house. If you have a soft black slate on your building, then you can be pretty sure it is from eastern PA. Yet some black slate from PA is hard. And some slate from PA is gray and hard. It is the hardness of the slate that we are interested in, as a hard slate roof can be restored when old, but a soft slate roof must be replaced. Many thousands of roofs have PA slate on them, but exactly where in PA those slates came from may be difficult to say. There are some exceptions to this, however.

The first, of course, is that if you live near a slate quarry, you are likely to have *that* slate on your roof. If you live in Chapman, PA, you probably have a Chapman slate roof. Same for the Peach Bottom area (Delta, PA, for example) — you have a Peach Bottom roof. In Bangor, PA, you probably have a slate on your roof from the Bangor area. However, all of these slates, and many others from PA have been shipped all over the eastern seaboard of the United States during the last century. And if you don't live near a slate quarry and you think your roof is a PA slate roof, how do you identify it?

Peach Bottom roof slate is easily identifiable because it is quite black, remains *smooth* (non-flaky) with age, and stays very hard. It also has a characteristic sheen to it that may not be evident unless you look at the back of the slate (assuming the front of the slate has been environmentally stained over time).

Chapman slate is easily identifiable because it is black, somewhat hard, and has diagonal bands (ribbons) across the full width of the slate that are evident from a distance — you can see them from the ground if the slates are on a roof.

Bangor black is a uniform black, smooth slate with no *apparent* ribbons, and is fairly soft, but can still last a good century. With age, these roof slates will appear to become flaky and will crumble. Many of the older Bangor slate roofs are reaching the end of their years at 75- 90 years old,

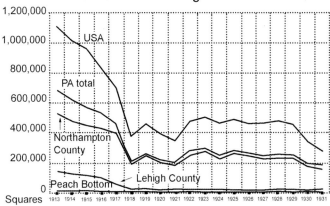

Production of Roofing Slate 1913-1931

The production of roofing slate in the United States dropped dramatically during WWI, never fully recovering. Pennsylvania has produced much of the roofing slate in the U.S., although relatively little Peach Bottom slate was ever produced.
[From *Slate in Pennsylvania*, p. 112]

Roofing Slate Sold in Pennsylvania -- 1890-1941

The production of roof slate peaked in Pennsylvania around 1902, with the bulk of the production occurring between 1896 and 1915. Production continues to this day, and the deposits remaining in the ground have barely had their surface scratched. Many homes were built during that period of peak production, and many had (and still have) PA slate on their roofs. [Source: *Mineral Resources of the United States* and *Minerals Yearbook* (dates unknown)].

SLATE ROOFING RECEIPT, 1876
FOR A PENNSYLVANIA BLACK SLATE ROOF IN PHILADELPHIA
Philadelphia, October 7, 1876, Levi Stokes, Dr. to True Blue Slate Co., for roofing house with slate. 3,444 ft @ 8 1/4 = 284.13, 41 gutter hooks @ 20 = 8.20 (total = $292.33), rec'd payment, True Blue Slate Company, Booth (?)

A. J. Williams of Williams and Sons Slate Co. in Wind Gap, PA, is a manufacturer of Pennsylvania's famous roofing slate. Williams also imports and sells premium roofing slates from selected foreign sources.

Photo by author.

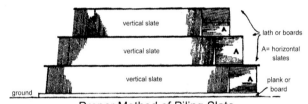

Proper Method of Piling Slate

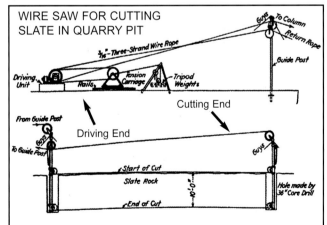

WIRE SAW FOR CUTTING SLATE IN QUARRY PIT

These saws became an essential part of the slate quarrying industry in Pennsylvania in 1928, yielding a more efficient quarrying process with less waste.
[From: Bowles, The Stone Industries, 1934, p. 256]

and many have already worn out. Bangor slate was shipped with a "Genuine Bangor Slate" label which some roofers left stuck to the slate when they installed the roof, so when the roof is removed 80 years later, the label is still on the back of the slate, allowing it to be readily identified.

Pen Argyl black slate may develop a light brownish surface color due to iron in the slate, and will also become flaky with age. In some cases, these slates can look remarkably similar to some old Vermont sea green slates to a layperson. People may point out old Pen Argyl slate roofs and mistakenly identify them as sea green roofs simply because of the "rusty" surface color that has developed over the years. When Pen Argyl slate gets old, however, it will be visibly deteriorating and individual slates will be seen falling apart on the roof. The pieces that fall to the ground will probably have soft edges, soft enough to crumble in your fingers. Old sea green slate, although changing to a brownish color, doesn't turn soft so quickly.

Some of the soft black slates from this region exude a chalky substance over time and may eventually display a whitish hue, especially around the edges.

Slate from the "gray beds" of eastern PA seems to be among the most admired for durability, second, perhaps, only to the Peach Bottom slate. They are fairly rare, "olive green in color" according to one published source, but to me look dark gray (but not completely black), and contain little carbon. When carbon makes up the bulk of the "ribbon" in ribbon slate, it greatly reduces the life of a piece of roofing slate, which may be one reason why gray slates last so long in comparison to their black cousins. In fact, the gray-bed slates of PA are said to resemble the green slate of Vermont in chemical composition.

The first slate quarry in Pennsylvania's Lehigh-Northampton district opened in 1812 at Slateford in Northhampton County, according to one source. Another source claims the first quarry was opened in 1832 by Samuel Taylor and actively operated from 1836 when James M. Porter joined him as partner. This occurred before railroad service existed into the area, and the slates were sent down the Delaware River on boats. This quarry only lasted about 30 years.

Another quarry near Slateford, the Snowden Quarry, opened in 1870 and operated until 1917, when it was deemed a non-essential war-time industry, and closed. During World War I, many quarry workers went to the steel mills for war production,

WHITE SPAR
BALDWIN
LISA WEEKS
JACKIE WEEKS
SALLY WEEKS
JENNY WEEKS
SMALL WASTE BEDI
SAM AND CELESE - HARD 24
LITTLE 5'
24 BED
GRANNIE RAINS + 3 FOOT
BIG 5'
SARAH BRUCH
HARD BIG RED
WASTE BEDI
ROUNDY BACK/GLDN. EAGLE
SAM BRAY
BLACK BED
WASTE BEDI
HANEY & ELLEN/HARD BED
THREE BEDI
WASTE BEDI
LITTLE BIG BED
BILL BUDGE
VAN JULIA/MILK & WATER
STRAIGHT BEDI/BOB CORY
JOHN WELTY & FAMILY
WASTE BEDI
WILCOX & JAN DENTITH
JOHNNIE NACK
CARBUNCLE & SILVER 24
LITTLE 18 BEDI
TWO 24 & 24 BED
WASTE BEDI OF GRAY
BLUE END
GRAY BED
WASTE BEDI
GENUINE 24 & RIBBON BED
2 LENGTH BEDI
GENUINE BIG BED
24 + HARD TWENTY
TWO COLOR RUN
FRONT BIG BED
BASTARD RUN
?

and many quarries closed and never reopened.

Slate was discovered in the Bangor area on a farm at what is now East Bangor, by Joseph Kellow in 1853. This quarry was operated from 1855 by fellows named Weidman and Derrick.

Bangor got its name from Robert M. Jones, a pioneer in the slate industry in the Bangor community who took the name from his hometown of Bangor, in northern Wales. Jones opened the Old Bangor Quarry in 1863, and the town that sprung up around it took the same name. Once again, the Welsh played the role of slate entrepreneurs, founding an industry in Pennsylvania from which entire communities emerged.

Some of today's PA slate quarries are huge, neatly cut holes that plunge vertically into the earth for hundreds of feet. The abandoned quarries of this type have vertical cliff-like walls dropping as much as 650 feet to a pool of water at the bottom. These can provide quite a hazard for persons out for a stroll in the woods unaware of the one-way drop awaiting them should their foot slip at the edge of one of the abandoned pits. Local folks tell of deer inadvertently dropping off the edges on a one-way free-fall to a watery grave. Fortunately, accidents are not common despite the unfenced holes (knock on wood).

One advantage to deep pits is that the temperature remains a steady 55 degrees Fahrenheit at the bottom where the workers toil daily to wrestle out the slate slabs. This provides a relatively cool work site in the summer and a relatively warm one in the winter.

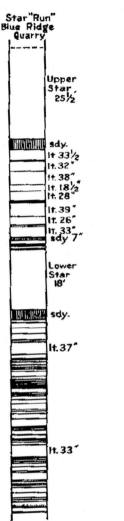

▲ Example of thickness measurements in beds in PA. Diagram represents 71' of depth. "lt." = light, "sdy" = sandy [From Slate in Pennsylvania]

CONVERTING SLATE BLOCKS INTO SHINGLES

The slate in Pennsylvania is still split into roofing shingles by hand, and the blocks are kept wet with water during the splitting process to enhance the ease of splitting. The stone is said to split *"with much greater ease if the quarry sap is not allowed to evaporate,"* an opinion echoed by slate workers in Wales and elsewhere in Europe, who insist that slate blocks left out on the surface of the earth to dry will not split as readily as freshly quarried slate.

The splitting of slate into the right thickness for roofing shingles may not entirely be the responsibility of the splitter, but may involve the block cutter and the blacksmith. The block cutter can split the incoming block to the proper thickness to allow for a predetermined number of slates to be produced from one slab. Since the standard thickness of a roofing slate is 3/16", the slab should be a factor of that thickness. In order for the block cutter to consistently split a block to the thickness desired, his chisel is conveniently made to the proper width and can be used as a measuring tool as well as a chisel. So the blacksmith is instructed to forge out a chisel that the block cutter can use to measure his cut. In this manner, the slate splitter (the guy who splits out the actual roofing slates) receives blocks that have

already been split to a convenient size, which he then splits by eye through the center, then in half again, then again to end up with the finished pieces of 3/16" roofing slate.

Naturally, this technique is not precise and some slates turn out to be thick or thin. However, the finished slates shipped out from the slate yards in the old days were amazingly uniform when the shingles were split from a good quality slate — the better the quality, the easier it is to split a thin shingle. When stacked, one can routinely expect 50 slates per foot of row of slates that are standard 3/16" thickness. This figure (50 slates per foot of row) is close enough that a person who has a stack of standard antique American slates need only measure the total length of the rows, in feet, and multiply by 50 to get a close estimate of the number of slates (10 feet of slate shingles stacked tightly on edge will amount to 500 slates, for example).

And slate *should* be stacked *on edge* on boards or slats, and not piled in flat, vertical piles like dishes, as the weight of flat stacks can weaken and break some of the slates over time. Roof slates are amazingly strong on edge, and can be thrown from

a roof onto a concrete driveway without breaking, as long as they land *on edge*.

The final trimming of the slate shingles was commonly done using a foot-powered trimmer with a straight, steel cutting blade about three feet long. The outer end of the blade was attached to a spring. Another type of trimmer had a rotary blade like an old-fashioned push-type lawn mower. Although these machines were once powered by foot or hand, today they're electrically powered.

Each slate shingle was laid against a steel bar on the trimming machine when it was trimmed, the bar being marked in inches so that the trimmer could trim the slate to the standard sizes, ranging from 6"x10" to 14" x 24". The person trimming the slate could easily estimate by eye the largest possible size for each untrimmed piece, and trim according-ly. Once trimmed, the slates were stacked in piles sorted by size, then handed to a person who had the job of punching the nail holes in the slate. The holes were punched into the back of the slate (the side fac-ing down against the roof) due to the countersinking effect of the hole punch, which then enabled the nail heads to sit flush into the slate. Two men on a good day can split and dress 12-14 squares of slate, if they have nice blocks to begin with.

CHAPMAN SLATE

It is worth mentioning the Chapman quar-ries, which are located (you guessed it) in Chapman, PA. The quarries are no longer in operation, but they were once considered the most productive producers of PA "hard vein" slate which are still serving thou-sands of roofs today. This black slate has a distinc-tive appearance because it is adorned with many rib-bons, or bands of various shades of black that cross the face of the slate at an angle, giving the slate a striated appearance. These ribbons are fairly hard and do not disintegrate as readily as the carbon rib-bons common to the "soft vein" slate of Pennsylvania. I have seen many hundred-year-old Chapman slate roofs still in fair shape, as these slates are one of the harder and more durable of the Pennsylvania black slate.

The Chapman Slate Company was one of the oldest producers of slate in Northampton County, with the Chapman Standard Quarry opening in 1860, and producing primarily roofing slate and slate slabs. Several other quarries operated in the Chapman vicinity. The "Chapman" Quarry consist-ed of two large holes separated by a wall of rock about 50 feet wide, the west hole being 450 feet long,

300 feet wide and 130 feet deep, while the east hole is about the same width but is about 1,000 feet long, and 150 feet deep, rivaling the largest quarries in the Bangor and Pen Argyl districts in volume. The old, abandoned quarry pit still lies just outside Chapman, PA, today, and probably won't be going anywhere soon. There was still some quarrying reportedly taking place there in 1924, and in 1931 the Chapman Slate Company was still listed as a slate producer in Pennsylvania. Today, however, no quarrying is taking place there and Chapman roof slates cannot be bought new.

On the other hand, Chapman slates are one type of black Pennsylvania roof slate that are recyclable, and care should be given to remove these slates from roofs of buildings being demolished or re-roofed so the salvaged slate can be used to keep existing Chapman roofs in good repair. Otherwise, if recycled Chapman slates are not available, almost any black Pennsylvania slate can be used to repair a Chapman roof, so long as the slate is sound, although no slate will match the roof quite as well as a genuine Chapman slate.

It should be noted that the slate in Pennsylvania seems to become harder and more durable as we move farther south in the state. When we arrive at the border between Pennsylvania and Maryland, we find a black slate so hard and durable that it is considered by many to be the best slate in the world, as we shall see in the next chapter.

▲ Slate blocks hoisted out of the Dally Slate Company quarry pit lay stacked and waiting in the modern slate mill. The edges have been sawed on a diamond saw. Some of these blocks will be used for structural slate, some for roofing.

▲ Chapman slate roof, good condition. Note "ribbons" running across the face of the slate at an angle, which is a distinctive characteristic of Chapman slate. Chapman ribbons tend to be hard, and don't undermine the quality of the slate. Chapman slates, first quarried around 1860, were the most productive of Pennsylvania's "hard vein" slates. If cared for, these roofs can last well over 100 years.

All photos this page by author.

▲ Boro Hall of Chapman, PA. The town is not much bigger, but it is home to Chapman slate, although the quarries are no longer in production. The "hall" has a Chapman slate roof.

▶ House near Chapman, PA, sided (and roofed) with Chapman slate.

▲ Slate block being hoisted out of quarry en route to the mill where it will be split down into roof slates.

▼ If roof slates are more expensive than the more commonly mass-produced, artificial roofing materials of today, then one sure reason is the hand-made nature of the material. Each piece of roofing slate is hand split from solid rock (below left), hand trimmed into the appropriate size and shape (below right), as shown at Dally Slate Company in Pen Argyl, PA, and then hand punched for nail holes (below center).

Photos by author.

A block of slate is being hoisted out of a slate quarry near Pen Argyl, PA, appearing as a light colored rectangle in the top center of the picture. The circle near the bottom of the photo indicates a quarry worker, dwarfed in this 350' deep, vertical-walled hole. The photo on the previous page shows the block emerging from the top of the pit. It will be taken into the mill where it will be cut and split to size, yielding PA's characteristic black slate.

Photo by author.

Chapter Seven
PEACH BOTTOM SLATE
Pennsylvania/Maryland Slate District

Professor Agassiz (late 1800s) "The Almighty might have made a more perfect fish than the trout, but he never did."
Mr. Humphrey (quarryman) "I say the same for the Peach Bottom slates."

What intrigued me most about Peach Bottom slate was its mystery. In nearly three decades of working on slate roofs I had never seen a Peach Bottom slate (or so I thought) and hadn't a clue about what one looked like. One day someone phoned me and told me he had some roof slates for sale. This was a guy just five miles down the road who described the slates as "really nice, beautiful Peach Bottom slate." I drove to his country home to look at the slate and sure enough, there lay a pile of really nice slate in his front yard, neatly stacked, with scalloped bottoms. They were a light gray slate with a reddish cast to some of them, which I recognized as sea green slate from Vermont. I didn't buy them because the guy wanted too much money, but I wondered how he came to the conclusion they were Peach Bottom slates. Apparently, he thought the color was peachy.

In my wanderings through the slate regions of Vermont, New York and Maine, I came upon old books about slate production in the United States, and Peach Bottom slate was mentioned as being of the highest quality, lasting hundreds of years. This was hard to believe as most of the slate from Pennsylvania is soft and relatively short-lived compared to the harder slate from Maine, Vermont, New York and Virginia. The notion that Pennsylvania, with all its soft slate beds, could also be the source of a slate so hard and durable as to deserve such high regard by authors of pre-1900 books only added to the mystery. This was compounded by the fact that

Peach Bottom slate was no longer in production and could not be bought new on the roofing market.

Furthermore, modern maps of Pennsylvania show no town named Peach Bottom. However, I happened to have an *old* map of Pennsylvania, and it did show a tiny little place named Peach Bottom just north of the Maryland border on the east bank of the Susquehanna River. So one day I couldn't stand it any longer, and I set out for the spot on the map, a full eight hours drive from my home in western PA. I was determined to find the truth about Peach Bottom slate once and for all. I dragged a couple of my kids with me to keep me company, and one July morning we took off. We drove all day until we reached a campground in Maryland just across the border from PA, where we pitched a tent and got a little bit of sleep under the starry night sky. The loud serenade of summer's nocturnal creatures kept us awake for a while. My fifteen-year-old daughter's comment was, "I wish those birds would shut up!" "Those are tree frogs," I said, "Go to sleep."

Early the next morning we drove the back

PEACH BOTTOM SLATE REGION

PEACH BOTTOM SLATE, FIRST USED 1734, IS THE OLDEST IN AMERICA. THE FIRST COMMERCIAL CUT HAVING BEEN MADE 1785 BY WORKMEN WHO WERE PRIMARILY WELSH. AT THE LONDON CRYSTAL PALACE EXPOSITION, 1850, PEACH BOTTOM SLATE WAS JUDGED BEST IN THE WORLD.

MARYLAND HISTORICAL SOCIETY

Road sign near Delta, PA, on the Maryland side of the PA/MD border.

Opposite: Peach Bottom slate graces barns as well as elegant masterpieces of architecture, such as the Castle Park Apartment Complex in the St. Louis, Missouri area. This roof was 120 years old in 2001, and the slates were nearly perfect. Photos by author.

roads to the dot on the map named Peach Bottom. We found one or two houses and a boat marina on the east bank of the Susquehanna, right across the river from the Peach Bottom nuclear power plant, which loomed on the opposite bank like a behemoth, its high-powered electrical transmission wires emanating from it, octopus-like, in every direction. This didn't look like a place for a slate quarry, in my opinion, but I did notice that the name of the road going uphill past the marina was Slate Hill Road. "Let's follow this road," I said to the kids, excited, like a bloodhound on a trail.

At the top of the hill, we passed an old farm with slate roofed buildings. We then came upon a small, old house, also with a slate roof, where an old man was roto-tilling his garden out back. I pulled over and shut off the car motor. The old fellow turned off his roto-tiller and we met in his side yard.

"I'm looking for the old Peach Bottom slate quarries," I said, after introducing myself. "Do you know anything about them?"

"Why sure I do. What do you want to know?"

"Where are they?"

The fellow turned around and pointed off across the green fields of corn and hay to a patch of trees on top of the hill we had just driven by. "Back there in those trees you'll find the old slate piles." I pumped the friendly man for more information, but he didn't know much about

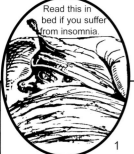

Read this in bed if you suffer from insomnia.

the quarries themselves. He did have a pile of slate in his back yard matching the solid black slate on the roof of his house, so I bought a few. "This house was built in the 1700s," he said. "See that roof? That Peach Bottom slate is just as good as the day it was put on."

The slate itself is coal black and very hard, with a sheen to it that I had not seen on any other black slate. I passed roofs with these slates on the drive to Peach Bottom, seeing more of them as I got closer to my destination and wondering if they were, in fact, the elusive Peach Bottom slate. I had not seen any slate like this back in western PA (although once I learned to recognize them, I did occasionally see them back home). The most striking characteristics of these slate roofs were the darkness of the black color and the shiny gleam.

The kids and I set out in the van for the wooded hill in the distance. We found a dirt tractor road leading up to the woods, and there we pulled off and set out on foot. The air was hot and muggy and puddles littered the lane, which was all but washed out from recent heavy rains. We were sweating by the time we got to the top of the hill, probably a half mile walk between corn fields and hay fields thick with huge butterflies. We stopped under a giant electrical transmission tower on a bluff overlooking the beautiful Susquehanna River. The nuclear power plant was now directly opposite us, connected to the tower by electrical lines hanging across the water, cracking and buzzing menacingly over our heads. My hair all but stood on end. From our vantage point I could see a mountain of slate rubble lurking in the trees across the corn field, so the kids and I wove through the rows of corn to get a closer look. When we reached the edge of the woods I plunged into a forest of poison ivy, wading through it to get to the slate pile. The kids were wearing shorts so I cautioned them to stay out of the woods and go back to the van and wait for me there.

I climbed the huge pile of scrap slate like a mountain goat and stood on top of it like some sort of conquering explorer, peering with some trepidation down the cliff-like bank of the river, on the brink of which I now perched. I had found the source of Peach Bottom slate! Or had I? Where was the pit? All old slate quarries had huge holes in the ground, usually filled with water, as well as huge piles of

ANALYSES OF PEACH BOTTOM SLATES FROM FIVE DIFFERENT QUARRIES (%)				
1	2	3	4	5
SiO_2 .... 55.88	.... 58.37	... 60.32	... 60.22	.... 44.15
TiO_2 .... 1.27	.... tr.	.... n.d.	.... n.d.	...... n.d.
$Al2O_3$ 21.85	.... 21.99	... 23.10	... 19.56	.... 30.84
$FeO (Fe_2O_3)$ .. 9.03	.... 10.66	.... 7.05	.... 5.24	.... 14.87
CaO 0.16	..... 0.30	.... ---.---	.... 3.87	..... 0.48
MgO 1.50	..... 1.20	.... 0.87	.... 2.30	..... 0.27
K_2O 3.64	..... 1.93	.... 3.83	.... 2.90	..... 4.36
Na_2O 4.46	.... ---.---	..... 0.49	.... 2.15	..... 0.51
CO_2 .... n.d.	..... 0.39	..... n.d.	.... n.d.	...... n.d.
FeS_2 .... 0.05	..... n.d.	..... 0.09	.... n.d.	...... n.d.
C 1.79	..... 0.93	..... n.d.	.... n.d.	...... n.d.
SO_3 .... 0.02	..... n.d.	..... n.d.	.... n.d.	...... n.d.
S .... n.d.	..... 11.0	..... n.d.	... 30.0	...... n.d.
MnO 58.0	..... tr.	.... n.d.	.... n.d.	...... n.d.

[From: Slate in Pennsylvania, p. 375]

This manse, built in 1954 next to the Chestnut Level Presbyterian Church, was probably one of the last houses to get a new Peach Bottom roof, although the slates may have been recycled. The nearby church, built in 1765, also has a Peach Bottom roof.

scrap slate. There was no visible hole here, although there had to be one somewhere nearby. I was reluctant to explore further through these dense woods, however, as the ground was covered with one of the thickest and most luxuriant growths of poison ivy I had ever seen, and my kids were now left unattended a half mile away along a rural roadside. The sweltering heat added to my desire to return to the van, so I picked up a heavy chunk of slate and carried it back to the road where I found the kids patiently waiting. "Let's stop at that farm we passed and see if we can photograph their slate roofs."

The middle-aged farmer, Allen Weikel, at that time a write-in candidate for Governor of PA, explained that much of the slate was actually quarried on the other side of the river. He was a very friendly fellow who took a good hour out of his busy day to make phone calls trying to locate someone who knew about the history of the quarries. Soon we hit paydirt. We had Roger Faill of the PA Dept. of Environmental Resources Geologic Mapping Division on the phone, and his advice was accurate and to the point: "The museum in Delta. That's where you should go. I've never been there, but I hear that they have some good information on Peach Bottom's history." So off we went to Delta, PA, right on the Maryland border on the west side of the Susquehanna, the heart of the Peach Bottom slate region.

By this time it was afternoon and we were all getting hungry. As soon as we reached the quaint little town of Delta, we stopped at the diner on the main street. I ordered their lunch special, and while waiting for my soup I asked the waitress if she knew anything about the old Peach Bottom slate quarries. "I know just the person you should talk to," she said. "Hold on a minute, let me see if he's still here." She disappeared into another room and soon returned with a wiry, elderly man named Harold Beucker, who pulled up a chair and began to fill me in on what he knew.

"I was the last person to haul Peach Bottom slate out of here," he said. "That was in 1956 or 57."

"Roof slate?"

"Yes, and I hauled it into the Pittsburgh area. Fox Chapel, if I remember right. Best slate ever produced. Best slate in the world."

"Why did they stop quarrying it?"

"Not economical. Too much waste, too hard to work, to split. Some of it went underwater when the river was dammed. After they quit making roof slate they continued to make slate granules for a few years. They used the granules for roofing, like what you see on shingle roofs."

"Where can I find out more about the history of the quarries?"

"The person you want to talk to is Ruth Ann Robinson. She lives just down the street. First house past the Welsh church. The red brick one. She's curator of the town museum. Museum's only open on weekends though. You won't be able to see it today."

After we finished eating, the kids and I set out for the Robinson house. No wonder I couldn't find an old quarry pit by the river, I mused as we walked, it was probably under water! The first thing I noticed when we reached the Robinson house, besides the Peach Bottom slate roofs on the church and the house, was a Welsh flag bumper sticker on the car parked in her driveway. Turns out that Ruth and her husband Don had both visited the slate areas of Wales. Don was making wooden slate-splitting hammers, like the ones used in Wales, in the workshop behind his house. Once again, a kind and friendly person, Ruth Ann took time out of her busy day to accommodate total strangers with absolutely no advance notice. She opened up the Old Line Museum and dug around for material to show us, and some of the historical information about Peach Bottom slate in this book is attributable to her.

One important fact that became clear to me during my investigation was this: Peach Bottom slate is from Pennsylvania *and* Maryland. The slate deposit straddles the state line and it is a resource shared by the two states. In fact, the slate was mined in Maryland longer than it was mined in Pennsylvania. It was all shipped out of Delta, PA, though, and therefore became known as a Pennsylvania slate. So now I'm trying to set the record straight. Peach Bottom slate is a Pennsylvania *and* Maryland slate. Marylanders are rightfully proud of their slate, but haven't been given credit where credit is due. In fact, I had to go to Maryland to photograph the highway sign, paid for by Maryland tax dollars, shown at the beginning of this chapter.

The Peach Bottom slate deposit extends about ten miles, running northeast to southwest in three parallel belts 75 to 120 feet thick, and crossing both the Susquehanna River and the PA/MD border near where the Susquehanna flows out of Pennsylvania. The slate deposit, locally known as a slate "ridge," extends from York County, PA, about 3 miles into Cardiff Township, Harford County, MD. Its total depth is estimated at 1000 feet, and its width is said to be one-fifth to one-half mile, although it seldom exceeds 200 feet in a quarry. About one and a half miles of the slate lay underneath the river until the construction of the Conowingo Dam submerged more of it. The quarries were located near Delta, PA, Peach Bottom, PA, and Cardiff, MD.

Slate splitters had some problems getting Peach Bottom slate to split to their satisfaction, and eighty-eight percent of the slate mined in the Peach Bottom district became waste.

Peach Bottom slate was exhibited in the Crystal Palace Exposition in England in 1850 by quarryman Roland Perry and was deemed the best roofing slate then known. It is perhaps this award more than any other that gained Peach Bottom slate its fame. Ironically, and despite its durability, Peach Bottom slate is no longer quarried. It's just too hard to produce finished slate products with this material at a price the market will bear. Perhaps this will change in the future and people will once again value durability over convenience and renew the markets for such fine slate products. Time will tell.

But don't hold your breath waiting. One recent summer a beautiful and rare Peach Bottom roof on an old historic inn in a small local town near me was damaged by a windstorm. I heard through the grapevine that contractors had been hired to tear the roof off and replace it with

---

" Peach Bottom " Slate.

The finest roofing slate in the world is that known in commerce as " Peach Bottom " slate, and which is taken **from** a ridge extending about seven miles from **Peach** Bottom, on the Susquehanna River, in **York** County, Pennsylvania, southwardly into Harford County, Maryland, about two miles of the ridge being in the latter State. The first quarries were opened at Peach Bottom, which gave the name to the slate. Since that time the industry has worked back along the ridge, until now three-fourths of the so-called Peach Bottom slate is taken from quarries in Harford County.

Stretched out for over a mile along the base of the ridge and hemmed in between it and Scott's Run is the thriving borough of Delta, which sprang up with the slate industry, and is dependent on it for existence. As the quarries moved toward the southwest the town followed them until it, too, has crossed Mason and Dixon's line and has extended into Harford, where, under the name of South Delta, it is now booming, if that progress may be called a boom which is the result of natural causes.

There are now in operation on the Harford side six quarries, giving employment to over two hundred men. The demand for Peach Bottom slate was never so great as now. The quarries in operation are being worked to their full capacity, new quarries are being opened and old ones re-opened, and as the supply of slate is practically inexhaustible, there is every reason to believe that, with proper management, the business of the quarries will continue to increase.

The Manufacturer and Builder, 1890

---

fiberglass shingles. I knew this would cost the inn owner many times the amount it would cost just to repair the roof, so I contacted the inn manager and informed her that the inn would be losing one of the best roofs in the world. She passed the word on to the owner, and it turned out that he didn't care, because his insurance was paying for the replacement roof. I then contacted the roofing contractor and offered to remove the slate, free of charge, so I could salvage them. I explained that these slates would still last hundreds of years, that they're no longer quarried, they're rare, and he would save a lot of work if someone removed the slate for him, etc. I told him that if he didn't like the idea of me working on his roof job, then he could take the slate off himself and I would buy them from him. At the very least, I said, I could be there when the slate was ripped off and gather them up before they went into a dump truck to be hauled to a landfill. The contractor said he'd get back to me. A couple of weeks later I drove by the inn to see the last of the Peach Bottom slate being smashed into a dumpster, while the last section of roof was having fiberglass shingles *stapled* to it. When I stopped and gave the contractor a piece of my mind, he didn't seem to understand what I was saying. Perhaps I was speaking the wrong language, but then I don't speak *Neanderthal* all that well.

I'm mentioning this incident as a warning to homeowners. *Contractors may not care about your roof!* Believe me, there are many contractors out there waiting like vultures to tear off your good slate roof, destroy the slate, and staple glorified tar paper on instead. They care about the money they're going to make, not about tradition, craftsmanship, durability, history and other things of character. Once your permanent roof is torn off and destroyed and you have it replaced with an inferior roofing product *guaranteed to fail* in 20 years or so, you are locked into a cycle of roof replacement that will continue to put money into the pockets of the contractors forever.

Furthermore, when insurance is paying for something, many contractors will push the homeowner to go for the most expensive option, no matter how wasteful. In other words, *if you don't protect your own slate roof, no one else will do it for you!* In this case I could say that the home owner deserved to lose the roof. But I saw this roof as an historic part of the community, and the building as one that may last for generations and have many future owners. It is disappointing when the owner of an historic building is not considerate of the community and not concerned about wasting good resources. Although this is hap-

pening every day in the United States, it is not legal in Europe to tear off a perfectly good, traditional stone roof and staple a tar paper product on instead. It is considered a breach of community standards and therefore prohibited. The entire community has a stake in the appearance of its buildings, and although this may be considered an unacceptable intrusion on personal liberties here in the USA, in Europe many quaint towns and villages continue to display the elegant charm they have demonstrated for centuries. Why? Because their building codes display a higher set of community standards which take into account such things as tradition and aesthetics. I fear that by the time we in the United States have evolved to that level of standards, there will be few, if any, traditional slate roofs left.

Peach Bottom slate was more expensive than the other slates available at the turn of the century (1800-1900), so it had a tendency to be used on more expensive buildings and homes. It is certainly found on common buildings *near the quarry district*, but farther away it was sought after by people who had the extra money to spend for the very best material available. Therefore, many ornate homes, churches and government buildings had Peach Bottom slate on them, and today if you have an older, fancy home or building with a hard, black slate on it, it may be Peach Bottom, unless the home is located near the Monson, Maine, or Buckingham, Virginia, slate districts (as these slates are also black and hard). For example, some buildings that had Peach Bottom slate on them by 1898 were the Pennsylvania State Capitol (roofed in 1820), Carnegie Library, Johns Hopkins University and Hospital buildings, Notre Dame Academy, the Biltmore Estate of George Vanderbilt, and the residence of William B. Astor on Fifth Avenue in New York City.

HISTORY

Peach Bottom slate was first discovered in 1734 by Welsh brothers who had settled in the area a few years prior. They used the slate to cover a roof, and the slates reportedly remained intact for 200 years until 1932 when the building burned. The roof was still in perfect condition. One history book tells the story like this:

"Two brothers, William and James Reese, natives of Wales, took up land in 1725 under patent of the English provincial Government, in what is now known as York County, PA, and during

The above advertisement is from a publication dated December, 1898: "The History and Characteristics of the Peach Bottom Roofing Slate."

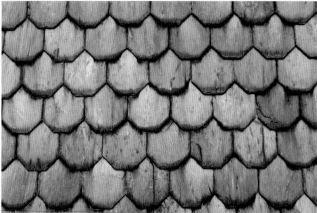

The octagonal roof (top, right) is being hoisted by crane onto the Caparosa house in Butler, PA. The main house already has an historic Peach Bottom slate roof (closeup, right). The tower will be slated with Buckingham slate from Arvonia, Virginia, which today most closely matches Peach Bottom slate both in appearance and durability. See photo of slated tower on page 91.

Photo at top by Kevin Caparosa, other photo by author.

the year 1734, when excavating for foundations for farm buildings, they discovered the slate rock from which they split the pieces necessary for roofing their buildings. This was the first use of slate in America." [From <u>Peach Bottom Roofing Slate</u>, 1898, p.5]

Those same pieces of slate were still in use in 1898 when the above historical account was published, on their *seventh* building, apparently having outlasted six others. *"They show no disintegration or decay* [after 163 years] *and still retain the strong, brilliant, blue-black color for which this slate has become famous by the name of Peach Bottom."* That seventh building was reported to be a hog pen located near Delta, PA, and still standing in 1930. It is reported in <u>The Stone Industries</u> (1934) that a sample of this slate was *"rescued from this lowly use and is now on exhibit at the United States Bureau of Mines in Washington, D.C. After nearly 200 years of service, it shows no evidence of deterioration."* Take a look at the photo of the one-inch-thick, 250-year-old gravestone illustrated on page 14 for an example of the durability of this slate.

One of the first slate roofs made of Peach Bottom slate was laid by Joseph Hewes, a signer of the Declaration of Independence. The first commercial quarrying began around 1785 by a fellow named William Docker, but quarrying didn't get into full swing until the Welsh stepped into the scene in large numbers around 1845. In 1848, about 30 Welsh people arrived in Philadelphia en route to Peach Bottom to become slate quarriers, splitters, blacksmiths and stone carvers. The Welsh introduced quarrying methods brought over from the quarries in Gwynedd, North Wales (see Chapter Four).

By 1858, 18 quarries were reported in operation west of the river and 11 east of it. At one time,

thirty Peach Bottom quarries or prospects were listed west of the Susquehanna River and twelve east of the river, including the remains of one I "discovered" with my kids. By 1929, only three Peach Bottom roofing slate operations existed in Maryland, and none in Pennsylvania, and a year later only two remained in Maryland. In 1997, there were no slate quarries producing Peach Bottom roofing slate.

To illustrate the strength of Peach Bottom slate, a compression test was conducted in 1893, indicating that the slate would crush at 385.6 tons per square foot of pressure applied parallel to natural cleavage, and 758.4 tons per square foot applied perpendicular to natural cleavage. Another set of tests (1892-94) showed an average crushing strength of 11,260 pounds per square inch, a figure typical of a good quality slate.

Because Peach Bottom slate had gained fame due to its award at the Crystal Palace Exposition in England in 1850, other quarries began to falsely refer to their slate as Peach Bottom as well. These imitations went by the names of Slatington Peach Bottom, Lehigh Valley Peach Bottom, and Peach Bottom of Lehigh Valley, and were decried by producers of "genuine" Peach Bottom slate as "cheap imitations named to deceive the public," which they probably were.

Today's lack of existing Peach Bottom quarries dictates the need to carefully salvage and reclaim Peach Bottom slates from buildings that are scheduled for demolition, or even scheduled for re-roofing. However, there is still one type of slate quarried in the USA today that somewhat resembles Peach Bottom both in appearance and durability, and which can be used with satisfactory success as a Peach Bottom substitute. This is the "Buckingham" slate of Virginia, which we discuss in the next chapter.

Old payment receipt for Peach Bottom slate.

1901 Advertisement

## Chapter Eight

VIRGINIA AND GEORGIA

Virginia slate (Arvonia-Buckingham slate belt) is black, very hard and durable, and is guaranteed by one manufacturer to last 150 years on a roof. It has a characteristic glistening sheen indicative of a high level of quartz, which sparkles in the sunlight when the slate is new. This slate is still quarried today in the little town of Arvonia, Virginia, deep in the rich heartland of the state.

This "lustrous" slate is said to be "blue-black," "deep black" or "oxford gray;" however, for our purposes "black" is close enough. It is said to have the lowest average absorption (less than .02 per cent) of any American slate when submerged in water, which is a very good indication of longevity. It is also claimed to have equal strength both with and across the grain, which is unusual for slate. A manufacturer also states that this slate shows the least deterioration of any other slate in corrosive tests. Add to this the fact that the slate does not change color with age, and you have, all in all, one of the best slates on the market today, rivaling Peach Bottom slate (which is not on the market), Monson slate (also not on the market, although the Glendyne Quarry in Quebec now offers a substitute), and the best green, purple and red slates of Vermont/New York. In other words, this slate should last the life of the building it is attached to. And if the building wears out before the slate does, the slate can be removed and used on another building. In some cases, as is done in some parts of Europe, the slate may be removed from an old building, the old roof boards removed and replaced, then the same slate renailed to the new roof boards on the same building with new nails. In this way a good slate roof can be preserved indefinitely.

The slate bed at Arvonia is 600 to 800 feet thick, 10 to 15 miles long, and 1 to 1 1/2 miles wide. It extends across parts of Buckingham and Fluvanna counties, and crosses the James River near the town of Bremo Bluff. Although often referred to as "Buckingham" slate in today's slate industry, that term is a trademarked product name, and the slate

itself is known geologically as "Arvonia" slate on state maps. The slate lies in three distinct veins, the first having such a fine grain that it can hardly be detected, the next having a coarser grain, and the third vein remains undeveloped.

The slate has an overburden of shale sitting on top of the working slate bed that is 40 to 60 feet thick. This overburden constitutes the only waste derived today from that quarry's open-pit operations as the slate refuse is crushed for road aggregate.

Virginia's slate quarries are in Buckingham County near Arvonia, in the center of the state.

Although many roof slates from this area are made from single beds, others show ribbons ("ribbing") indicating they were cleaved through a succession of beds, giving the slate an attractive striated appearance, but not affecting durability in this type of slate. Some thin beds only a few inches thick contain fossils of crinoids, brachiopods and trilobites considered to be of Ordovician age (500 million years old). Other fossils including clams have been found. The deepest quarry in the district was reported to be over 350 feet deep and depths of over 200 feet are not uncommon. One of the largest single

Opposite: Many government buildings in Washington D.C. are covered with Buckingham, Virginia, slate roofs, such as the Smithsonian Institution Building, known as the "Castle."

Photo by author.

▲ David Crummette, of Buckingham Virginia Slate Corporation (1995), Arvonia, Virginia, demonstrates the use of a modern electrically powered slate trimming machine with a "guillotine" style blade. Untrimmed pieces sit to his left, foreground. The trimmed Buckingham "blue-black" slate is shown below and on page 92.

Photos by author.

slabs of slate from the Arvonia-Buckingham slate belt trimmed to seven and a half feet by eleven and a half feet in width and length, and was able to be split to only one and a half inches in thickness!

Slate deposits are also found in the Esmont area in Albemarle County, in the Warrenton area of Fauquier and Culpeper Counties, and in the Quantico slate belt in Stafford and Spotsylvania Counties, all of which are inactive.

Here again, as we have seen throughout the other major slate producing regions of the United States, Welsh people essentially started the slate quarries. The name "Arvonia" for example, comes from "Arvon" which is a derivative of "Caernarvon," the name of the county (and city) in North Wales from which many a slate worker emigrated. A "caern" is a "fort" in Welsh, and the name *Caernarvon* was shortened to just *Arvon* when the Welsh settled in Virginia, eventually to again be altered to its final form, *Arvonia*. A principal town in Caernarvon, Wales, is Bangor, and *that* name was given both to Bangor, Maine, and Bangor, Pennsylvania, by Welsh immigrants.

Early reports of the use of slate in Virginia are sketchy, and by all accounts the first slate roofs came to Virginia from England. In 1662, the Virginia General Assembly authorized the building of 32 houses at "James City" and specified slate roofs for these buildings. At that time, however, slate was not being procured from Virginia's own deposits. Likewise, in 1706, slate was again specified for the roof of the Governor's house in Williamsburg, authorizing its purchase from England. During the mid 1700s, the *Virginia Gazette* published notices of shipments of slate from Germany, Ireland and Algiers.

By 1755, slate roofs were being described in print, along with lead, shingle and tile roofs, on Virginia houses, and although Virginia's slate deposits were known then and were probably worked to some extent, it is not clear what percentage, if any, of these roofs were made of Virginia slate. In 1794, a contract was awarded to a roofing firm for the purpose of slating the roof of Virginia's Capitol with slate from a Buckingham quarry, stipulating that the job had to be completed by 1796. Again, for reasons that are not clear today, that contract was never honored, and the Capitol was instead roofed with wooden shingles.

It is known that at least one quarry was in operation in Buckingham County by 1812. Unfortunately, just about that time, Thomas Jefferson, noted for (among other things) his architectural genius, decided that "slate is a bad covering because [of its] constantly getting out of repair." He preferred roofs covered with tin instead, approving of a roof design that only rose vertically a half inch in each foot of horizontal run, which is practically flat by today's standards. So, for a while, major roof contracts were given to tin roofers rather than slate roofers, and Jefferson advised *"to cover with tin instead of shingles. It is the lightest and most durable cover in the world. We know that it will last 100 years, and how much more we do not know."*

Well, Jefferson's advice was clearly off target and, unfortunately, disastrous. The tin roofs lasted only ten years, and when they were replaced, the owners specified slate, one person noting in his diary that he had *"commenced taking off roof of the house to be replaced by a new one to get rid of the evils of flat roofing and spouts and gutters, or in other words to supercede the Jeffersonian by the common sense plan."*

By 1835, the tide was turning for Virginia slate. Geologist W. B. Rogers reported to the Virginia legislature that *"in texture, density, and capacity of resisting atmospheric agents [Buckingham County slate] can scarcely be excelled by a similar material in any part of the world."*

By 1851, the slate was receiving "honorable mention" at the London World's Fair in England, followed in 1876 by a gold medal at the Philadelphia Exposition, and yet another medal in 1893 at the World's Columbian Exposition.

Perhaps the greatest factor involved in getting Virginia's slate industry moving was the Welsh influence. In 1867, as Virginia still smoldered from the Civil War, immigrants from North Wales began to move into central Virginia in search of opportunity in the slate quarries. Some of the Welsh people came down from Vermont where they had settled for the same purposes. By 1881, one Virginia Welsh quarry employed 100 mostly Welsh workers, when a week's shipment of roofing slate may have equaled a hundred tons in weight and traveled as far as San Francisco by rail. The advent of the railroad into the Virginia slate district in 1885 opened the final door to the markets, and the slate industry flourished, booming between 1900 and 1910, approximately the same time slate production peaked in Pennsylvania.

Once again, World War One had a bad impact on the slate industry, and of eight Virginia slate quarrying companies existing before the war, only four continued afterward. By 1928, there were only three companies in operation, but with the introduction of earth-moving equipment, production actually rose to new highs. Still, the Great

▲ Preservation expert Bryan Blundell squirms through a trap door in the Smithsonian Institution Building known as the Castle, surrounded by Buckingham County, Virginia, roofing slate.

▲ "Buckingham" slates from Buckingham County, Virginia, sold by Buckingham Slate Company, are often referred to as "blue-black" slates. Here they are being installed on an octagonal tower by Barry Smith at the Caparosa residence in Butler, Pennsylvania (above).

Photos by author.

▲ Worker at Buckingham-Virginia Slate Corporation in Arvonia, Virginia, trimming the famous "Buckingham" slates.

Photo courtesy of Buckingham Virginia Slate Corp.

Depression caused quarry operations to virtually cease in the early 1930s, when, in 1932, slate production was as low as it had been in 50 years. By 1940, the three companies had recovered sufficiently to employ about 420 workers, but then along came World War Two, grinding operations to a halt, after which operations resumed at one third the pre-war level.

By 1962, however, 100,000 tons of finished slate products were being shipped from Buckingham County, Virginia, to all parts of the United States.

Today, the immense deposits of slate in Virginia have hardly been touched. Sources of Virginia's high quality slate are listed in the back of this book.

GEORGIA

The state of Georgia was never a big producer of roofing slate, nor could it begin to compare to Virginia in either quantity or quality of slate produced. Yet, slate mines did flourish for a while in this southernmost slate producing state. Deposits suitable for roofing are found in the northwestern corner of the state near the Cohutta, Silicoa, Pine Log and Dug Down Mountains in the southern Appalachian range. The quarries were localized in two areas: at Rockmart in Polk County, and at Fairmount in Bartow County. Both quarry areas initially produced roofing slate, although both subsequently manufactured only crushed slate products (aggregate), until eventually all production died off in the 1970s (Fairmount) and 1980s (Rockmart). Slate deposits have also been reported in Murray, Gordon and Fannin Counties. Lindsay (1974), reports that the principal slate producing states in America in 1897 included Georgia, and Lindsay's *History of the North Wales Slate Industry* strongly implies that Georgian slate was exported to Europe at that time.

ROCKMART

Georgia's two quarry regions produce distinctly different types of slate. The Rockmart slate, debatably of Paleozoic age (between 230 million and 600 million years old), is known as black, or "blue-black" (actually gray), and is very similar to the slate of eastern Pennsylvania's Lehigh-Northampton "hard vein" district. Some Rockmart slate remains quite hard and durable a century later, and appears similar to Peach Bottom slate, while other varieties

▲ The Rockmart Presbyterian Church, built of slate, displays the gray Rockmart slate on the bell tower.

(Above right) Gravestone at the old Methodist Church cemetery in Van Wert, marking the grave of a soldier who died in 1861. The slate is still in very good condition, although many of the old slate gravestones in this cemetery have eroded beyond recognition, demonstrating that Rockmart slate is both hard and soft, depending on the quarry site.

▼ Typical Rockmart slate roof.

Photos by author.

▲ An example of Rockmart's harder slate, still in good condition despite its age.
Photo by author.

seem to be softer, becoming rather soft after a century. Apparently, the different grades of Rockmart slate depended on their original location within the actual quarries, the harder slate coming from one pit or mine, and the softer from another.

Much of the slate quarried has been ground up for aggregate and used as a concrete filler, asphalt shingle coating, etc., due to the lack of a smooth-textured, finely grained slate suitable for splitting into shingles. As a result, there is not an abundance of Georgia slate roofs in the Rockmart area at the time of this writing, and many of those that still exist tend to be, unfortunately, in poor repair.

According to the Geological Survey of Georgia (Bulletin 27, 1912), Rockmart slate is a "very bluish-gray color and of slightly rough and lusterless surface." The survey adds, "This is a mica slate of the fading series [will change color somewhat with exposure to the weather], although some obtained from a tunnel at one of the quarries is reported to have kept its color for many years. Its fissibility [ability to split] is fair."

Once again, the development of the slate industry in the Rockmart area was attributed to the Welsh (pronounced "Welch" by the locals, with a hard "ch"), although the first quarry was opened around 1850 by a Mr. Blance, who discovered slate on his own plantation in 1849. Mr. Blance was born on May 8, 1799, in Savannah, Georgia, of French parents, and moved to the Rockmart area in 1847. One story describes him stumbling over an outcropping of slate as he was tying up his family cow, thereby making his memorable discovery. Another version describes Mr. Blance's discovery as a byproduct of a road-grading project he was doing at the time (maybe the cow was pulling the grader). In any case, he subsequently established the Blanceville Slate Mines and invited the Welsh to step in and lend a hand. No doubt the Welsh were lured by stories of *"the largest, most inexhaustible deposit of slate south of Pennsylvania"* reportedly discovered near the village of Van Wert (an extension of Rockmart). One notable Welshman was Colonel Seaborn Jones, credited as one of the originators of the Rockmart slate industry. Colonel Jones established extensive land holdings and became a major benefactor of the city of Rockmart.

The average value of farmland in Polk County in 1850, when slate was being discovered, was about $10 an acre, and most people earned their living growing cotton, corn, peaches and other fruits, or operating dairy farms. Some worked in cotton mills. That all changed as the slate industry expanded and became a major employer of men in the area.

Although the Civil War interrupted the development of Rockmart's slate industry as the men took up arms and engaged in battle, in 1863 some mining was done to provide roofing for military purposes. The quarries nevertheless remained at a standstill for fifteen years after the war, reopening around 1880. They then produced their greatest yield during the following decade, with 5,000 squares of roofing slate (enough to cover 420 houses) valued at $22,500, being produced in 1894. According to the U. S. Geological Survey, Georgia produced 38,097 squares of roofing slate from 1879 until the beginning of World War I, valued at $165,918, although these figures are thought to be incomplete. It is reported that the total roof slate production of the Rockmart district was over 50,000 squares, enough to cover about 4,200 average size homes.

Roofing slate production dropped off after 1900 and was replaced by the quarrying of slate to make light-weight aggregate. The aggregate, known as "Galite" (a contraction of Georgia lite), was made by heating crushed slate to 2,000 degrees F., which causes it to expand. The resulting "expanded shale aggregate" was added to concrete to reduce the concrete's weight, but not its strength, and is incorporated into such structures as the Golden Gate Bridge and the White House. The Rockmart Slate Corporation quarry in Rockmart, Georgia, was planning to renew the production of ground slate in 1997 to be used as a horticultural mulch material called "Slatescape."

FAIRMOUNT

The slate near Fairmount, thought to be Cambrian in age (600 million years old), is green, similar to the sea green slate of Vermont, but with more color variety, showing shades of blue, buff, brown and red.

Fairmount slate at quarry sites appeared to the author to be quite hard, although not a single Fairmount slate *roof* could be found by the author and a colleague, who spent a couple of days in 1997 searching in vain for slate roofs in the Fairmount area. It is reported in historical accounts that these slates did not "possess the hardness of the best Rockmart slates," and due to the lack of existing roofs, the author had no way to visibly and with certainty determine the longevity of Fairmount slate roofs.

There weren't many local folks who remembered the operating slate quarries either, and those that did only remembered the slate being ground into aggregate. No one was old enough at the time to remember slate ever being quarried for roofing,

although one older farmer did remember a hotel on Slate Mine Road that had Fairmount slate on the roof, but which had burned to the ground. It appeared that, although Fairmount slate was once quarried for roofing, it was only done in small quantities and the roofs were eventually lost to neglect or ruin. There are some roofs in Atlanta, and presumably in other areas of Georgia, with green slate roofs which may have originated in Fairmount, but the origin of the roofs is pure speculation due to the similarity in appearance to green and sea green roofs of Vermont.

The green slate belt was first prospected by Mr. G. W. Davis, a slate worker from Pennsylvania employed in the Rockmart district. The exact date is uncertain. The belt extends through Gordon, Bartow and Murray Counties. Huge water-filled pits remain hidden among the forested hills about four miles south of Fairmount, bearing evidence of the old slate industry. From these pits emanate large tunnels for a total distance underground of 12 miles, according to quarry owner Ray Sullivan (see photos, page 96).

▲ Slate Mine Road near Fairmount, GA, is reportedly a site of old Georgia slate mines, but only traces now remain there. The huge quarry holes pictured on the next page are several miles away, on the other side of town.

Photo by author.

▲ Abandoned slate quarry pit near Fairmount, GA. Note person standing near top, just right of center (circled).
▼ These huge tunnels extend 12 miles underground, now accessible only by boat.

Photos by author.

## Chapter Nine

MAINE AND CANADA

The slate from the Monson region of Maine is black, smooth and will last for centuries on a roof. The people of that area will say what is commonly heard in all slate regions in the US, that their slate is "the best slate in the world." Unfortunately, at the time of this writing, Monson slate is no longer being quarried for roofing. Quarries stopped producing roofing slate in Monson for the same reason they stopped elsewhere throughout the world: increased cost, market fluctuations and decreased demand.

Older Monson slate roofs are, nevertheless, excellent roofs very much worthy of restoration efforts. People who put their money or efforts into restoring their Monson roofs will benefit greatly in the long run, both financially, by keeping a permanent roof over their heads, and aesthetically, by preserving a traditional, natural and beautiful stone roof.

The restoration of Monson roofs requires ample supplies of replacement slates, and since the slates are no longer being manufactured, old roofs that are being removed (as on buildings to be demolished) should be carefully salvaged and stored until needed. This careful recycling of Monson slate will ensure that many of the old roofs will be kept intact, as one of the most necessary ingredients in slate roof restoration is replacement slates that match the slate on the roof. Recycling of roof slate is discussed in greater detail in Chapter 16.

Monson is a quiet, quaint little village, founded in 1822, just two years after Maine gained statehood. Nestled on the shore of Lake Hebron, near Monson Pond, it is just a few miles south of the beautiful Moosehead Lake in north central Maine.

Life changed for the residents of Monson in the summer of 1870, when William Griffith Jones, a Welshman, discovered an outcropping of slate while riding his horse. Welshmen at that time were considered the foremost authorities on excavating and cutting granite, marble and slate, and apparently Mr. Jones understood that slate had commercial value, so he bought the slate-bearing land from a hotel keeper and started the development of his Eureka slate quarry within days.

Mr. Jones discovered only the tip of an ice-berg when he saw that first slate outcropping, for it was only a tiny section of one of Monson's two immense slate belts. The village belt was 200 feet in width, and became one of two belts known as the Monson-Maine Co. vein and the Portland-Monson Co. vein, both of which ran east to west.

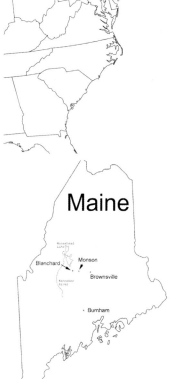

Soon after Jones' discovery, slate quarries sprung up all over the area. The first quarry after Jones' was called the Hebron Pond Quarry, started by a Mr. Chapin. It suffered an unfortunate accident in 1881, when two men were injured and a third man killed. In 1872, a Mr. Norris opened a quarry and installed steam power, a new-fangled addition to the Monson area quarries at that time, but he didn't have much luck with it — his buildings burned to the ground five years later.

In the meantime, many people of English, Irish and Scottish descent immigrated to the Monson area from New Brunswick and Nova Scotia to develop and work the quarries. William Thomas, Commissioner of Immigration, brought Swedes to the area in 1870, and later Finnish people arrived to help work the slate mines.

In 1873, Fred Jackson founded the Cove Quarry, which utilized some of Mr. Norris' machinery, thereby sparing its loss in the later fire. The same year, Dexter and Portland Companies founded

▲ Monson, Maine, slate roofs with the characteristic smooth sheen of the legendary Monson slate. Photos by author.

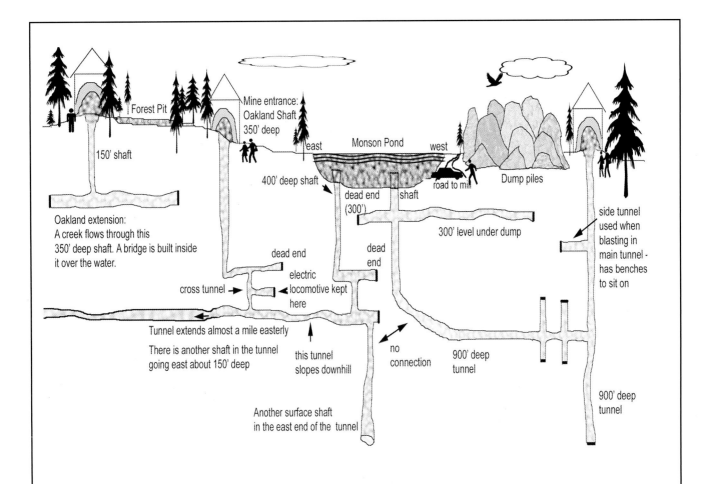

A Monson, Maine Slate Mine
Diagram of Mine Showing Shafts and Details
From a crude, hand-drawn map by Arthur Day, slate quarry worker for Monson, Maine Slate Co.
drawn at about the age of 80, from memory (special thanks to Berwin Storer).

Arthur Day recorded the layout of the mine where he worked by a crude sketch on a piece of paper, from memory, before he died. The only notes associated with his mine diagram stated, "The west end tunnels do not connect with the east end tunnels. There is no heavy flow of water in any of these tunnels and shafts, just a few trickling springs. This is the reason they never found any of the dye they put in the dump. The water in the pits neutralizes trickling spring water so no flowage occurs."

the Dirigo Quarry, followed by the Forest Quarry (opened by people named Forest) in 1874. Then the Underground Quarry, operated by John Folsom, sprung up in 1875; then John Tripp's Oakland Quarry in 1877; the Monson-Maine Quarry in 1880 (by Allen Williams); the Kineo Quarry and the Burmah Quarry in 1882; the West Monson Quarry (W. M. Jones) in 1895 (the buildings burned January 16, 1897); the Matthews Quarry in 1902; the Portland-Monson Quarry in 1906; the Farm Quarry and the Wilkins Quarry in 1910; and the Eighteen Quarry in 1919 (abandoned in 1922).

Some of these smaller quarries were eventually bought up by the larger ones, and by 1882, 12 years after Mr. Jones discovered the first slate outcropping, $150,000 had been invested in the operation of five quarries in the Monson area. Six hundred and twenty railcar loads of roofing slate, equivalent to 2,500,000 square feet of roofing, were being manufactured annually, requiring 200 men and 35 horses, as well as twelve teamsters to haul the slate to the railroad. Two thousand five hundred cords of wood were used each year, and about $6,000 was being generated annually in local wages by the quarries. By 1885, a newspaper was being published out of Bangor, Maine, called "The Monson Slate."

By 1922, a scant forty years later, there were only three quarries in operation — the Portland-Monson, the Monson-Maine and the General Slate Co. The Monson-Maine Quarry outlived the General Slate Co., but was eventually abandoned in 1943. The Portland-Monson Slate Company was run by the Coleman family until April 1, 1965, eventually to change hands to the Tatko family, who renamed the company the Sheldon Slate Products Co. in the early 1990s. Sheldon Slate Products of Monson, Maine, the only slate quarry in operation in that area today, produces countertops, floor tiles, flagstones etc., but not roofing slate.

On the opposite page is a map of the Monson-Maine Co. mine shafts and tunnels as recorded by Arthur Day, a quarryman who drew the map for posterity in the 1980s at about age 80. He indicates that some of the tunnels descended to 900 feet, and that an electric locomotive eventually came into use in the tunnels, presumably to move the slate along tracks laid on the tunnel floor. He also shows where a side tunnel was carved from the main tunnel to allow men a safe place to duck, complete with benches to sit on, when blasting took place in the main tunnel. Other quarrymen of the area tell of mine tunnels 1,100 feet deep, and of a shaft under Monson Pond, two of which are evident in Mr. Day's diagram.

In addition to the quarries in the Monson area, the neighboring town of Brownsville, twenty miles to the east, had quarries that produced much of the famous roof slate known as Monson slate. Also, a small town called Blanchard a few miles to the west of Monson quarried slate, as did Burnham, about 45 miles south of Monson.

The quarry business was not without its dangers, as we have seen in previous chapters, and Monson was not immune to them. Numerous accidents took place over the years in addition to the loss of buildings to fire. Not only were workers killed or injured in the mines, but people drowned in abandoned quarry pits that filled with water, and others died after falling over the cliff-like sides of the open slate pits.

Perhaps the most well-known use of Monson slate was for the Kennedy Memorial stone marking

▲ A slate splitter at Trinity Slate in Newfoundland using a pneumatic splitter. (Right) The purple and green slates that are characteristic of Newfoundland grace an older home. These slates have a reputation for being quite durable. Photo at left courtesy of Newfoundland Slate, Inc.; photo at right by author.

The Glendyne Slate Quarry in Quebec, Canada, produces the sought after "North Country Black" roofing slate.

Photos courtesy of Glendyne Slate Quarry.

CANADIAN SLATE QUARRYING AREAS

One of the first known Canadian slate quarries opened in **Rimouski** in 1728.

Glendyne Slate Quarry at **St.-Marc-du-Lac-Long** is an internationally known source of high quality black roofing slate, known by the trade name of "North Country Slate" in the U.S.

The durable purple and green slates of Newfoundland are quarried on the shores of **Burgoyne's Cove**, off Trinity Bay, on the remote east coast of Newfoundland.

The **Kingsbury, Richmond, Danville and Melbourne** area of Quebec once quarried a black slate similar to that of eastern Pennsylvania. A slate museum is now located in Melbourne.

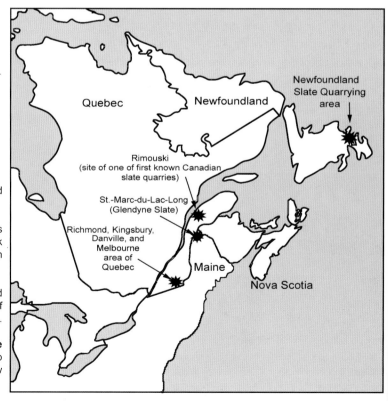

President John F. Kennedy's grave at Arlington National Cemetery. Saint Patrick's Cathedral in New York City also has a Monson slate roof.

(Much of the historical information about Monson's quarries comes from *"The History of Monson, Maine - 1822 - 1972,"* published by the town of Monson, Maine).

CANADA

"Newfoundland possesses one of the great roofing slate deposits in the world."
Professor G. B. Walcott, Director of the Smithsonian Institute (late 1800s)

One of the first known slate quarries to be opened in Canada was near Rimouski on the south bank of the St. Lawrence River. Opened in 1728, the slate was found to be of a poor quality and the quarry was abandoned in 1733. Another formerly productive Canadian slate district is located about 60 miles north of Newport, Vermont, near Richmond and Danville, Quebec. Here, slate was first mined in 1854 at the Steele Quarry (later called the Bedard Quarry), about 3 kilometers southeast of Richmond. The first profitable quarry opened in 1860, less than 2 kilometers northeast of the village of Kingsbury, and was known as the Melbourne Quarry (also called Walton Slate). A square of slate at that time sold for $3.50 Canadian, while the quarry wage 30 years later still only hovered around eighteen to twenty-three cents an hour. The slate from this region was black and similar to the slates of eastern Pennsylvania.

In 1878, the Melbourne Quarry shut down, and the New Rockland Quarry, which began in 1868, then became the only important slate quarry in the region, monopolizing 80% of the Canadian roofing slate market. The New Rockland Quarry employed 150-300 workers in the late 1800s, many of whom were Welsh and Cornish, and who were localized in the village of New Rockland. The quarry continued until 1924, after which the inhabitants of the village moved away and now no buildings remain there. By the time the New Rockland Quarry was in its waning years, it produced no more than a third of the slate consumed in Canada, while much was instead imported from the United States.

Today, one may visit Melbourne's "Slate Interpretation Center," a slate-roofed museum at 5 Belmont Street, Melbourne, Quebec, just across the river from Richmond via an old steel bridge (10 miles from Kingsbury, or about 60 miles north of the Vermont border). The center was incorporated in 1992, and is dedicated to the preservation of slate roofs in the Kingsbury-Melbourne-Richmond area of Quebec.

Slate was discovered in the mid-1800s on the eastern seacoast of Canada in Newfoundland, the slate being of Cambrian age, meaning it had been highly metamorphosed into a very hard and durable stone, and is therefore categorized as S1 (hard) slate. This type of slate can be expected to last hundreds of years on a roof, with proper maintenance. It is identical to the slate of Wales, particularly from the Penrhyn, Dinorwic, and Dorothea Quarry areas, yielding shades of purple and green.

The first quarry in the Newfoundland region was opened by the Welsh in the mid 1850s and most of the slate quarried at that time was used locally or shipped overseas. The quarry shut down in the early 1900s, as did many others throughout North America, due to economic hardships and market fluctuations. However, the Newfoundland quarry *reopened* in the mid-1980s, and in 1991 the ownership changed and the operation expanded into a new, large mill building with new equipment. Roofing slate was subsequently produced there under the trademarked brand name "Trinity Slate," using fully automated equipment including diamond wire machines in the quarry, laser guided saw systems for cutting blocks, pneumatic hammers for splitting the individual roof slates, and pneumatic trimmers for trimming the slates to their final size.

In the late 1990s, Newfoundland claimed to have the largest production of natural roofing slate in North America with distribution in the UK, Europe, Japan, Australia, New Zealand, and North America. Newfoundland Slate, Inc. with its quarry site at Burgoyne's Cove, off Trinity Bay, laid claim to being "the only slate quarrier in North America with the prestigious ISO 9002 Certification," an international standard for quality assurance. However, at the end of the 20th century, Newfoundland slate was no longer in production, and with the vagaries of slate manufacturing such as they are, the reader may have to do some independent homework to keep up with the changes in the industry. The latest indication is that, in 2003, roofing slate was again available from the Newfoundland quarries.

Roof slate is still being quarried at the Glendyne Quarry at St-Marc du Lac Long in Quebec, which is about a hundred miles north of Monson, Maine. This quarry began operation in 1995 and yields a slate similar in appearance to Monson slate.

## Chapter Ten

INTERNATIONAL SLATE

Germany, Italy, France, Spain, China, Scotland and the U.K.

This is by no means meant to be an exhaustive description of international slate. There have been and still are slate quarries in numerous countries not elaborated upon here, such as South Africa, Australia, Switzerland, India, Belgium, Brazil, the Czech Republic, Norway and others. Consider this as only an *introduction* to the slate of the world, and please understand that any omissions and deficiencies in the information presented herein are simply a reflection of the author's limited knowledge.

GERMANY

German slate roofing expertise is renowned across Europe and around the world. German slating styles and skills have been perfected over hundreds of years, if not thousands. Archeological evidence in Germany indicates that roofing slate has not only been used there since 300 BC, but a unique style of slating known as "Altdeutsche" (Old German) has been carried on since that ancient time. Germany is the source of many fine slate roofing tools made by such well-known firms as the Freund Company and the Carl Kammerling (CK) Company. The world's second largest roofing slate distributor, Rathscheck, also resides in Germany, in the beautiful Mosel River Valley.

Germany's characteristic black roof slate is still mined in Germany at four locations — the Mayen area in the Mosel Valley of western Germany (known as "Mosel Schieffer," or Mosel Slate); the Bad Fredeburg area in the pine-covered mountains of north central Germany (known as Magog Slate); the rural hills south of the Mosel Valley in the Bundenbach area; and Lehesten, where Schmiedebach (Thuringia) slate is mined in an open quarry.

Germany also has at least four roofing schools, one in the Bad Fredeburg area, one in the Mayen area, one in Lehesten (Thuringia) and one in

Schlema in the Erzgebirge in the Saxony Ore Mountains (See Chapter 14). These schools, funded by tax dollars, teach the various old traditional slate roofing skills and techniques as well as tile roofing, sheet metal work, artificial slate and even low-slope roofing applications. It seems that in Germany, like in the United States, roofing is almost strictly a man's trade (to the relief of most women, no doubt), as no females were seen studying at either of these schools during the author's visits. The schools were

SLATE PRODUCTION IN EUROPE

Slate is also produced in Norway (see Industry Resource Guide in the back of this book).

SLATE IMPORTS TO THE U.S. September 1997 (customs values in dollars; all slate products)	
Italy	896,626
India	729,295
China	526,890
Brazil	377,432
UK	193,075
Canada	156,122
Spain	130,000
South Africa	67,211
Portugal	12,764
Mexico	11,576
Philippines	7,149

SLATE EXPORTS FROM THE U.S. September 1997 (customs values in dollars; all slate products)	
Bahamas	199,419
Belize	170,172
Canada	155,389
Japan	80,065
Venezuela	55,351
Cayman Islands	44,317
Mexico	30,455
South Africa	25,939
Ireland	9,636

Source: Stone World, January 1998; originally from U.S. Dept. of Commerce

Opposite page: Building at the Castle complex in Edinburgh, Scotland, built in 1630, with graduated Scottish slate roof.
Photo by author.

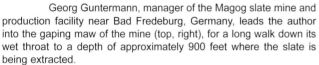

Georg Guntermann, manager of the Magog slate mine and production facility near Bad Fredeburg, Germany, leads the author into the gaping maw of the mine (top, right), for a long walk down its wet throat to a depth of approximately 900 feet where the slate is being extracted.

Rathscheck Company in Mayen, Germany, maintains a stock of roofing slates from around the world (right), here inspected by Dr. Uwe Dittmar. Dr. Dittmar examines Rathscheck's own slate (below), deep in their Katzenberg Mine near Mayen, a mine with 27 km of tunnels and a record of slate production since 1362. See Chapter 14 for more information about German slating styles and techniques.

Photos by author.

teeming with eager, hard-working young men, however, who were being trained in a respected European trade. Master level training was also offered for those roofers who had already gone into the roofing industry and gained sufficient experience on the job.

Perhaps what most sets the Germans apart from the rest of the world in the realm of slate roofing is the unique style of slating seen everywhere on Germany's charming Tudor-style dwellings. Altdeutsche and its very similar modern cousin, Schuppen (fish-scale) slating is unheard of in the United States, although such roofs can be seen throughout Europe wherever German slaters have been employed to display their incredible craft. The slating style is unique in that the slate is in the shape of a parallelogram rather than a rectangle, and laid on the roof with one side and the top overlapped in such a manner that only one corner is exposed. This corner is always cut into a curved shape, usually by hand with a slater's hammer and stake on the job site (in the Altdeutsche method), or manufactured as such in the Schuppen method. Each piece of roofing slate is nailed in place with three nails onto a solid board roof deck rather than the lath roofs more common in the UK and other parts of Europe. The result is a totally unique style that enables the roofer to dispense with much of the exposed flashing metal common to American roofs and instead simply wrap the slates over the valleys and dormers in a smooth, sweeping style that is both artistic and functional. Needless to say, this German slating technique requires specific training and time to master, which is perhaps why it is not more widespread throughout the world when compared to standard lap rectangular slating — a much easier, simpler, but less exotic system of roofing.

With 2,300 years of tradition behind them dating back to the Romans, the Germans have kept alive and perfected a unique style of slating that is incredible and beautiful. Examples of the fine art of German slating can be found in Chapter 14.

CHINA
Special thanks to John Wright, IEL International, Hong Kong (www.iel.com.hk/slate); and to John Xue (johnchina@china.com).

Chinese roofing slates have become a common sight in American and European roofing slate markets, as China continues to grow as a major producer and world supplier of roofing slates. Although Chinese slate is available in many colors ranging from black to red, yellow, green, gray, blue and purple, one of the most widely used and respected Chinese slates is a black variety, which can be very hard and can perform quite well under ASTM testing. As in all slate-producing areas of the world, a range of slate qualities can be found in China. Some of the slate quarried there has a record of longevity observable on the older buildings still standing there today. Other quarries are so new that no track record for their slate has yet been established. Quality control can also vary significantly from producer to producer. It is always wise to buy slates from a known source (quarry) that has a good business reputation, and this is especially true when buying slates from abroad; otherwise what you order and what you get may be two entirely different things! For best results when buying in quantity, it is important either to visit the slate mills yourself or have a reliable person available to you in China who knows what's going on.

Commercial development of Chinese roofing slate for western markets only began in the 1980s. Until then, Chinese slates had been used mostly in remote villages. As the Chinese economy develops, State Owned Enterprises (through which most business was previously transacted in China) are allowing more private companies to use their quarries or lease the mineral rights. This has caused more Chinese slate to become available on the world market. Some new slate deposits have been discovered accidentally during construction of large infrastructure projects in China.

There are numerous sources of slate in China, including the following provinces: Beijing, Hebei, Hubei, Jiangxi, Shaanxi, Yunnan and Sichuan.

Currently, 50% of the export market originates from Hebei, which is the province that includes Beijing. Beijing slate from Men Tou Gou has reportedly been used for roofing for over 500

▲ A Chinese black-slate quarry in the North of Jiangxi province in full operation is busy with workers, one of whom splits individual roof slates from a block (below).

CHINESE SLATE

Back-breaking work greets Chinese slate quarry workers (top left) as they move an extremely heavy piece of stone. Slate is being punched for nail holes (top right), trimmed (above) and cut from blocks (above right) at Chinese quarries.

The finished product awaits shipment (right).

All China photos provided courtesy of
John Wright, IEL International, Hong Kong
(www.iel.com.hk/slate).

years. About 80% of the production is now exported for flooring, walls, roofing, etc. The color varies from black to black/gray to purple, green, jade green, etc. The green slate from Fang Shan is very popular, especially for use in billiards tables.

Zhu Xi and Zhu Shan areas in the Hubei Province are supposed to have the largest reserve of Cambrian, Ordovician and Silurian slates. Colors include black, gray, bean green and green. Unfortunately, the overburden is quite large, averaging 50 meters.

The Shaanxi slate vein runs for over 200 km east-west, averaging 50 km wide. Most of the slate was formed during the Ordovician and Cambrian periods, 400 to 600 million years ago. Here you will find the beautiful, high quality black slate, which comprises 50% of the national production capacity. People from all over the world have come to visit the area for sourcing, purchasing and investment in slate.

Many Chinese quarries are served by poor transport systems which makes getting the material to port costly. Exporting quarries are often located close to a main navigable river such as the Yangtze or Yellow River or near rail lines. During the Chinese New Year in late January or February (depending on the lunar cycle), some quarries may close for a month, meaning production can come to a standstill during this period.

China is almost exactly the same size as the U.S. with similar climate differences. The best time to visit is in the spring (March/April) or autumn (Sept/Oct). Winters can be very cold and summers extremely hot. Most major cities have air transportation and the rail system is efficient, although even 1st class can be basic. Roads are variable, ranging from highways to muddy tracks. Consequently, it can take up to 5 days to reach some quarries. The two main transport hubs are Hong Kong and Beijing. Any business person planning to visit China will either need to speak Mandarin or have an interpreter. English is widely spoken in Beijing, Xian, Shanghai and elsewhere in tourist areas, but as soon as you get off the beaten path, you will have great difficulty communicating without some knowledge of the language.

ITALY
"Here is where slate was born."

A visit to the Italian slate region will take the traveler over rugged, mountainous terrain in a jungle-like coastal setting with very winding, narrow roads. Ancient villages clinging to steep hillsides or nestled in misty valleys are dotted with pastel stucco houses. The slate or tile roofed abodes overlook spectacular views of the Gulf of Tegullio where palm trees dot the coastline.

The author had the pleasure of being introduced to Italian slate by Vittorio Terzo Arata (founder and president) and Sophie Boulay (export manager) of Euroslate in Orero, Italy. Vittorio is a wiry, fit, 70-year-old, with dark glasses and a bristly, gray crew cut, who speaks English fairly well. An animated, energetic man for his age, Vittorio was dressed in work boots with clothes that were obviously being used for hands-on work in the mill. You couldn't have picked him out from any other worker by his clothes, despite the fact that he owned and operated three slate factories. Vittorio is not a suit-and-tie type — he apparently gained his success by working side by side with the others in his company. His grandfather, Vittorio I, started the first slate mill there in 1880.

Euroslate's primary product is billiards slates, with the claim to being the largest billiards slate factory in the world, producing 100 billiards slate sets per day, pocket holes and all, ready to ship out. It's all done on an automated factory line designed by Vittorio. They also hand split slates for that natural cleaved look for table tops and flooring, and they do make roofing slates. The ones in stock at the time were with all sawn edges. Their roofing slates tend to be larger in size, 1 cm thick

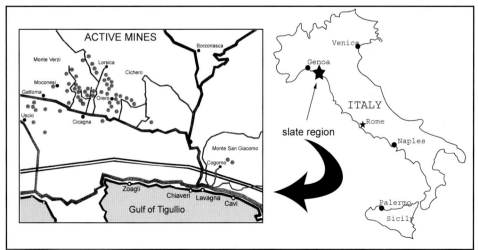

(about 1/2 inch), and a typical size is 57x13 cm (about 12x24 inches), with no nail holes. Vittorio insists that hooking them in place with stainless steel hooks is the best installation method, and they even have the machine that makes the hooks from a roll of wire. Perhaps with this large, thick slate, a hook would be preferred.

The slate here is black, mined underground, and easily split into large sheets. It reminds me of the slate in eastern Pennsylvania (USA). However, it has a large amount of calcium in it and much of it turns white upon exposure to the weather. Many of the older roofs in the area are made of this slate, which seems to hold up pretty well, especially when used with a half-inch thickness. Some of these older roofs are fairly low slope, perhaps 4:12, but installed with a head lap that reaches down to the top of the third slate below, a technique known as "triple slating" or "triple covering" (see Chapter 14).

There are reportedly 50 or more slate mines operating in this area of Italy, called the Val Fontanabuona ("valley of the good spring"), just east of Genoa, plus another 10 or 12 on the French border at the Arma di Taggia exit off the highway. The word for slate in Italian is "*ardesia.*"

According to Vittorio, the Romans found that slate was being used in this area 20 centuries ago. They called slate "tegmen," and the name of the Gulf of Tegullio is derived from this word. He also reports that the history of slate in this region goes back to the bronze age, 4,000 years ago, according to archeological digs in the village of Chiavari, nearby. Therefore, says Vittorio, "*Here is where slate was born.*" Italians have made billiards slates for three centuries.

Vittorio's mine is underground with access through a tunnel that begins at the very top of a mountain reached via a winding, narrow, overgrown road. You can drive right down into the throat of the mine and see the men extracting the slate blocks deep inside. The blocks are covered with a layer of mud and then a sheet of plastic before being taken out of the mine and to the mill, in order to preserve the natural moisture and therefore the best splitting qualities. The mine is characterized by huge, solid

Vittorio Terzo Arata of Euroslate, and Orion Jenkins, the author's son, gaze across the rugged terrain of northern Italy (below), where the distinctive black slate is quarried. Sophie Boulay and a worker at Euroslate inspect the roofing slate before palleting (center). Italian slates are known for their "weathering" characteristics which can create a beautiful roof (bottom).

Photos by author.

slate pillars that are left in place in order to support the roof, each about 30 meters square. For more information about Italian slate, see Chapter 14.

FRANCE

Angers is a city in northwest-ern France, capi-tal of the depart-ment of Maine-et-Loire, located 191 miles southwest of Paris at the confluence of the river Loire and the Maine. With a reputation for being one of France's most pleasant cities (pop. 136,038 in 1982), Angers is known for its wine and the famous Cointreau liqueur — and for its slate.

France is the world's largest *consumer* of roof-ing slate, utilizing not only their own locally mined slate, but also imported slates, especially the black Spanish slates, which are similar to French slates (Spain is currently the world's largest *producer* of roofing slates). The French black slate (also described as "blue-black") is mined near Angers at the Fresnais Mine in Trelaze in the beautiful Loire River Valley. This "Grade A" material has been used as roofing for over 500 years, and today has a repu-tation for being very consistent in thickness. A tonne of this slate equals approximately 3 squares (300 square feet of roof coverage). There are five or six thousand companies doing slate roof installation in France today, a country with a population of about 60 million people. Angers is known as "Black Angers" or the "Black City" for the black slate roofs that cover the entire area. There is a roofing school in Angers known as the Ecole Superieure Couverture at Rue Darwin, Belle-beille.

The evolution of the slate industry in France parallels that of the UK, with the most progress accomplished after the mid 1800s. Viollet-le-Duc's Dictionnaire contains accounts of slate roofing in Angers from the eleventh century, and slate became commonplace in France by 1200. By the end of the 1800s, according to Planat's Encyclopedie, French quarries were located in Angers and Poligny, where gray and dark blue slates were mined.

French slate quarries were some of the most

Opposite page: Castle in Angers, France with local slate on the roof. Photo by author.

dangerous in the mining industries. In 1892, six men out of 794 employed in the Angers mine were killed, or a proportion of 7.56 per 1,000, while in the same year the proportion in coal mines was 1.18 in 1,000, and, in all the mines and quarries of France, 1.09 in 1,000.

A Scientific American Supplement of 1894 describes the making of roof slate in Angers in this manner: *"The working of the slate for roofing must be done in situ before the material has dried, and, in losing its water, has lost also its fissility. The block is first slit into thick plates by means of a chisel. Then the slitter, seated under a wind screen, his feet incased in thick sabots and his legs wrapped with rags to prevent him from wound-ing himself, places the fragment of schist between his legs, and, provided with a very thin chisel and a mallet, strikes the edge in order to detach sheets from it. All that has to be done then is to give this sheet its proper form and size. This is done by means of a plane iron–very heavy iron knife, whose extremity is shored into a ring and which turns down against the edge of a block. It is estimated that an outside laborer earns from fifty-five to seventy cents a day, and a pitman from sixty to eighty cents."*

"In 1893, the slate quarries of Maine-et-Loire gave employment to 1,034 inside and 2,185 outside labor-

Partial reproduction of an old article about the Angers slate mines.

Application of the Electric Light at the Angers (France) Slate Quarries.

In 1882 the Société des Grands Caneaux undertook the first subterranean exploitation of the important slate deposits of Marne-et Loire. After working these over fifty years, the quarries, with their galle-ries running in every direction, have reached an im-mense depth. The only circumstance that has been found relatively disadvantageous in these under-ground works is that of the darkness which prevails therein, and which renders the work of the laborers less advantageous and a surveillance of the walls quite difficult. This inconvenience was remedied at first by the substitution of gaslight for the miners' smoky lamps formerly used, and more recently by the application of the electric light, which, indeed, appears to completely solve the problem.

The first gas works were erected in 1847 at the Fresnais quarry, for lighting chamber No. 1, and consisted of one cast-iron retort and one gasometer. When the possibility of sending the gas down to a depth of 650 feet was discovered, and when the supe-riority of this mode of lighting the subterranean chambers was recognized, the works in the course of the year 1855 were enlarged and equipped with two

(not continued)

ers — say altogether 3,219, of whom 1,511 were slitters. In 1891, a total of 2,749 laborers were divided into 2,121 for the center of Angers (Trelaze), 109 for the Foret and 135 for Misengrain. During the year these 2,749 laborers produced 159,820,047 slates, worth $710,650."

Slate is still actively mined in the Angers area of France today.

THE UK AND IRELAND

IRELAND

Roof slate was quarried in Ireland as late as the 1930s. Slate mined in Ireland included a fine slate similar to Welsh slate near Newtownards, Co. Down; a coarse and heavy slate near Kilrush in Co. Clare, and slate in Killaloe, also in Co. Clare. A blue-black slate was quarried in Valentia Island (Co. Kerry); a softer, greenish gray slate near Ross, Wexford and Wicklow in Waterford County; a low-quality clay slate near Westport was wrestled from the ground in County Mayo; and more slate was removed from the earth at Clonkilty in Co. Cork. Extensive quarry operations also took place at Curraghbally, Co. Clare and Portroe, Co. Tipperary where 10,000 tons per year were being produced in the 1840s. Bluish-gray slates were also quarried near Ashford Bridge, Co. Wicklow.

ENGLAND AND WALES

For an in-depth look at Welsh slate history, please refer to Chapter 4. The UK still has 36 operational, natural slate quarries, according to a year 2000 publication by Historic Scotland. These are located in three main slate producing regions of the UK including Wales, Cornwall and the Lake District. The largest of these are in Wales and are operated by Alfred MacAlpine Ltd., in north Wales at Bethesda. Other Welsh quarries are located at Blaenau Ffestiniog, including Cwt-y-Bugail, Ffestiniog, Gloddfa Ganol Slate Mine and Nantile.

England's main quarries outside Wales are in Cornwall (primarily the Delabole Quarry) and in the Lake District (Burlington and Kirkstone quarries). Burlington slate from Kirby-in-Furness, Cumbria (the Lake District) of northern England, is either a "Blue-Grey" slate or a "Westmoreland Green" slate. These popular slates have been used for over 300 years.

Also in the Lake District, the Buttermere and Westmorland Green Slate Company Limited is now England's only underground slate mine (as opposed to an open quarry). Westmorland Green Slate is taken from beneath the hills in eleven miles of tunnels. Quarrying was taking place here by the 1750s, and from 1833, development expanded with underground mines as well as open quarries. The Buttermere Green Slate Company was established in 1879, but slate mining and quarrying ceased in 1986, then restarted in February of 1997.

British slate quarries reached their peak in the late 1870s with an annual production of half a million tons, 90% of which came from the 670 Welsh quarries operating at the time. By the year 2000, nine Welsh quarries were producing 40,000 tons annually.

Despite the UK's long history of slate production and slate roofing, today natural slate accounts for only 7% of total roofing products used there, which includes locally quarried material and recycled slates as well as imported slates. Concrete tiles now make up the bulk of the roofing at 72%, artificial slates make up 14%, and clay tiles make up 6%. The balance (1%) is made up of metal roofing or other shingles.

SCOTLAND

Scotland has its own slate roofing history dating back to the mid 1500s. Production levels rose to 25 to 30 million slates per year at its peak at the end of the 19th century, employing thereby 1,000 to 1,500 men. By 1937, this had declined to 10 million slates and 370 men. By World War II, roofing slate production was negligible, and by the mid-1960s slate production in Scotland had come to an end.

Scottish slates tend to be rough, course, thick, and extremely durable, perhaps owing to the great geological age of the slates from the west coast of Scotland — at 650 million years, the oldest in the UK. Scottish slates are traditionally center nailed (single nail at the top center of the slate) onto a fully

Slate Producing Areas of the UK (★)

boarded roof deck, as described in Chapter 14. Most of the slate found in Scotland comes from the Grampian Highlands, bordered on the south by the Highland Boundary Fault, or Highland border.

The four main slate producing localities were:

1) Ballachulish, which produced a dark gray/blue slate with a distinctive, well-marked grain referred to as lineation;

2) Easedale Islands, producing a dark blue/gray slate occasionally showing a brown weathering, with a marked grain;

3) The Highland Boundary Fault, producing a variety of colors including blue, gray, green, purple and mottled, generally lighter than the slates of the west Highlands. These slates have the finest texture in Scotland, almost entirely without grain.

4) The Banff-Aberdeen Slate Belts in the northeast of Scotland, including the Macduff quarries, producing a rough, course, dark blue/gray slate without grain.

The Scottish government has shown a great interest in the historic preservation of its nation's antiquities and has published

Delabole slate roof in Cornwall, England (top, right). Welsh slate on a cottage in western Wales (right). Lake District slate on a barn in northwest England (below).

All photos by Dave Starkie.

▲ Easedale slate on a Scottish house built in 1785.
▼ Working woman at the Franvisa slate mill, northwest Spain.

Photos by author.

comprehensive books on Scottish slate quarries and Scottish slate (see reference section). These are available from the Publications Department, Historic Scotland, Longmore House, Salisbury Place, Edinburgh EH9 1SH, Scotland (www.historic-scotland.gov.uk).

SPAIN

Last, but not least, Spain is the big boy on the world's roof slate production block, producing more roofing slate than any other country in the world, with slate quarries scattered across the nation and particularly concentrated in the northwest in the provinces of Orense and Leon. The most common Spanish slate seen in the U.S. is the smooth-grained, jet black variety that is also prized throughout Europe.

Spanish slate had begun to develop a questionable reputation in the U.S. at one time, due to imports of poor quality materials that were beginning to give Spanish slate a bad name (a similar situation occurred with Chinese slates too, by the way.) Apparently, American importers were just buying the cheapest slate they could get without researching the quarries of origin or the characteristics of the slate. As a result, they were receiving slates with high pyrite levels that would develop unsightly red rust spots in only a year, slates that were too thin to nail onto a roof, and/or slates with no nail holes at all. Many roofing contractors in the United States do not realize that slates produced for European markets are quite different from those preferred in the U.S. market. For example, European (especially French and German) slates tend to be thinner than American slates, and European slates are frequently produced without nail holes as Europeans often don't nail slates — they install them with slate hooks instead. Things have changed somewhat over the years, however, and now a high degree of quality control at the Spanish end and reputable suppliers at the American end have introduced high quality Spanish slates to the American market.

Some Spanish quarries produce slates particularly *for* the American market. These slates tend to be thicker than the European variety and they're punched with nail holes for a 3" or 4" headlap. There are quite a few companies marketing roof slate in Spain today, some of which are listed in the Industry Resource Guide at the back of this book.

SPAIN

slate producing
provinces

Province	Slate Produced
Lugo	gray, green, multicolor
Leon	black, gray
Guipuzcoa	gray
Barcelona	multicolor
Segovia	gray, multicolor
Badajoz	black
Zamora	black
Orense	black, gray
La Coruna	gray

Source: Piedras Naturales, Anuario 1997

▲ The rugged mountains of northwestern Spain yield a high quality black slate produced by (among others) Franvisa slate company (Pizarras Franvisa). Franvisa's quarry is at the top right corner of the photo on the mountainside. Pilar Cubelos Martin (left) and Francisco Vime Losada, of Franvisa, inspect the pallets of roofing slate awaiting export. Top: A typical Spanish slate mill.

Photos by author.

THE SLATE ROOF BIBLE
SECOND EDITION
Part II

Installing Slate Roofs

Bats were rudely evicted, live wasp nests were destroyed, and feet were baked — all in a summer morning's work on a slate roof. At 95 degrees Fahrenheit in the shade, roof soot clings to sweat as evidenced by three slaters taking a merciful lunch break.

Photo by author.

Foreword – Part II

WHAT'S IT LIKE TO WORK ON AN OLD SLATE ROOF?

A friend once asked what it was like to work on slate roofs. He worked in an office and was thinking about quitting his job and was even considering taking up slate roofing as a trade. I was certain that he would immediately change his mind if he had any idea of what slaters go through on an average summer day, so I decided to describe the work in terms he would probably understand:

Imagine you're in your office one fine summer day, sitting at your desk, comfortably typing away at your computer, listening to the hum of the air conditioner while your wonderful co-worker graciously brings you a freshly brewed cup of coffee.

Then the air conditioner dies. You continue typing until you realize that not only has the air conditioner stopped, but someone has turned *on* the furnace! The temperature rises, passing 100° F while sweat beads on your brow. Then you become aware that now someone has turned on a *humidifier*. The heat is so intense and muggy that soon you start ripping off clothing, flinging your suit coat and tie to the floor, along with your shirt. Sweat is now pouring into your eyes, blinding you, so you grab a rag and wrap it around your forehead. Your co-worker informs you that this is now your new work environment, and you must do your day-to-day tasks under these sweltering conditions, whether you like it or not. She hands you a jug of drinking water, which you desperately chug.

Stunned by this new development, you pull open your desk drawer, hoping to find something to fan yourself with, when a cloud of black dust flies in your face covering your sweaty skin. You now look like a coal miner as you take a tissue and wipe the dirt granules out of your eyes so you can continue to work. You pull your drawer open further and a swarm of yellow jackets flies out mad as hell as they go for your face. Some lodge in your hair as you frantically swat at them. In a panic, you fall backward in your chair, scrambling on the floor to get away.

You run for the closet to get a can of hornet spray, and as you poke around looking for the spray, pounding the buzzing bees in your hair, a bat flies out of nowhere, startling you. You duck as the bat skims past your head chirping insanely, then you watch it assume a circling pattern in the office as it flies around looking for a place to land. Fortunately,

you're not afraid of bats or hornets, but when you grab the can of spray you wince and drop it to the floor. The temperature in the room has risen to 135°F and the can is literally too hot to touch, so you put on some gloves, despite the heat, grab the spray, sneak back to your desk and spray the hell out of the yellow jackets, while muttering some vengeful obscenities. Then you fling open a window to get some air and, luckily, the bat flies out.

Now that most of the bees have been killed, you sit back down at your desk, being careful not to touch anything metallic without insulating yourself from the hot surfaces. Only now you find that your typewriter is gone, and in its place are steel and iron hand tools, including a formidable looking hammer and a long tool that looks like a sword. Your co-worker informs you that your job henceforth will be to beat on the sword as hard as you can with the hammer. "And don't cut yourself on the sharp edges!"

"Sharp edges of what!?" you ask.

"Of the slate and sheet metal, of course!" she impatiently replies. "They can be sharp as razor blades, dummy!"

Sure enough, you nick your finger the first time you impatiently beat on the slate ripper, so you stop and wrap the bleeding appendage with a bandage, wondering if maybe you should have washed your filthy hand first. When you resume your labor you realize that your desk is covered with a thick layer of bat droppings, and you've been grinding the guano into the skin of your arms. Just when you've stopped to ponder this new discovery, an amazing, unexpected and terrifying thing happens: the floor of the room suddenly heaves into the air, tilting at a 45 degree angle! Everything in the room goes flying downhill! Your desk somehow remains attached to the "floor," however, so you cling to it for dear life.

Your co-worker yells up from the bottom of the heap shouting that she has a hook ladder you can hook to the top of the room so you have something to stand on. She carefully slides the ladder up the slanting floor past your desk and somehow it hooks at the high end of the room. You carefully stand on the ladder, crouched low and leaning to compensate for the 45 degree angle of your new work surface. "Back to work!" she yells up, "but here, take these." She throws up a pair of rubber-soled shoes. You slowly take a seated position on the hook ladder, and remove your wing tips. The soft, lightweight shoes give you better traction.

Back to work!? OK. But you find that the tilted "floor" is so hot you can't touch it and you can't even rest your foot on it to stabilize yourself, because it burns the soles of your feet right through the shoes. You must keep your entire body perched only on the aluminum hook ladder, which holds you an inch or two off the floor surface, now radiating heat like a heat lamp. You take off your headband and wipe your sweaty forehead, clinging with one gloved hand to the ladder. Then you replace the sweatband and grab the iron implements. The last thing you're worried about right now is bats or bees, although you don't want to be startled under these conditions and risk falling. So you move slowly and deliberately as you once again begin beating as hard as you can on the iron tools in your hands.

But you can't relax yet. Suddenly, the entire work environment lurches forty feet in the air and all the walls and ceilings drop away, leaving you out in the beating sun on a steep slope with your hind-end flapping in the wind. The heat, dirt, bats, bees and humidity remain with you, however, and now you have the blasting sun beating down on you too. Although there is an occasional breeze, it doesn't seem to do anything but blow the heat around. With both hands clinging to the ladder, you look over your shoulder and see your co-worker standing way down there on the ground. "No more pushing paper for you," she yells up. "You need to come down here and carry up these slabs of stone!"

"What! How many?"

"Oh, a ton or so," she says, without exaggerating.

You are now asking yourself if your miserable working conditions can possibly be any worse. You can't imagine what else can go wrong when suddenly you smell something, and it doesn't smell good. "Oh god, what's that?" you ask yourself out loud. "What's what?" replies your co-worker in a gruff voice. You're astonished to discover that your gracious, gentle, attractive co-worker is now clinging to the same hook ladder you're clinging to, and she's had a horrible sex change operation. *He* now bears a striking resemblance to a species of lower primate. Worse, he has developed a notable proclivity for the use of expletives in his language. The smell? As luck would have it, your co-worker is now listed in the Guinness Book of World Records as the world's greatest source of natural gas!

Just at that moment, a thunderstorm blows in out of nowhere and light-

ning begins to flash. You realize you've suddenly become a human lightning rod, so you scramble down the hook ladder and hasten down the ground ladder, fearing for your life as you rush to get off the roof. Rain begins to pelt down and the wet roof surface turns as slick as ice. Your co-worker has enough sense to get down ahead of you, so you join him on the ground where you seek shelter together. The pouring rain has cooled things off, washed some of the dirt off your face, and now you've got your feet back on the ground. Your prayers of gratitude are suddenly interrupted with a shocking thought — you forgot to close the gaping hole you left in the roof and now it's raining buckets!

You'd give anything for a nice, cushy office job again — wouldn't you?

[Unfortunately, the above scenario is not much of an exaggeration. Interestingly enough, it's a true story. The office worker was apparently a masochist at heart, because he did quit his middle management position at the factory where he worked, despite my earnest discouragement; and now, seven years later, he is a very successful slate roof restoration contractor. Because he started out in the slate roofing business on the author's crew, he managed to get his photo in this book in more places than one, including slating an octagonal tower, shown in Chapter Eight.]

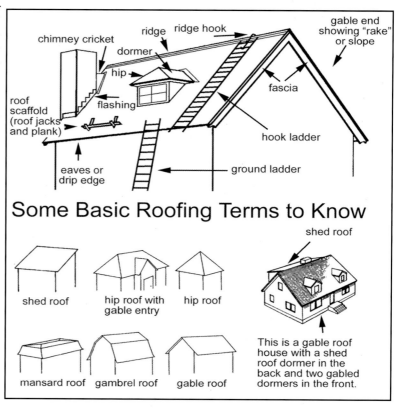

Some Basic Roofing Terms to Know

Chapter Eleven

SAFETY FIRST

Most old slate roofs are in need of repair, and ninety percent of what's involved in repairing them is simply a willingness to climb. If a slate roof could be laid down on the ground, almost anyone would be able to climb on it and repair it. But it's not down here — it's up there. This is where the element of risk comes into play, and safety precautions must be taken seriously.

You can actually walk out the front door of your house, trip on your shoe string, fall, hit your head on a rock, and die. If you fall from a height of six feet, same story. If you fall from a roof top . . . well, let's not imagine the consequences, but let's be aware of the possibility and take measures to prevent any sort of accident. The author, at the time of this writing, has worked on slate roofs for nearly 35 years and will try to pass on some insight and experience related to safety when working on the high, steep, slippery surfaces characteristic of slate roofs.

But first the obligatory disclaimer. I'm passing on this information for educational purposes, but I can't be held responsible for the actions of others. The author cannot control the attitude of the reader, nor sharpen his or her senses. You are responsible for that, and if you trip on your shoestring on the way to the ladder, don't blame me. When working at heights, you must proceed with utmost care and caution, and *you* must take full personal responsibility for *yourself*. Read all instructions carefully, and follow the recommendations of the manufacturer of any roofing tools and/or equipment. Be aware of safety codes and regulations in your area, and do not proceed with any roofing work unless and until you, and you alone, will take full responsibility for your actions.

Perhaps you heard the true story of the man who wanted to work on his roof but feared for his safety. He tied a stout rope to the rear bumper of his car, flung the rope up over the peak of his house, then climbed up a ladder on the other side of the house and tied the rope around his waist. While he was working happily away, his wife had a sudden and unexpected urge to go shopping. Unfortunately, the man had neglected to inform his wife of his ingenious safety precautions. She hopped in the car and took off without a second thought. This story was reported in the newspaper, and although the man was yanked off the roof and dragged down the street, he reportedly survived, though his final condition was unclear.

Even though this man had safety in mind, he went about it in the wrong way. This illustrates a case where a person violated all the rules of safety — he had the wrong attitude (too careless), didn't understand the potential hazards (in this case, not informing his wife), and he didn't use the proper equipment (ropes wrapped around one's waist are a mistake).

1) PROPER ATTITUDE

First, and perhaps most important, safety is a mental attitude. You must be careful and cautious. You must realize that *you* are responsible for your own safety, and if you fall or hurt yourself *you*, and perhaps your family as well, must suffer. If you don't take care of yourself, no one is going to do it for you. You probably understand this when driving a car or when chopping vegetables at the kitchen counter; you *must* understand this when climbing on a roof. The man in the incident above could have at least opened the hood of the car to indicate that it shouldn't be moved (or he could have put a note on the steering wheel). You don't want to be telling yourself what you *should* have done — instead, think ahead!

Never rely on someone else's judgment about whether a situation is safe or not (unless you're an apprentice working alongside a *master* roofer). Use your own judgment. If you don't feel safe in any given situation, don't proceed. Stop and do whatever is necessary to improve the safety of the situation to *your* satisfaction, then continue.

1. KEEP A SAFETY ATTITUDE.

2. KNOW THE HAZARDS.

3. USE PROPER EQUIPMENT.

Never be in a hurry when working on a slate roof. It only takes one slip and you're history. This issue about attitude cannot be overemphasized. Keep your eyes and ears open at all times when working at heights.

Always make certain the correct equipment is being properly used and *always* be aware of the potential hazards. Find them before they find you, and prevent accidents before they happen.

2) BE AWARE OF THE ROOFTOP HAZARDS

There are a number of hazards that exist on old slate roofs that you must memorize and look for *every time* you get on *any* roof. They can be summed up as: BEES, BATS, CHIMNEYS, WIND, LIGHTNING, ELECTRICAL WIRES and CO-WORKERS.

BEES

Bees are one of the worst hazards on old slate roofs. One recent summer in a local town, a man was found at the bottom of a ladder riddled with bee stings, and dead. He had probably set his ladder up in front of a nest of yellow jackets, which may not have been apparent to him, and when they attacked him he panicked, fell off the ladder and thereby made his last mistake. You must *always* look for nests of bees during the summer months on any house, rural or urban, before even putting up a ladder. There are generally seven types of common bees: yellow jackets, honey bees, paper wasps, bald-faced (white-faced) hornets, bumble bees, carpenter bees and mud daubers.

The first and worst are *yellow jackets*. These are small, mean and nasty wasps that live in highly populated paper nests sometimes shaped like footballs, but the nests are usually *not* visible — they're in the wall or roof of the house. Sometimes the nests are in the ground and yellow jackets are sometimes called ground bees (there are other types of ground bees, too). Sometimes the nests are exposed under the eaves of the building anyplace from down on the porch to the very top of a three story house. Yellow

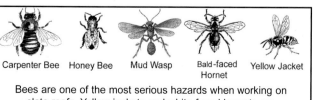

Carpenter Bee   Honey Bee   Mud Wasp   Bald-faced Hornet   Yellow Jacket

Bees are one of the most serious hazards when working on slate roofs. Yellow jackets and white-faced hornets are especially aggressive.

jackets look sort of like honey bees, and many people can't tell them apart, but honey bees are fuzzy and yellow jackets aren't. Also, yellow jackets die off in the winter *except for the queen*, who hibernates and starts a new nest in the spring. Therefore, yellow jackets don't nest in the same spot from one year to the next, but they *will* nest in the same house. Honey bees, on the other hand, usually stay put in the same nest year after year.

One reason yellow jackets are such a hazard is because they're just plain *mean*. They'll go out of their way to sting you. I've had them fly from under the eaves clear over the peak of the house to the other side of the roof just to try, successfully, to sting me. When you're hanging on a ladder and perched precariously at some considerable height, you don't want a bunch of yellow jackets flying in your face. It's a natural reaction to recoil, swat like hell and run when attacked by bees. You can't do that when on a ladder or a roof, so make sure you don't put yourself in that situation. If by chance a bee does fly at you, it is critical that you hang on to whatever it is you're holding on to, and don't panic! Leave the area as quickly as possible without being reckless. Don't let a puny bee force you to meet your maker.

I might add that some people are allergic to bee stings, and yellow jacket stings are among the most venomous, according to my observations. A friend of mine was stung in the forehead by one, and her head swelled up like a pumpkin. She looked like something out of a horror movie for about two days. My neighbor was stung by one and his whole body swelled up, and he had trouble breathing for hours. I have worked on slate roofs since 1968 and have never been stung by any bee except yellow jackets and honey bees, and only once or twice. I am not allergic to them and was once stung by 17 yellow jackets as a teenager (not on a roof) with no adverse reaction. But now I spot them before they spot me, and I avoid them if I can. If I can't avoid them, I spray them with hornet spray *if* I can get close enough to spray the nest itself. It does no good to just spray the bees because there are thousands of them, but if you can knock out the nest the remaining live bees will usually leave you alone (they become demoralized and disoriented after the queen has died).

Being a honeybee keeper, I can safely say that the best defense against harmful bees is awareness of their habits, patterns and weaknesses, avoidance of them, or eradication if all else fails. I don't recommend hunting down and killing every yellow jacket nest you can find, because then you're just asking for trouble, but if you do need to kill them, here are some tips:

Most of the bees will be out working on a sunny day. That's how you can locate the nest — look for the insects flying toward and away from the house. When you see insects flying back and forth like that, you can trace their flight path back to the house and either find the nest, or you may see a small hole under a window sill, between some clapboards, in a chimney, or even on the roof itself where a slate is missing. You'll see the bees flying in and out of the hole. Once you've located the nest (almost *every* old farmhouse and many city houses have them), stay away from it. *Do not* put a ladder up in front of a bee flyway. Remember that bees, like some people, are friendliest on sunny, dry days, and meanest on cloudy and muggy days. This is a fact. So on a sunny, dry day you may be able to work closer to a bee nest than on a bad weather day, but *don't count on it.*

Because the bees are working during the day, the ideal time to spray the nest is at night when they're all home. Unfortunately, night time is the hardest time to spray because you can't see anything, so dusk or dawn may have to do. Otherwise, you can spray them during the day and expect to see bees returning to the nest all day and swarming around the door to their nest, wondering what the heck happened to their home. When they realize their home and queen have been destroyed they probably won't attack you in a mad frenzy, as they're most defensive when protecting their queen. Once the queen is dead, the bees don't seem to know what to do. However, they may still sting, so watch out.

The yellow jacket nest starts out in the spring with only one bee — the queen. She builds a small paper nest about the size of a golf ball, then starts laying and hatching eggs. By fall, the nest will be the size of a football and have thousands of bees in it. Obviously, a good time to destroy a nest is in the spring when it is tiny. The longer the nest is allowed to expand, the more bees it will have and the more effort it will take to get rid of it. Of course, the bigger a hidden nest gets, the easier it is to locate, because there are more bees flying in and out of it. Fall is the worst time for roofers where bees are concerned, because the nests are bloated with bees, and in September and October, the roof worker needs to be especially vigilant.

Do not try to kill yellow jackets or any other bees with general insect spray — it doesn't work and it just makes the bees mad. Use only wasp and hor-

net spray. Finally, if you have to approach a nest closely to spray it, wear coveralls, gloves, and a bee veil. A prudent roofer will always keep these essential materials in his work truck. They may save you from bee stings many a time. If the nest is in the ground, you can pour a flammable liquid down the hole and light it, unless the hole is near a house or other object that you don't want to burn to the ground!

Honeybees aren't as mean as yellow jackets, but will defend their nest just as violently if need be. Their nests are rarely exposed and are made of wax cells where the bees store honey and hatch out young. You can locate their nests in the same manner as yellow jacket nests — by looking for their flyway. You will see them flying in and out of a hole in the building.

Again, their behavior is affected by the weather, and they don't want you anywhere near their nest when rain is impending and humidity is rising, probably because their nests are vulnerable to rain. They, too, are friendliest during sunny, dry weather. Honeybees (and yellow jackets) tend to attack black and rough surfaces first. That's why they go for your hair. Maybe they evolved in such a manner that they think everybody is a bear trying to steal honey, and so they dive for that part of you that looks like a bear. Knowing this, you would never approach a bee nest wearing dark, rough clothing. Beekeepers wear smooth, white coveralls. Honeybees generally will not go out of their way to sting you like yellow jackets will, but if the weather's bad and you resemble a bear — look out!

I have never destroyed a honeybee nest and probably never will — they're beneficial insects. It's easier to wear protective clothing around them than it is to try to kill them. A full grown honeybee nest will contain tens of thousands of bees that will pollinate many flowers, as well as produce much honey and beeswax. But they do live in the walls of old houses where they remain year after year, although they're not as common as yellow jackets. The solution to honeybees is to avoid them. If you can't, then wear protective clothing and work around them when they're in a good mood.

Bald-faced (white-faced) hornets are mean like yellow jackets. They live in paper nests which can get rather large (football size or bigger) and hang down from the eaves of a house, containing as many as 10,000 hornets. I have seen their nests hanging on almost anything, including tree branches and brush, and saw one nest sandwiched between two windows, thereby exposing the entire inner workings of the hive. These nasty critters can't be trusted. They'll come after you aggressively and *should* be destroyed.

Paper wasps are not mean, they just look like it. They're a relatively benign stinging insect. They *will* sting you though, if you disturb their nest. You can get quite close to a paper wasp nest and they won't come after you unless you vibrate the nest by banging near it. It is recommended to spray their nests when you have to work near one because a number of roofers do get stung by these fellows.

One of my workers was stung by a paper wasp while tying a ladder to a rain gutter. The nest was up under the slate at the eaves, and he stuck his hand practically right in it, then developed an allergic reaction when stung. One minute he was normal and the next he was covered with hives, his face was swollen, and he was lying on the tailgate of the truck trying to breathe. I watched another fellow also get stung in the hand by paper wasps for the same reason — reaching under the eaves of a dormer (a favorite spot for paper wasps) without looking first. Another roofer I know fell off the peak of a roof after exposing a paper wasp nest while removing ridge iron. The wasps came flying out at him and he backed off the roof in panic, falling nearly thirty feet off the gable end. He was lucky — he only broke his leg! You must *always* look under eaves and in any area that may hide a nest of bees before putting your face or hands in it. Remember to find them before they find you! Paper wasps love to build nests under ridge metal, so always remove ridge metal and other flashings carefully and have a can of hornet spray handy in the summer months.

Bumble bees and carpenter bees look very similar. They're round, black and yellow and fuzzy, and they're usually not mean. Bumble bees live in colonies often in the ground, while carpenter bees live in holes they bore in wood. There are large carpenter bees and small ones. The small ones seem to live in groups, while the large ones are loners. You'll often see the large carpenter bees when on the roof of old buildings, especially barns, because they live there, too. They're very territorial and will hover in the air like a tiny helicopter staring at you from a couple feet away, trying to intimidate you. If any other insect happens to fly past at that moment, the carpenter bee will chase after it until it leaves its territory, then it will come back to you as if to ask "What the heck are *you* still doing here?!" I've never known anyone to be stung by one, although I've had to swat a few with my slate ripper just to get rid of the persistent little devils. Their danger lies in their ability to startle you, a danger you can't afford when working at heights.

Finally, *mud daubers* are common on old roofs. They're long, thin, purplish wasps that nest in clumps of mud they build on or in chimneys, under flashings, and especially in attics. They're not mean and are usually not a problem.

BATS

Bats come in various sizes and shapes and are very common around, in and under old slate roofs. They are not a danger except that they may startle you if they fly out at you, and your instinctive recoil can knock you off the roof. However, if you're aware of their presence and expect them, you won't be startled if one does appear.

Bats love to hang out under ridge metal, under chimney flashing, in old chimneys (sometimes between the bricks), under loose slates, and especially in the attics of old houses. They can get into the roof through almost any hole, and are almost impossible to keep out. Again, these are beneficial creatures that eat a lot of bugs, so the solution to bats is not to kill them but to learn to live with them. I've seen people try getting rid of bats by putting mothballs in their attics (not very effective), and by using electronic pest repellent (might work). Some people swear that you only need to keep a light on in the attic to keep them away, but this is hearsay and I can't vouch for it.

Bats hibernate in the winter, then become active again when it warms up. In the fall when the days are cold, bats, like wasps, move in slow motion and are easily captured. They do have a lot of teeth and might bite, so I don't recommend handling them without gloves. Of course, everyone says they carry rabies, but this threat may be exaggerated.

You can usually locate bats easily because first, their droppings give off a characteristic odor, and secondly, the bats themselves give off a familiar high-pitched squeak or chirp when agitated. Furthermore, you can hear them scrambling and scratching under ridge metal when you've climbed on the roof and disturbed them. On one roof, I shoved my slate ripper under a broken slate to remove it, and when I pulled the ripper out a live bat came out with it! If you have an old house with a slate roof, you probably have bats. If you expect them and look for them, they won't startle you and cause an accident.

CHIMNEYS

Never grab onto a chimney for support unless you're quite sure that it is solid — most aren't. Many of the old chimneys on slate roofs have soft mortar, and the bricks can easily be lifted apart. This becomes a hazard when someone grabs onto a chimney to steady himself when moving along a ridge, and the bricks give way. Also, when taking chimneys apart, remember that bats do live in them, and sometimes bees do, too. Find them before they find you. Finally, chimney swifts (birds) also live inside chimneys in shelf-like nests fastened to the side of the bricks or stone. When you peer down into the top of the chimney, a startled bird may fly out in your face. Be prepared.

WIND, LIGHTNING, HEAT AND RAIN

Wind is an obvious hazard when working on roofs. It's much windier on top of a roof than on the ground, and on very windy days it may be a good idea to stay completely off the roof. The danger of wind is magnified when you're carrying a hook ladder up onto a roof, across the roof, or back down off the roof. The wind will catch the ladder and push it, which may cause you to lose your balance. Also, wind will readily knock a ground ladder off a house, so *erected ground ladders must always be tied to the house* to prevent this from happening. Aluminum ladders are light and easily blown down onto a person, car, or other property, not to mention the damage to the ladder, or the embarrassment to the stranded roofer.

Roofers are human lightning rods. *Always* vacate the roof as soon as any lightning becomes evident. Thunderstorms have a way of coming out of nowhere, so this can be a real hazard, especially when aluminum ladders are being used, as aluminum is a great conductor of electricity. Don't stand on the ground near a standing aluminum ladder during an electrical storm, either.

Heat can be terrible on a slate roof during the summer months. I have personally measured a temperature of 135⁰ Fahrenheit on the working sur-

face of a slate roof, and have read that 140⁰ is not uncommon. People do keel over from heat stroke, and if you are susceptible to this sort of thing, stay off a slate roof during the heat of a summer day. Furthermore, sunburn can lead to skin cancer in later years, so protect yourself from excessive, chronic solar exposure. Rain, dew, and of course frost make the surface of slate roofs wet and slippery. As a rule, *never work on a wet or frozen slate roof.*

ELECTRICAL WIRES

Electrical wires are a very serious hazard. If your aluminum ladder happens to hit a high-voltage wire while you're touching the ladder, you're dead — instantly. Electric lines may contain as much as 7,200 volts, or may be stepped down to 2,400 volts until the lines reach the transformers located on a pole near the house, where the voltage is reduced to 110/220 volts. The "high voltage" lines are certain death, and the 110/220 lines can also kill, so *always* be aware of the location of electrical lines and keep your aluminum ladders away from them.

When electrical lines enter the house from a public electric service source, there will always be at least two lines, and they will attach to the house on an insulated mount, usually leading to an electrical meter. This differs from telephone lines because telephones only need a single line (although if the house has several phones there may be several phone lines), and phone lines usually don't lead to an insulated mount, and never to an electrical meter. If you see lines entering the building you want to work on and aren't sure what they are, avoid them.

Electricity causes muscles to contract. If you're holding on to a ladder and you hit an electrical line with it, the electricity will cause your hand muscles to contract forcing you to remain holding on to the ladder. If you have to touch a surface to see if it is conducting electricity (don't try this), touch it with the *back* of your hand, then, if the surface is "hot" it will contract your muscles and throw your hand away from the surface. Better yet, if you have any doubts about whether a metal surface on the outside of a house is "hot" or not (it happens), call the electric company and let the experts take care of it.

I have heard several accounts of people getting electrocuted while working on construction jobs. In one case, a man was carrying sheet metal roofing up onto a barn roof, and the wind caught the metal and blew it against a high voltage line while the guy still had hold of it (he died instantly). One

excess sun   lightning   rain or wet   wind or gusts   ice or frost

WEATHER HAZARDS ON SLATE ROOFS

SLATE ROOFS ARE SLIPPERY WHEN WET!

highly publicized case involved three men trying to take down a metal flag pole in the front yard of a town hall. The pole hit a high voltage line and the men stood in the yard in the center of town welded to the flag pole literally frying to death in front of a crowd of horrified onlookers for a full half hour until the electric company could shut off the power (the three men all died instantly).

Interestingly, when I contacted my local electric company (West Penn Power) to get accurate information on voltage levels of power lines, they routed my call to four different people (receptionist, engineer, public relations officer and safety manager) before finally stating that they wouldn't give me that information. They told me to look it up at the library. So much for the "experts." Go figure.

CO-WORKERS

Co-workers can be an unintentional hazard. Like the uninformed wife who took off in the car without first ensuring that her actions wouldn't adversely affect her husband's safety, co-workers cannot always be relied on to maintain safe working conditions. As stated earlier, don't rely on the judgment of co-workers when your own safety is at stake. Co-workers can drop things on you from above, throw or drop things on the ground without looking, bump into you carelessly on the roof causing you to lose your balance, knock you while carrying a ladder on the roof, put up a ladder recklessly, attach a ladder hook to a ladder carelessly, install a roof bracket incorrectly, fail to tie ladders in place securely, overload roof scaffolds with too much weight, etc. This is not to suggest that co-workers are bad, but accidents can and do happen, and you must keep an eye on your co-workers when working at heights as if they were hazards themselves, because they *can* be. Conversely, co-workers can greatly enhance your safety, but accidents occur when people get tired, cranky, impatient, over-worked and careless. So consider this "a word to the wise."

3) USE THE PROPER EQUIPMENT

You can safely do almost any job on a slate roof with the following equipment: ground ladders, roof ladders (also called hook ladders), roof jacks (also called roof brackets), planks and ladder jacks (also called ladder brackets). Sources of tools and equipment are listed in the next chapter and in the back of this book, although ladders, roof jacks, ladder jacks and planks are available at almost any lumber yard. The only specialty tool is the ridge hook, which is used to make a hook ladder (available from several sources, including jenkinsslate.com). In severe cases, a "harness" may be necessary, which is a safety belt that wraps around the worker and ties on to something to prevent a fall. In 35 years of working at heights, I have had to wear a harness on only a few jobs, and I generally avoid anything with ropes on a slate roof because ropes impede mobility and create a tripping and snagging hazard. People who are not so accustomed to working at heights may want to use a harness more often. Some people use them all the time, and in some cases they're required by government safety codes.

GROUND LADDERS

A ground ladder is one that stands on the ground and is used to get up to the roof. This is the typical ladder that most people are familiar with. Aluminum ladders are recommended because they're light in weight and easy to handle, although aluminum ladders do conduct electricity and are very dangerous around electrical wires. Ladders must be put up properly or they can be very hazardous. Ladders will blow off the house if not tied on, and are typically tied to a rain gutter, or a nail or hook driven into the fascia. They must not be set up too steeply, or not steep enough (at too great an angle), and must be put up plumb (perpendicular to the force of gravity — not leaning sideways). Although smaller ladders can be put up by one person, longer ladders should be put up by two people, for safety's sake (see next chapter).

HOOK LADDERS AND RIDGE HOOKS

Once you get up to the eaves of a roof, you generally must use a hook ladder to get to the peak. Hook ladders enable the roofer to gain access to most spots on a roof safely, without damaging the slate. Don't try to climb up a slate roof by walking on the slate, hanging on a rope. Not only does this damage the slate, but a loose slate can slip out when you step on it, and you will suddenly know what it's like to ride a skateboard forty feet in the air.

Good ladder hooks (ridge hooks) are inexpensive and will attach to almost any ladder. Typically, you would separate an aluminum extension ladder into its two halves, then attach a hook to one of the halves in order to make a hook ladder. The ladder is simply slid up the roof until the hook drops over the peak and secures itself. Usually, the hook is facing upward when the ladder is slid up the roof, then the ladder is flipped over when the hook reaches the top. Better hooks have little wheels attached to make it easier to roll them up the roof.

The hook ladder must be long enough to reach from the peak of the roof to the eaves in order to be able to climb onto it from the ground ladder. It is critically important that the ridge hook be secured tightly to the hook ladder, as several contractors have told me that their ridge hooks have come off at the most inopportune of times. This is somewhat surprising to me, as I have used hook ladders since 1968 and have never had a hook come off a ladder. But then, I do keep a safety attitude, and check my equipment regularly.

A greater danger when using hook ladders arises on old barns, and some old buildings. Such roofs tend to have a hump in the middle, halfway between the peak and eaves. This causes the bottom of the hook ladder to be lifted off the roof several inches, and if you put your weight on the bottom of the ladder, the top will pop up (the see-saw effect), the hook will disengage, and you and your ladder will quickly plummet earthward. *This is a very serious hazard when using hook ladders on old barn roofs, and you must be very careful that the bottom of the hook ladder is lying tightly against the roof when the top is firmly secured.* Otherwise, a block of some kind (a piece of 2x4 or 4x4) must be tied to the bottom and underneath the hook ladder to prevent the seesaw effect from ruining your life. Also, a co-worker can sit on the top of the hook ladder to make sure it doesn't disengage.

Hook ladders can also be hazardous when they have to be carried around on a roof, especially long hook ladders. Typically, the hook ladder is carried along the ridge like a tight-rope walker carries his balancing pole along a tight-rope. This is the easiest way to move a ladder around on a roof, but it requires a good bit of balance. However, a ladder will catch the wind much more readily than a pole will, and a gust of wind can knock a person off balance when carrying a hook ladder along a ridge. Alternatively, the person can sit on the ridge and drag the hook ladder along the roof, but this actually requires a lot more effort (but little balance).

Many contractors will use what are called "chicken ladders" instead of hook ladders. These are home-made contraptions, usually made of wood, that function in the same manner as hook ladders. They are not as safe or convenient as an aluminum ladder with a ridge hook attached to it though, and therefore are *not* recommended.

HOOK LADDERS ARE SAFER THAN EITHER ROPES OR "CHICKEN LADDERS"

Swiveling angle-iron end piece

Ladder hook (also called ridge hook) shown attached to ladder. Hook ladders allow for relatively safe access to the top of roofs. This type of hook is best suited for slate roofs because it has a hinged, flat piece of metal contacting the slate, which prevents breakage. This hook is easily removable and can be attached to any pair of rungs on the ladder, allowing for the working length of the ladder to be adjusted. **The manufacturer of this hook specifies that they be used in pairs (two hooks on the end of a ladder instead of one),** but two hooks on one ladder are sometimes heavy and hard to handle, thereby posing a hazard themselves, and some roofers use only one per ladder.

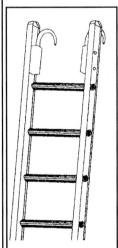

This is a common type of hook ladder <u>not</u> recommended for use on slate roofs, because the hooks are pointed and will damage the slate.

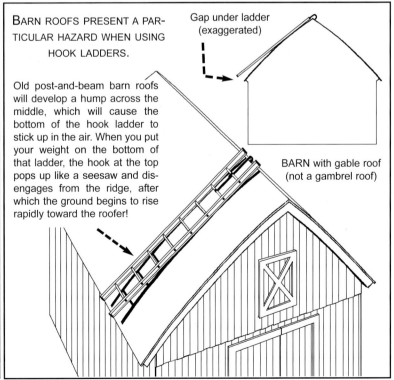

BARN ROOFS PRESENT A PARTICULAR HAZARD WHEN USING HOOK LADDERS.

Old post-and-beam barn roofs will develop a hump across the middle, which will cause the bottom of the hook ladder to stick up in the air. When you put your weight on the bottom of that ladder, the hook at the top pops up like a seesaw and disengages from the ridge, after which the ground begins to rise rapidly toward the roofer!

Gap under ladder (exaggerated)

BARN with gable roof (not a gambrel roof)

LADDER JACKS, ROOF JACKS
AND PLANKS

Again, these tools are discussed in greater detail in the next chapter, but are mentioned here because you cannot safely work on slate roofs, *at times*, without them. Hook ladders only gain access to roofs that have horizontal ridges, which not all slate roofs have. Some roofs come to a point like a pyramid,

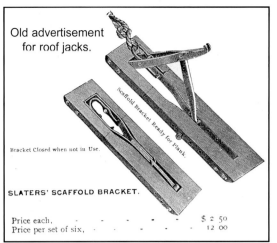

Old advertisement for roof jacks.

Scaffold Bracket Ready for Plank.

Bracket Closed when not in Use.

SLATERS' SCAFFOLD BRACKET.

Price each, - - - - - $ 2 50
Price per set of six, - - - - 12 00

some have flat tops, while others, such as gambrel roofs, do have horizontal ridges but they can't be easily reached with a hook ladder. When you can't get a hook ladder up to a ridge, then you have to use roof jacks and planks in order to get on the roof.

The terms roof *jacks* and ladder *jacks* are old-fashioned roofers' terms. Manufacturers call these things roof brackets and ladder brackets. They will generally be referred to as roof and ladder jacks in this book for tradition's sake.

Roof jacks usually nail to the surface of the roof, *right through and on top of the slates*, although the tongue can be slid underneath a slate before nailing if necessary. They support horizontal planks which are used as working platforms. A ladder can then be carefully set on these planks and laid flat on the roof to get up higher. Roof jacks, when properly nailed in place and used with a good quality, proper size plank, are quite safe and indispensable for slate roof work. They can be nailed into a slate roof and removed from the roof without damaging the roof (this procedure is described in the next chapter).

A *ladder jack* does not attach to the roof, but instead attaches to a ladder. They must be used in pairs, much like roof jacks, and they have the same purpose as roof jacks, namely to support planks so as to provide a work platform. They come in handy on

slate roofs when two hook ladders are positioned on the roof next to each other, and a ladder jack is attached to each one, then one or more planks are laid across the ladder jacks. This allows for a solid roof platform to be quickly and easily constructed without the need to nail anything into the roof. This "roof scaffold" system is particularly useful when working on chimneys, as a hook ladder can often be positioned on either side of the chimney, and planks can be run across from one ladder to the other just below the chimney in a very short period of time.

Let's take another look at the guy who tied himself to his car and got yanked off the roof. He obviously had a ground ladder or he wouldn't have been on his roof at all. If he had had a hook ladder he would have been able to climb onto his roof without the need for a rope. Two hook ladders would have enabled him to get practically anywhere on his roof (one to go up and down from the ground ladder, and one to move around the roof by walking along the ridge, which is the "sidewalk" on the roof). If he had the sort of roof where a hook ladder wouldn't work, he could have nailed a pair of roof jacks to his roof, working from his ground ladder, then laid a plank across them and climbed onto this roof plat-

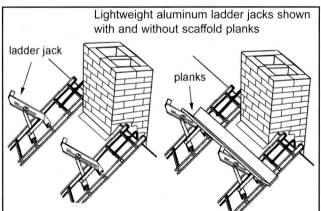

Lightweight aluminum ladder jacks shown with and without scaffold planks

ladder jack

planks

Two hook ladders with ladder jacks are positioned on either side of a chimney (left). 2"x10" planks are laid across both the ladders and the jacks to make a safe and easily installed roof scaffold (right). Double-wide planks can be laid on both the ladder and the jacks to make a platform with greater capacity. If this scaffold is expected to bear heavy loads, then use two ridge hooks per ladder. Access the other side of the chimney with a set of roof jacks and planks, or use two more hook ladders with longer planks. Be careful not to overload the roof scaffold, as too much weight can mean disaster.

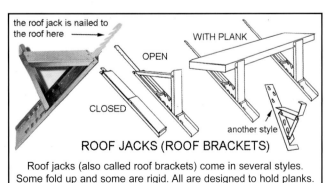

the roof jack is nailed to the roof here

WITH PLANK

OPEN

CLOSED

another style

ROOF JACKS (ROOF BRACKETS)

Roof jacks (also called roof brackets) come in several styles. Some fold up and some are rigid. All are designed to hold planks.

form. He could then have set another ladder onto this platform to get to the spot that needed fixed. If he still couldn't reach the spot, he'd have to install more roof jacks and planks until he got to where he wanted. Then if his wife got a sudden urge to run to the store, he'd survive unscathed.

OSHA

The Occupational Safety and Health Administration (OSHA) under the U.S. Department of Labor, promulgates safety standards for the construction industry that apply to all employees and subcontractors on any construction job. These regulations are detailed and extensive and require a thick volume all to themselves. Persons subject to OSHA regulations should contact the U.S. Department of Labor and obtain a copy of the OSHA regulations as they apply to the construction industry.

OSHA regulations require that *"no contractor or subcontractor shall require any laborer or mechanic... to work in surroundings or under working conditions which are unsanitary, hazardous, or dangerous to his health or safety."* They further require that *"each employee on a walking/working surface with an unprotected side or edge which is six feet (1.8m) or more above a lower level shall be protected from falling by the use of guardrail systems, safety net systems, or personal fall arrest systems."* For steep roofs, OSHA includes, *"Each employee on a steep roof with unprotected sides and edges 6 feet (1.8m) or more above lower levels shall be protected from falling by guardrail systems with toeboards, safety net systems, or personal fall arrest systems."* OSHA promulgates detailed regulations on every aspect of construction workplace safety from the type of extension cords that can be used, to fire protection, sanitation, illumination, scaffolding, safety belts, ladders, tools, written safety records and just about anything else you can think of.

According to the National Roofing Contractors Association Safety Manual (2000), roofing slide guards (roof jacks and minimum 2"x6" planks) can be used as alternative means of fall protection on residential steep slope roofs, as long as the roof has an 8:12 slope or less and a rafter length of 25' or less. Such slide guards must be installed along the entire length of the eaves and then every eight feet up the roof. Roof brackets must be supplemented with guardrails and toeboards or with personal fall arrest systems.

Nothing enhances the safety of working at heights quite like frame scaffolding, especially when securely tied to the building and otherwise used properly — with safety rails, clamp-on platforms, self-leveling feet, and a hoist for lifting heavy material to the top. The author is shown, above left, preparing to install new Vermont sea green slates on the Staples home in Sandy Lake, Pennsylvania. Brent Ulisky, the author's step-son, erected the scaffolding and is preparing to send up slates (above right).   Photo at left by Brent Ulisky, at right by author.

## Chapter Twelve

BASIC TOOLS AND EQUIPMENT

Don't read this chapter until you've read the previous Chapter Eleven on safety, which introduces you to some of the equipment required for slate work and explains some of the hazards one can, and *will* run into when working on slate roofs. This chapter will show you how that equipment, as well as hand tools, are meant to be used.

Slate work is a specialized endeavor requiring some unusual tools and equipment not readily found at a hardware store. If you don't use the right tools, you'll have a very hard time doing any work on a slate roof. The tools you will need *are* available, however, and it is strongly urged that if you intend to do any work on any slate roof, you invest in a few tools appropriate for the job. Slate tools are not prohibitively expensive and are well worth the investment.

SLATE RIPPERS

There are several hand tools necessary for working on slate roofs, and the most important is the slate ripper. The ripper is a long, sword-like steel tool that is inserted under a slate in order to hook the nails holding the slate and then pull or rip them out. They do not cut the nails as many people believe, and they should not do so, as the remaining cut-off nail shaft left underneath the slate creates an irritating obstruction when trying to slide the new slate into place. The nails should be completely removed, and that is the job of the ripper, which is inserted *underneath* the slate to be removed. The nail shaft is hooked by the end of the ripper and the nail is pulled out, usually bent in half, by vigorously pounding on the ripper handle with a hammer.

Once both nails are pulled out, the slate can be removed (some roof slates have more than two nails). This procedure is necessary when replacing slates, as the ripper enables the worker to remove the old, broken slate, which must be done before a new slate can be slid into place. Slates are also removed

from existing slate roofs when flashing metal is replaced, or when holes are cut into the roofs to install (for example) skylights. The ripper is as important to the slater as the hammer is to the carpenter.

Although slate roofs are unique in many ways, they have one particular quality that separates them from most other roofs: they can readily be taken apart and put back together. A slate roof should be seen as an entity consisting of thousands of parts, each removable and replaceable. Those parts are the slates themselves, and they're removable with the help of a slate ripper. Any individual slate can be removed from the field of the roof, and replaced, without hurting the roof in any way. This is also done whenever flashing is being replaced, all of which is routine when the proper tools are used.

Like an automobile, slate roofs need to be "tuned up" now and then. A car needs its spark plugs replaced, while a slate roof needs a few slates replaced occasionally. When you consider that an average slate roof may have 3,000 or more slates on it, and that many slate roofs after a hundred years have only had maybe 30 slates replaced, then you can see that 99% of the roof has *not* needed repaired in a century's time. Nevertheless, many people will convince themselves that a 1% breakdown after a

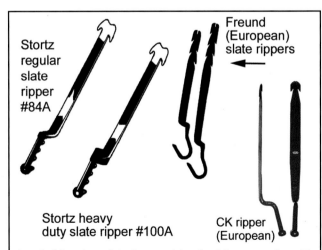

Stortz regular slate ripper #84A

Freund (European) slate rippers

Stortz heavy duty slate ripper #100A

CK ripper (European)

A typical American slate ripper weighs about three pounds and is about 30" long overall. The end with the hooks is inserted under the slate to pull out (not cut) the nails. Sources of slate rippers include: Stortz; North American Bocker; AJC Hatchet Co.; ABC Supply Co., Inc.; and Jenkinsslate.com.

Opposite page:
Old building in Angers, northwest France, displays
the characteristic black slate of the region.
Photo by author.

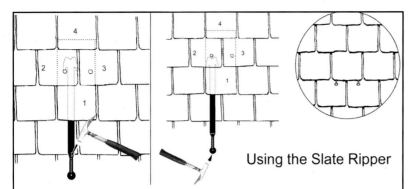

Using the Slate Ripper

The diagram above shows the nails holding slate #1 in place (slates 2, 3 and 4 are overlapping slate #1). In order to remove slate #1 from the roof, the two nails must be pulled out, one at a time, with the ripper, as shown. The ripper is inserted underneath the slate to be removed. A hammer is used to pound downward on the handle of the ripper in order to force the nails out, while the worker pulls on the ripper with the other hand. Once the slate is out, a new slate can be installed (see Chapter 17). Sometimes it is difficult to get the ripper under the slate and hooked on the nail, because the slates are too "tight." This problem can be alleviated by gently sliding a couple of nails under the bottom edge of the slate to wedge it up a bit (shown at right, above), and/or by pounding on the slate nail with the tip of the ripper before trying to remove it, as shown above, middle.

century means it's time to replace the roof. This is similar to saying that when the spark plugs of your car wear out, it's time to replace the car!

The reason this may be hard to understand is because slate roofs follow a time line that is not on a human scale. Humans typically last maybe 80 years, while slate roofs may last hundreds. So when people see an 80-year-old roof, they jump to the conclusion that its time is up. This misperception is compounded by the fact that America is a throw-away culture where people take for granted that when something breaks, it is simply thrown away and replaced. Slate roofs, however, don't break. Like cars, they wear out, but unlike cars, it's usually over a period of time which may be centuries, depending on the type of slate. And like cars, they can not only be tuned up, but also completely overhauled.

A standard overhaul of a slate roof requires replacement of such things as the metal flashings, chimney tops, broken slates and possibly some roof sheathing (wood). A total and drastic overhaul may involve removing the entire slate roof, replacing the wood underneath, then re-nailing the same slate to the new wood with new nails. At this point in time (early 21st Century), many American slate roofs do need overhauled somewhat if they're going to last another century. However, like a mechanic who can't get your spark plugs out without a spark plug wrench, a slater can't take your roof apart without a slate ripper.

Like any craftperson's tool, the ripper is a tool that becomes more versatile as the user gains more experience. It can pull nails out, push nails

out, and cut nails by both pushing and pulling. It can be used to feel around under slate to find obstructions, old repairs, hidden nails under tarred slates, etc. It even makes for a suitable bee swatter in an emergency. It should never be used to pry with, however, as it bends easily and is not designed for prying.

Rippers are designed to pull out roofing nails and not some of the larger nails (such as 8 penny and 16 penny nails) that some roofers will incorrectly use to fasten replacement slates. A quick way to break a ripper is to hook it onto an eight penny nail and try to pull it out by beating on the tool. Broken rippers can be welded back together, however, and used again for quite some time.

Sometimes a ripper will bend the nail but not pull it out, and even though the old slate will come out, the old bent nail or piece of nail remains behind and blocks the replacement slate from sliding in. The solution is to use the ripper to bend the nail back and forth until it breaks off. That's one reason the point of the ripper is indented — so nails can be pushed with the tool as well as pulled. Sometimes there is absolutely no way to get a ripper hooked on a nail, and then the nail must be cut by using the point of the ripper and pounding upward on the end of the handle. This is only recommended as a last resort because you risk pushing the nail up under the slate where it won't come out at all (without removing more slates). Alternatively, when a nail nub is left under the slate because the ripper cut the nail rather than pulled it, and the nub interferes with sliding the new slate into place, it is often easiest to simply remove more overlying slates, expose the culprit, and pull it out or pound it down with a hammer.

The pointed end of the ripper can also make a suitable chisel for scraping old roofing cement off the surface of a slate. Many tarred slates are cleaned up this way only to reveal that there was no reason whatsoever for the slate to be tarred in the first place — no hole, no crack, no exposed nail!

Hand tools can get very hot on slate roofs during the summer months, and rippers are particularly notorious for this. They can actually get too hot to handle, and gloves must be worn when using them. When the ripper is not being used, the point is slid far enough under a slate on the roof so that the ripper will remain safely lodged, out of the way, and easily accessible until it is needed.

SLATE CUTTERS

Contrary to popular opinion, roof slates are not difficult to cut using a slate cutter, a simple hand tool which is similar in principle to a paper cutter. The slate cutter has a long arm with a cutting blade on it; the slate is worked through the blade with one hand while the cutting arm is operated with the other. New slates, however, are quite a bit more difficult to cut than most old slates.

Slate roofing became an art long before electricity came into common usage, and usually no electric tools are needed to work with roof slate. Occasionally, a diamond blade cutter or a masonry blade may be needed to cut very thick slates, or very hard, new slates, especially when delicate cuts are required. Generally speaking, though, standard thickness (3/16" to 1/4") roofing slates need nothing but a hand-held manual slate cutter. The disadvantage of using a diamond blade to cut slate is that the saw cut leaves a square edge, whereas most roof slates have beveled edges. The hand-held slate cutters leave a beveled edge on the slate, matching the original slate much better than a cut from an electric saw. Also, electric cutters are much more dangerous than manual ones, especially when used on the roof.

The best hand-held slate cutters can cut both convex and concave curves in slate, and punch holes in the slate, as well as make a simple, straight cut. Many commercial slate cutters today don't have hole punches on them and can't cut concave curves. Concave curved cuts are necessary when the slater is making a scalloped cut in a square slate to match the scalloped design on an existing roof.

Old slate tools can occasionally be found at flea markets, auctions, garage sales and antique stores. Old tools can sometimes be superior to new ones, and are often used by professional slaters. You may still find an old Pearson cutter collecting dust in someone's basement or attic, as well as an old Belden, Pecto or Red Devil ripper or slater's hammer, but not for long — these old tools are disappearing fast.

A professional slater will carry a cutter with him while on the roof, so it must not be a cumbersome or awkward tool. Most cutters are designed so they can be mounted on a board, which is fine when using the cutter on a flat surface, such as on the ground. But when up on a roof, a board is not necessary, as the slate is being cut by a roofer usually crouched in an awkward position, and the bottom half of the cutter is simply tucked into the roofer's tool belt as the slate is cut. The cutter is then opened

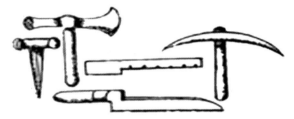

Randle Holme 1688: slater's tools
Courtesy of www.stoneroof.org.uk

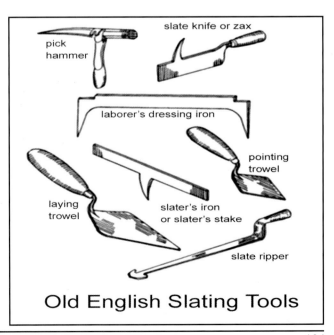

slate knife or zax

pick hammer

laborer's dressing iron

pointing trowel

laying trowel

slater's iron or slater's stake

slate ripper

Old English Slating Tools

up and laid over the ridge of the roof for safe storage, or propped in a rung of a hook ladder until needed again.

Nifty little slate cutters are the "one-hand" cutters, which are small (about 10" long) and can be carried in a tool belt. Made in France and Germany and sold in the U.S. by Stortz and Bocker, they will cut roof slate much as a pair of tin snips cuts sheet metal.

SLATE HAMMERS

A slate hammer is a unique tool that is very handy for a slater because it not only drives nails, but also punches holes in slate. Traditional installation hammers are also used to *cut* slate. If you intend to do any significant amount of work on slate roofs, a slate hammer is an essential tool.

Slate hammers generally fall into two loose categories: *installation* hammers and *restoration* hammers. The installation hammers are distinguished by their slate-cutting shafts. They're available as right-handed or left-handed hammers depending on which side of the shaft the cutting edge is on. Some, such as the Gilbert and Becker slate hammer, have cutting edges on both sides and are therefore suitable for either right-hand or left-hand use. This type of slating hammer will cut slate on the job site by chopping the roof slate while it's laid over a slater's stake. A stake is a T-shaped piece of iron that is driven into the ground or into a board or block of wood and used as a back support for a piece of slate while the slate hammer chops the slate to size (see next page). Instead of a stake, though, almost any straight edge, such as the edge of a ladder or even another slate, may do. Installation hammers also tend to be lighter than the others — the amount of weight needed to pound slating nails is not great. Notable installation hammers include the German-made Stortz hammer, the American-made Gilbert and Becker hammer, the German-made Freund slating hammer, and a variety of other European brands.

Restoration hammers, on the other hand, tend to be heavier hammers which are more useful for beating on slate rippers, pounding larger nails, and doing the various demolition and reconstruction tasks that restoration work dictates. Although these hammers do not include a slate-cutting shaft, they do have a slate punch end which can also cut slates in a pinch by perforating along a line, then breaking. Notable restoration hammers include the American-made Estwing Latthammers and the German-made CK and Freund hammers. Europeans refer to these hammers as "carpenters' hammers," but Americans tend to refer to them as "European roofing hammers." An American carpenter wouldn't be caught dead with a hammer designed to punch holes in roof slate, so the term "carpenters' hammer" wouldn't make sense in the U.S. In Europe, on the other hand, slate roofs are so common that even carpenters want their hammers to be able to punch a hole in a roof slate.

Few people realize that a hole can be punched easily in roof slate by using a sharp object — slates of standard thickness do not need to be drilled. Traditionally, holes are punched in roof slate

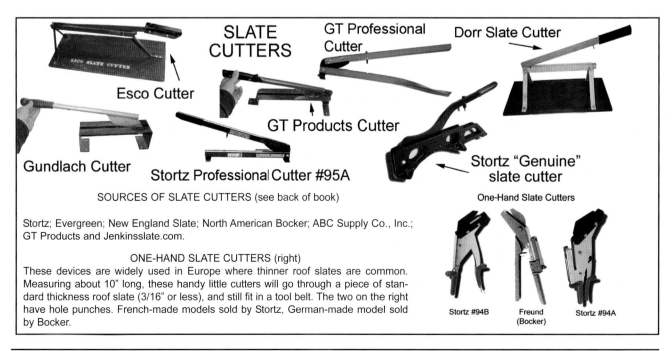

SLATE CUTTERS

GT Professional Cutter

Dorr Slate Cutter

Esco Cutter

GT Products Cutter

Gundlach Cutter

Stortz Professional Cutter #95A

Stortz "Genuine" slate cutter

SOURCES OF SLATE CUTTERS (see back of book)

Stortz; Evergreen; New England Slate; North American Bocker; ABC Supply Co., Inc.; GT Products and Jenkinsslate.com.

ONE-HAND SLATE CUTTERS (right)
These devices are widely used in Europe where thinner roof slates are common. Measuring about 10" long, these handy little cutters will go through a piece of standard thickness roof slate (3/16" or less), and still fit in a tool belt. The two on the right have hole punches. French-made models sold by Stortz, German-made model sold by Bocker.

One-Hand Slate Cutters

Stortz #94B   Freund (Bocker)   Stortz #94A

Slate Roofing Hammers

Estwing Hammers: 34+ ounces with a 21-ounce (600 gm) head with cross-hatch face. Overall length is a standard 13" (330mm).

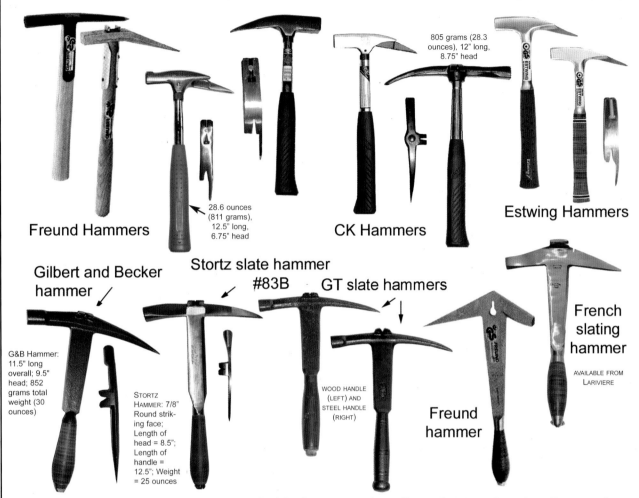

805 grams (28.3 ounces), 12" long, 8.75" head

28.6 ounces (811 grams), 12.5" long, 6.75" head

Freund Hammers

CK Hammers

Estwing Hammers

Gilbert and Becker hammer

Stortz slate hammer #83B

GT slate hammers

G&B Hammer: 11.5" long overall; 9.5" head; 852 grams total weight (30 ounces)

STORTZ HAMMER: 7/8" Round striking face; Length of head = 8.5"; Length of handle = 12.5"; Weight = 25 ounces

WOOD HANDLE (LEFT) AND STEEL HANDLE (RIGHT)

Freund hammer

French slating hammer

AVAILABLE FROM LARIVIERE

Top row of hammers is more suitable for general roofing, slate roof restoration, and roof construction. The bottom row of hammers is more suited for slate installation.

The classic German-made Stortz slate hammer shown above is a "right-handed" hammer. The cutting edge, on the right side of the shaft, as shown, cuts roof slate with a chopping action when the slate is backed with a support, such as another slate, or a slater's "stake" (see below). The pointed end punches holes in slate. The handle is wrapped in leather. The leather-handled Gilbert and Becker slate hammer is an American made classic with a double-edged shank that can be used either right- or left-handed. Freund and CK hammers, also German imports, include the European "roofing" hammers (called "carpenter's hammers" in Europe), which are ideal for slate roof repair and restoration due to their heavier weight. Most of the top row of hammers are heavier hammers having no slate-cutting shank (as the ones below have), although all have a slate punch. The heavier hammers are more suitable for beating on a slate ripper (for example) than are the lighter slate installation hammers. The American-made Estwing hammers, like some of the Freund and CK hammers, are called carpenter's hammers in Europe, but are slate roof restoration hammers in the U.S.

SOURCES OF SLATE HAMMERS (see rear of book): John Stortz and Son; North American Bocker; Carl Kammerling and Co.; ABC Supply Co., Inc., GT Products and Jenkinsslate.com.

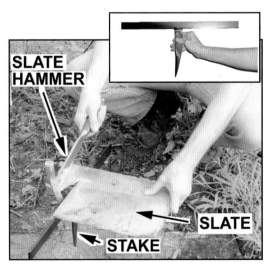

SLATE HAMMER
SLATE
STAKE

Slater's Stake (approx. 20" long) — to support a roofing slate while being trimmed with a slate hammer.

using any slate hammer, all of which have sharp points for that purpose. If the hammer doesn't have a slate punch on it, it's not a slate hammer.

A crude way to *cut* slate is to use the pointed end of a slate hammer (or a hammer and nail) and perforate holes along a line on the slate, then break the slate on the dotted line. Once the slate is cut, the jagged edge is tapped straight-on with the hammer to remove the roughness. In this manner, even ham-

▲ John Stortz and Son, Inc., Philadelphia, have provided for over 150 years some of the highest quality slate roofing tools found anywhere. Their standard (USA) ripper is widely considered the best available on the market and indispensable to the slate roofing trade. Above, John C. Stortz puts the finishing touches on one of the rippers. Stortz also manufactures/distributes slate hammers, cutters, stakes, tongs and many other tools for roofing and masonry applications. Photo by author.

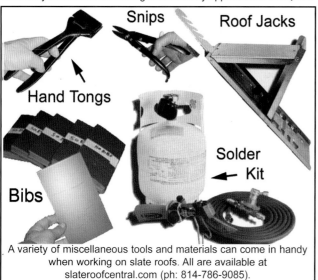

A variety of miscellaneous tools and materials can come in handy when working on slate roofs. All are available at slateroofcentral.com (ph: 814-786-9085).

mers without cutting shanks can be used to cut slates. Also, this slate-cutting technique can be performed with *any* hammer and a nail, so that even the average do-it-your-selfer can cut roof slate without special tools.

MISCELLANEOUS HAND TOOLS

Other hand tools commonly required for work on slate roofs and frequently carried in a slater's tool belt are tin snips (aviation snips) for cutting flashing and sheet metal; utility knife for general purposes; tape measure (25' length); chalk line (for chalking lines when installing new roofs or valleys); utility pencil (carpenter's pencil or other marker); flat pry bar (also called "wonderbar"), for removing face-nails, prying up slates, removing ridge iron and flashings, scraping tar off roofs, etc.; hand tongs for bending sheet metal; thin chisel for flashing work; and a nail punch, such as a 1/2" bolt about 4" long, for setting the nail when replacing a slate, and for punching nails down through the slate.

An additional set of tools is required for masonry work on chimneys, and that is discussed in Chapter 19.

GROUND LADDERS

Not much work would be done on slate roofs without ground ladders — they're critically important pieces of equipment that have a degree of versatility to them and, with experience, can be used to gain access to even the most unlikely places. The recommended type of ladder is either a medium or heavy duty commercial aluminum ladder. Be sure to read the section pertaining to ladders in the safety chapter first, and don't electrocute yourself when putting up or taking down an aluminum ladder.

The first task when using a ladder is putting it up. This can be done by one person by simply lodging the base of the ladder against a solid object such as the side of a house, and "walking" the ladder up. To walk the ladder up, simply start at the top of the ladder, hoist it overhead at arms length, and start walking toward the stabilized base of the ladder, moving your hands forward rung to rung until the ladder is standing straight up. Then pull the base of the ladder back away from the building so the ladder is standing vertical, extend the fly until it reaches the eaves, and lean it against the building. If you're walking up an extension ladder, do *not* extend the ladder *before* walking it up; instead, extend it after it has been stood up in a vertical position (this may take some practice for the inexperienced person). The ladder is taken down in the same manner — the extension is retracted, the base is wedged against something solid and stationary, and the ladder is walked down backwards.

Once a ladder is standing vertically and not leaning on something, it's very unstable unless it's kept perfectly plumb (straight up and down). So it's extremely important to hold the ladder plumb when moving it around, such as alongside the house. Better to leave the top of the ladder leaning slightly against the building when moving it along, if possible.

It's easier to handle a ladder with two people. To put up a ladder with two people, one keeps a foot on the base of the ladder while the other walks it up. The ladder is turned on edge during this procedure to make it easier to keep a foot on the ladder. The foot must not be taken off the ladder until the ladder is vertical, otherwise the base will fly up, the top of the ladder will plummet to the ground, and loud, unpleasant noises will come from the guy who was trying to walk it up.

Once the ladder is up in a vertical position, one person stands behind the ladder (between the ladder and the building) and supports it with two hands, while the other extends the fly (moveable) section of the ladder, even climbing the ladder if necessary. The person supporting the ladder must be strong enough to hold it nearly plumb, but leaning a little toward the house so it doesn't fall backward or sideways with the other person on it.

If a person is putting a ladder up alone and must extend the fly, the ladder is held plumb with one hand, a foot solidly holds down the bottom rung of the ladder, and the other hand extends the fly either by pushing on it, or by pulling on the ladder rope, until the fly reaches the eaves of the building, if possible. Then the person can climb the ladder and extend the fly to the desired height by pushing the fly upward as the ladder is climbed, being careful that the ladder is leaning against the building with enough slope to keep it from falling backward. Once the ladder has been extended, the base can be moved back from the building to its proper final distance, which should leave a slope on the ladder that is not too steep or too shallow.

The ladder top should *always* be tied to the building after the ladder has been erected. Rain spouting often comes in handy as a place to tie on to, but when no rain spouting is available, something else must be used. A 16-penny nail driven solidly into the fascia and bent into a hook will make a quick anchor to tie to when nothing else is available.

Many aluminum ladders come equipped with an attached rope meant to be used to extend the ladder. Some people swear by this rope, others take the rope off right away and never use it at all. Originally, extension ladders were made of wood and were quite heavy, and ropes weren't sufficient to extend them, so many older roofers developed the habit of extending the ladder by climbing up the rungs and pushing the "fly" ahead of themselves. This requires two people when a ladder is long or must be extended very far, but can easily be done by one person in most situations.

Any time a ladder is stood up vertically and the base placed in its permanent location, the ladder will not stand plumb if the ground on which it is standing is not level. Before the ladder is leaned against the building, it must be plumb in order for it to be safe, otherwise it may slide in one direction or another along the eaves or along the rain spouting, causing the ladder climber to lose balance. To check whether a ladder that is already leaning against a building is plumb, stand on the bottom rung and pull the top of the ladder away from the building slightly. A plumb ladder will drop right back against the building in the original place. The top of an incorrectly erected ladder will move to one side or another. It's always wise to make certain a ladder is plumb before you climb it, especially if someone else put it up. You don't want to get to the top and realize the ladder hasn't been put up properly and the top is now sliding sideways, with you on it!

To make the ladder plumb on an uneven surface, *don't* prop one of the legs on top of something if you can avoid doing so. Instead, *dig a hole* and drop the other leg into it. *It is much more stable and safer to dig a hole for one ladder foot than to prop the other foot on something.* A hole can be dug quickly in soil with the claw end of a claw hammer, and easily restored after the ladder is moved. This can't be done, of course, when on pavement, or setting a ladder up on a roof (porch roof for example), and in these cases a prop must be used to level the ladder's feet. A solid plank with one end propped on a block of some sort usually does the trick.

Most ladders have feet that swivel backward or forward. These come in handy for leveling the base of the ladder when the surface is just slightly out of level. Many ladders can be lowered slightly on one side if the foot on that side is swiveled backward, and can be raised slightly if the foot is swiveled forward. Some ladders do just the opposite, so you'll have to experiment with yours. It's routine to climb

Ladders have a top and a bottom. The bottom has the feet. The moveable part of the ladder is called the "fly."

MAKING A GROUND LADDER PLUMB VERTICAL

Moving a ladder foot backward or forward will drop or raise that foot slightly thereby helping to plumb the ladder.

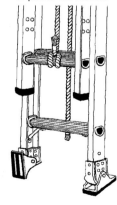

a ladder when its feet are bent (swiveled) backward or forward.

The more extended the ladder is the more top-heavy it will be, and the less stable it will be when trying to move it. *The rule of thumb is to retract the fly before moving a ladder, then extend it again when the ladder is where you want it. The other rule of thumb when moving a ladder is to keep it as plumb as possible when moving it. Often it's best to walk the ladder down, move it, and walk it back up.*

Most of the work commonly done on slate roofs is done on hook ladders (roof ladders), not on ground ladders, unless working along the eaves or drip edge of a roof. If both hands are required when working at the top of a ground ladder, then one of the worker's legs should be wrapped around a pair of ladder rungs as shown in the illustration on the following page.

A good all-around ladder for working on a two-story house is a 32-foot, medium duty commercial ladder. A 40-foot ladder is necessary for higher homes, but a 32-foot ladder will easily reach well beyond the eaves of most homes, and it's a relatively easy ladder to handle, especially for one person.

HOOK LADDERS

A hook ladder is any ladder that has a ridge hook attached to the end, so the ladder can be hooked over the ridge of a roof. Hook ladders are one of the most important pieces of equipment for anyone wanting to work on a slate roof. They make a difficult and dangerous job relatively safe and easy, and they're not hard to come by because you don't have to buy a special ladder with a hook

already attached to it. Instead, you buy the hook separately and attach it to one of your own aluminum ladders. Hook ladders are discussed in the previous chapter, and if you haven't read about them there already, now is a good time to do so.

Hook ladders enable a roofer to climb on slate roofs without putting weight on the slate itself, thereby preventing damage to the slate. They also keep the roofer off the slate when the roof is too hot to touch, preventing the roofer from becoming roof-burned (it happens). *Never use ropes to climb on slate roofs* because they force you to walk on the slates, which are subject to breakage under the weight of a person's feet. Slate roofs are not sidewalks and should not be walked on unless absolutely necessary. One of the main reasons slate roofs go bad on porches and other low-slope roofs is because people tend to walk on them. Generally speaking, the steeper and higher a roof, the better condition it will stay in, since people stay off steep roofs. In addition to ropes, home-made chicken ladders (wooden hook ladders) don't make any sense when steel ladder hooks are so inexpensive and relatively safe.

A good ladder hook has a flat, swiveling piece of angle iron on the end which prevents the slate from breaking under the pressure of the hook. It should also be easy to clamp onto a ladder. Although the hook is typically attached to the top two rungs of the hook ladder, it can be attached to any pair of rungs lower down on the ladder in situations where the available hook ladder is too long to be practical.

The better ladder hooks have a rolling wheel that enables the hook ladder to be rolled up the roof. A rolling hook should have a wheel that only rolls in one direction (up and down the roof), and not sideways. So if you have a hook with a *swivel* wheel on it, you should take the wheel off before an accident happens.

It should be noted that the manufacturers of some ladder hooks specify that two hooks be used at the same time, side by side, on the same rungs of each hook ladder. If you do any amount of slate work, however, you'll soon learn that a hook ladder with two steel hooks is very top-heavy and hard to handle and therefore a hazard of its own. Two hooks are likely to be required, however, when a heavy weight will be loaded on the ladder hooks, such as a roof scaffold holding masonry materials.

The disadvantage of a hook that can be fastened on and taken off the ladder easily is that the hook may come off when you don't want it to, like when you're sliding the ladder up the roof to hook it

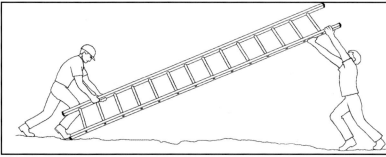

It is much easier for two people to put up a long or heavy ladder than one person alone. The fellow on the left keeps his foot planted <u>firmly</u> on the ladder until it is walked up <u>all the way</u> by the other fellow. He also pulls on the ladder to help lift it as the other fellow walks it up. The ladder is turned on edge to allow for a secure place to set one's foot. The procedure in reverse will safely bring a ladder down.

Always watch for electrical wires when putting up or taking down a ladder!

on the ridge. The author has *never* had this happen, but other contractors have had their hooks fall off. The solution to this potential problem is to *always* make sure the hook is attached square, snug and centered on the rungs of the ladder and the wing nut is firmly secured. Obviously, if you put the hook on the ladder hastily, you're asking for trouble. The top edges of ladder rungs are slanted in one direction so as to allow for a level surface to stand on when the ladder is angled against a building.

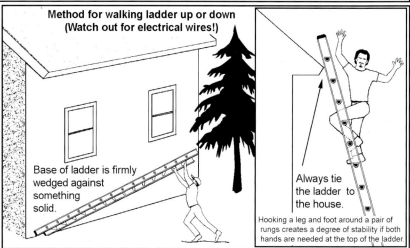

Method for walking ladder up or down (Watch out for electrical wires!)

Base of ladder is firmly wedged against something solid.

Always tie the ladder to the house.

Hooking a leg and foot around a pair of rungs creates a degree of stability if both hands are needed at the top of the ladder.

Hooks are designed to clamp onto those rungs and are made to accommodate that slant, so they *can* be put on backwards by someone who's not paying attention, and if so, they *won't* attach firmly. When you're standing on a hook ladder on a high roof, the hook itself is the only thing protecting you from the force of gravity, a force to be reckoned with when working on steep slopes. That hook should be highly respected and used properly, with utmost care.

There are other styles of ladder hooks, such as pointed ones that are used in pairs and attach to the side rails of the ladder; these are used by people who only need to get up and down the roof (such as firemen) and don't need to actually work on the roof itself. They are of no use to slate roofers.

In order to get a hook ladder up onto a roof, first the ground ladder must be set up properly. Then the hook ladder is "walked" up (often by propping it against the base of the ground ladder) and leaned against the ground ladder, leaving enough room beside it to allow the worker to climb the ground ladder. The ground ladder is then climbed until the climber reaches about 3/4 of the way up the hook ladder. The worker then holds the hook ladder against his shoulder while grasping one of its rungs tightly, and climbs up to the top of the ground ladder carrying the hook ladder. At the top, the hook ladder is raised up far enough to tilt over the edge of the eaves and lay on the roof, after which it is carefully (so as not to damage slate) slid up the roof,

hook facing up, until the top of the hook ladder reaches the peak. The hook ladder is then flipped over and hooked on the ridge. The reverse procedure will get the hook ladder down. If the hook ladder needs to be moved sideways one way or the other, it can be "rolled" from the ground ladder (turned on its back, then onto its front in a rolling motion).

It takes some practice to get hook ladders up and down safely, and it can be a dangerous job, especially on high, steep roofs requiring long hook ladders, or on windy days. Perhaps the most important trick in carrying hook ladders up to and down from roofs is to learn where the center of balance of the hook ladder is (hook ladders are top-heavy because of the hook). Grab the ladder just above its center of balance when carrying it up or down a ground ladder. It will then tilt over the eaves and lie on (or come off) the roof easily. Make sure your body's chest level is just above the pivot point at the eaves of the roof when you tip the hook ladder on or off the roof. Once again, remember the rules of safety, and if it doesn't feel safe to you, don't do it — you can always practice with a shorter, lighter ladder first.

A professional slater will often use several hook ladders at the same time on a roof. One may be used just to get up and down from the ground ladder to the roof peak, while another may be used to work the other side of the roof. Smaller hook ladders may be used to get around chimneys or to work on dormer roofs, and often hook ladders will be paired to make a roof scaffold around a chimney. So it's always a good idea to have more than one hook handy, and more than one hook ladder available. Fortunately, any extension ladder can be taken apart and one half (or both) put to use as a hook ladder, then the extension ladder can be put back together when the hook ladder is no longer needed. This is one reason to remove ropes from ladders when not needed — then the ladder can be taken apart and put back together more easily.

ROOF JACKS (BRACKETS)

Roof jacks are important tools for any roofer, especially slate roofers. Also called roof brackets, they're used to support planks on the roof and thereby create roof scaffolds, which allow a safe place to stand or to set another ladder on the roof surface. It's very important to know how to attach roof jacks to a slate roof safely without damaging the roof, as roof scaffolds will make many a seemingly impossible roof job vastly easier and safer. Speaking of safety, roof jacks are discussed in the previous chapter, and that information should be reviewed before attempting to use roof jacks.

Many roof jacks fold up for storage, and this can be a hazard when the jacks have not been correctly latched open before being nailed on a roof. You don't want to get up on a roof scaffold and have it suddenly collapse because a roof jack wasn't locked open. Always check them *before* putting your weight on them, or else use fixed (non-collapsible) jacks.

Roof jacks are typically nailed on *top* of the slate roof, in the *slots* between the slates where the slates abut one another side to side, and are nailed *through* the underlying slate. In the same way that a hammer and a nail can be used to punch a hole in a slate, a nail can usually be driven through a slate without cracking it (new slates or thicker slates may

need to be drilled). This is especially true of older slates, although some of the very hard slates, such as the New York red slates or Peach Bottom slates, may tend to crack when nailed through even after a century of age, and may require drilling first with a masonry bit. For the most part, though, a nail can be driven through an old slate roof with impunity. Alternatively, a slate can be removed using a slate ripper and the roof jack can then be nailed to the roof without damaging any slates. The removed slate is later replaced using a slate hook.

Roof jacks will always have at least three holes through which to nail them. Many have two sets of three holes. There is a good reason to have three holes on a roof jack — that's how many you should use when safely attaching the jack to the roof. One nail is not enough, two *may* do, but three is the right number. When nailing a roof jack to a roof always use three nail holes or slots and make sure all three nails hit something solid. If you're not hitting something solid with your nails, move the roof jack and try again (and be sure to cover your old nail holes with flashing slid *under* the slate, if you punctured the roof!).

Furthermore, on steep roofs use 16d common nails (3.5" long) when nailing roof jacks through the slate. When doing so, it's critically important to make sure the nail *heads* have properly hooked the roof jack, and aren't driven in so far that the roof jack has nothing to hold on to (it's the nail *heads* that keep the jack from sliding off the roof, so use stout nails with good, large heads). When nailing jacks to a felted roof deck when no slate is present, standard 1.5" roofing nails will usually suffice.

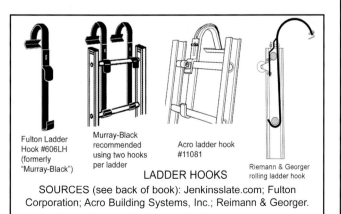

Fulton Ladder Hook #606LH (formerly "Murray-Black")

Murray-Black recommended using two hooks per ladder

Acro ladder hook #11081

Riemann & Georger rolling ladder hook

LADDER HOOKS

SOURCES (see back of book): Jenkinsslate.com; Fulton Corporation; Acro Building Systems, Inc.; Reimann & Georger.

VERSATILITY OF HOOK LADDERS

The longer hook ladder in this illustration is slightly angled across the roof, which is often necessary to gain access to difficult areas. The shorter hook ladder has the hook attached two rungs down from the top to shorten the working length of the hook ladder in order to access a section of roof that would otherwise be difficult to reach. The hook ladder can be angled across the roof only when the ladder hook has a swiveling piece of angle-iron on the end that contacts the roof.

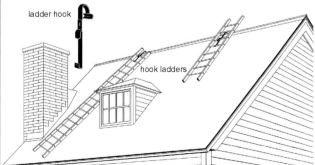

ladder hook

hook ladders

The jacks themselves only *hook* onto the nails, and they're readily removable with an upward tap of the hammer on the bottom of the jack, although removing the nails is often another matter altogether. The nails must be removed from the slate roof or else covered with hidden flashing after the work is done. Sometimes 16 penny nails can be pried from the roof with a hammer using a slate ripper or the jack itself as a backing to prevent the pressure of the hammer head from breaking the slate. If this won't work, then the nails are simply driven into the roof using a nail punch (a 1/2" x 4" bolt will do fine), and left there.

After removing the roof jack, you'll have three nail heads or nail holes (depending on whether you pulled the nails out or drove them in) situated in a slot on the roof between two slates, and these will leak. This is not a problem, however, as a piece of metal flashing called a *bib* is simply slid *under* the slot and *over* the holes to leave the area leak-proof. The metal should be non-corrosive (copper is ideal, but aluminum, especially *brown*-painted aluminum, works well too), and should be bent lengthwise slightly so as to be force-fit under the slate, thereby preventing it from sliding back out. A lengthwise bend in the middle of the metal not only wedges the metal in place, but helps the metal ride over the nail heads, if there are any. The insertion of this metal flashing is almost always assisted by a slate ripper, which pushes the flashing into place.

The metal flashing should be at least 4" wide and long enough to cover the nail holes with at least an inch or two of overlap on the bottom, and 2" of underlap under the next course of slates above. A length of 7" works well on many roofs, although sometimes longer and wider bib flashing is required. *Don't use shiny flashing or white flashing, as it looks unsightly in the cracks between the slates.* Using the above technique, roof jacks can be safely attached to and removed from slate roofs without creating leaks.

SCAFFOLDS

Roof scaffolds are either typically made with roof jacks and planks, as already mentioned, or with ladder jacks (on hook ladders) with planks. Planks can be either wood or aluminum. The roof scaffold is only as good as its weakest part — even if roof jacks are nailed in place with three 16 penny nails each, and (if collapsible) firmly latched open, a bad plank will make the scaffold unsafe. A 2"x10" plank free of large knots, splits, checks or other flaws makes a good scaffold plank. Typically, two roof

jacks are sufficient for a plank that's eight feet long, although three roof jacks over a span of eight feet are much safer when a lot of weight is involved. If you doubt the strength of your roof scaffold, you can always add a central roof jack to beef it up.

Other types of scaffolding are also useful when doing work on slate roofs, and these include ground scaffolding (pipe or frame scaffolding) which sits on the ground and stacks one stage on top of another; pump jacks, which attach to vertical 4x4's; and planks (either wood or aluminum) that sit on ladder jacks positioned on ground ladders below the eaves of a building. Ground scaffoldings usually won't be required when doing general slate restoration work, although they can come in handy at times. Frame scaffolding especially can greatly enhance the safety of high jobs, as well as new installations, eave and gutter work. Frame scaffolding can readily be rented, and the rental agencies will often erect the scaffold for an extra fee.

When using frame scaffolding, *always* tie the frames to the building every three stages of height, minimum. If the building is masonry, drill holes into the mortar joints and slip in lead sleeves, then screw sturdy eye-hooks into the sleeves to tie to. These hooks will be readily removable when the scaffold comes down. *Always* use a complete set of safety rails at the top of the scaffold, which includes vertical posts and a double set of horizontal rails. To make it easier to get materials up to the top of the scaffold, a well-wheel and pulley system made specifically for the scaffold will be necessary. Otherwise, a much more expensive electric ladder hoist system works well, too. And finally, *always* (in my opinion) use aluminum planks at the top of the scaffold, the kind that are made specifically for the scaffold and that hook safely into place. Do not use wood planks. Aluminum scaffold platforms are not that expensive and safety should never be compromised to save a buck. Wood planks may be cheap, but they're narrow, can slip or break, and just aren't worth it. You can always tell a good contractor or craftsman by the condition of his or her equipment. Wood planks on scaffold frames are sub-standard and should be avoided. I guarantee that when you read about people dying from falls from scaffolding (and you will), it's because they did not use proper aluminum scaffold planks, proper safety rails, or proper tie-off procedures.

Furthermore, pipe scaffold must be set up exactly plumb. Adjustable screw feet make this job infinitely easier. Do not pile blocks up on the ground to try to create a level surface and then set

ATTACHING ROOF JACKS TO A SLATE ROOF

On standard thickness (3/16" to 1/4") slate roofs, the roof jacks can be nailed on TOP of the slates in a slot between the slates. On thicker slates such as on the bottom of a graduated roof, the roof jacks should be nailed in the same position but UNDERNEATH the top slate. In both cases, the nails penetrate the slate through the overlying slot. Use three 16 penny common nails, making sure they all hit something solid. To remove, knock off the bracket, then pry out the nails using the bracket as a backing to prevent breaking the slate, or pound the nails down into the roof using a hammer and a bolt. Install a copper or brown-aluminum bib flashing under the slate, but over the holes.

Alternatively, first remove the slate where the roof jack is to be nailed, then nail the jack directly to the roof sheathing in the space between the underlying slates. When done, remove the jack and replace the slate (this is the most common roof jack installation procedure on ceramic tile roofs).

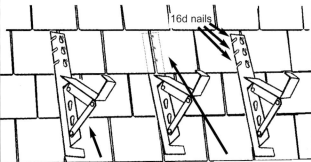

On standard thickness slates (3/16"-1/4"), nail roof jacks on top of the slates through slot between the slates. Alternatively, remove a slate and nail the roof jack to the roof in the space between the underlying slates.	Slide roof jack *underneath* very thick slates if nailing on top of slates. Otherwise, remove the top slate and nail the jack to the roof deck between the underlying slates.

After removing the roof jack, slide a piece of non-shiny metal "bib" flashing under the slate to cover the holes, (or nail heads) as shown. Use the point of a slate ripper to push the metal under the slate. Bend the metal lengthwise to make it easier to ride over the nail heads, and to force-fit the metal so it won't slide back out. Use copper or brown-painted aluminum flashing so the metal won't rust.

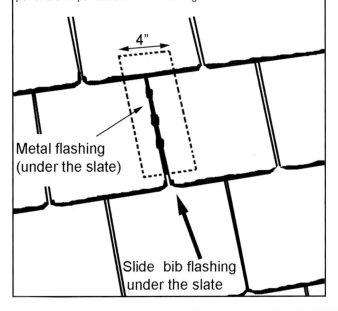

your scaffold on top of them; if you don't have screw feet, then buy some. Finally, avoid electrical wires (and lightning) when using metal scaffolding.

SLATE HOOKS

Slate hooks are widely used by slate roofers to attach replacement slates, instead of using the nail and bib flashing technique discussed in the Chapter 17, and no list of slate equipment would be complete without the slate hook. The slate hook is a simple copper, galvanized, or stainless steel hook that is nailed into the roof after a bad slate has been removed. The new slate is slid into place and the hook keeps it from sliding back out. Copper hooks can be difficult to use on roofs that have hardwood sheathing because they may bend when you try to nail them into place. Galvanized slate hooks will rust in time and leach rust stains onto the slate roof. Stainless steel slate hooks are recommended.

NAILS

The common nails for nailing standard thickness (3/16") slate are 1½-inch-long, 11 gauge copper roofing nails (see Chapter 13). One way to calculate nail length is by using the following formula: nail length = thickness of slate X2 plus one inch. For a standard quarter inch slate, the nail length, then, would be 1.5 inches.

Although copper roofing nails are the preferred nails specified by professionals, 99% of old slate roofs do not have copper nails — they have *hot-dipped galvanized* nails, or even cut steel nails, both of which are far less expensive than copper and will usually last at least a century. In fact, many century-old nails of these types are in good enough condition to be used over again. So even though industry professionals insist on using copper nails, if you're using a good, *hot-dipped galvanized nail* to fasten the slates to a roof (especially salvaged slates), you'll find that they work quite well. Copper nails, however, *are* superior to hot-dipped galvanized, and stainless steel are superior to copper, so these are the nails to use when installing new slate. Aluminum nails should be avoided as they don't last long enough on a roof. Longer and heavier copper nails are required for thicker slates or for nailing copper ridges, ceramic tiles, etc.

Never use *electroplated* galvanized nails, which are cheap nails made for asphalt shingles, and should not be used on a slate roof. Electroplated nails are marked "EG" roofing nails (electro-galvanized), whereas hot-dipped nails are clearly marked "hot-dipped."

The traditional nail for fastening galvanized ("tin") ridge iron to a slate roof is the eight penny (2 1/2" long) hot-dipped galvanized nail. After this nail is pounded into place, the head is covered with a dab of caulk (lifetime durability clear silicon is recommended). Some roofers insist on using gasketed nails on ridge iron, but a standard eight penny nail with a caulked head works best. Copper ridge, of course, requires copper nails. When nailing copper flashings, valleys or ridges, one should always use copper or brass nails to prevent galvanic action. Galvanic action occurs when dissimilar metals are placed in contact with each other, leading to the deterioration of the electropositive metal. This is discussed further in the Chapter 18.

Stainless steel roofing nails should be considered whenever installing a new slate roof onto old, hard roof sheathing. Southern yellow pine or northern hardwoods, although they can last for centuries on a roof, become quite hard when dry and may bend copper roofing nails. Switch to stainless steel roofing nails to solve this problem.

ROOF CEMENT AND CAULK

Remember these two rules about roof cement, also known as "mastic," or "tar" — a black, plastic material that is either trowel or brush grade:

1) *Roof cement should never be visible on the surface of a slate roof.* This goes for both installations and for repair work. Granted, there is one type of slate ridge installation which requires exposed nail heads (an installation procedure that I do not recommend, shown on the Smithsonian, p. 27), but these nail heads can be caulked with lifetime silicon caulk rather than the more unsightly roof cement. 2) *Roof cement should not be routinely used underneath slates during installation* as is sometimes recommended by printed installation directions. The incorrect practice of cementing eave, valley and ridge slates into place was, unfortunately, published as an accepted procedure in the 1926 book, Slate Roofs, now extensively reprinted and therefore perpetuating an un-ending plague to those of us who actually have to repair or maintain slate roofs. Fortunately, most slaters of old either never saw that book or else had sense enough to know not to cement the slates into place, as 99.9% of all old slate roofs were not installed in this manner. Slate roofs are ingenious roofing systems because they can be taken apart and put back together, *but not when they have*

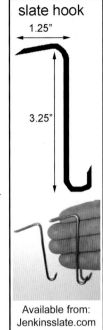

slate hook

1.25"

3.25"

Available from: Jenkinsslate.com and from slate suppliers (see back of book)

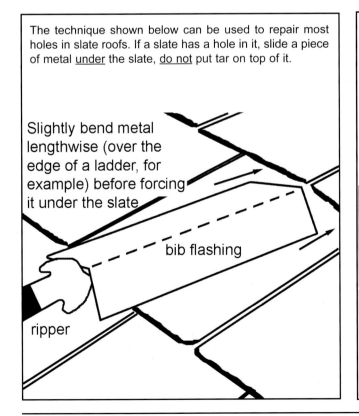

The technique shown below can be used to repair most holes in slate roofs. If a slate has a hole in it, slide a piece of metal <u>under</u> the slate, <u>do not</u> put tar on top of it.

Slightly bend metal lengthwise (over the edge of a ladder, for example) before forcing it under the slate

bib flashing

ripper

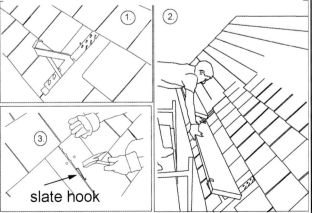

USING ROOF JACKS
WITHOUT NAILING THROUGH SLATES

1) Remove a slate by pulling it out with a ripper. Nail a roof jack into place by nailing directly into the roof deck in the slot between the slates underneath.
2) Stage the roof with adequate roof scaffolding.
3) When done, remove the roof jack by knocking it off the nails and then either pry the nails out or punch them down. Then, install a stainless steel slate hook where the slate is missing and hook a slate in place. The slate can always be removed again later should the unlikely need for future roof scaffolding present itself.

slate hook

been cemented into place! Slates are best installed with nails, and no roof cement is needed. The only exception to this rule is when small pieces of slate are being installed in an unusual circumstance, such as on a steep hip or small finial, or when edge slates are exposed to abnormally high winds and likely to be blown off — then some cement or caulk *under the slates* is recommended.

Roof cement comes in two standard varieties: standard or "wet surface." Wet surface roof cement can be used to seal leaks on damp surfaces, such as during a rainstorm, and it's a good idea to keep some handy if you're a roofer, although standard roof cement has more body and is the preferred material for general roof repairs. Roof cement is also available in caulking tubes.

A good all-around caulk for slate roofs is clear, lifetime silicon caulk. It really sticks to just about anything, holds tightly, lasts a long time, and is waterproof immediately (although it can't be used on damp surfaces). It's used to seal exposed nail heads, to glue small pieces of slate into place when necessary, and to fill mortar joints on chimneys after they've been reflashed.

LADDER JACKS AND STAND-OFFS

Ladder jacks (also called ladder brackets) are used to support planks and create working platforms on ground ladders or on hook ladders. There are many different types of ladder jacks, and they're most often used often by contractors who attach them to ground ladders for the purpose of working on the siding, gutters, soffit, fascia or eaves of a house. Ladder jacks are most useful to slaters when attached to hook ladders to create a quick and sturdy roof scaffold without nailing anything into the roof. As such, the best ladder jacks to use are lightweight, aluminum ladder jacks, which can be carried up and down a ladder with ease. They require no tools to attach to the ladder, and simply clamp onto the ladder by design. They must be used in pairs, and therefore require pairs of hook ladders. They're especially handy when working on chimneys. Hook ladders must be heavy duty if the roof scaffold attached to them is going to support heavy weight, such as bricks.

Planks are laid across the ladder jacks in order to create a work platform. Aluminum planks span longer distances and can be much safer than wooden planks when the ladder jacks are attached to ground ladders, especially as aluminum planks can be clamped to the ladder jacks and can be used with safety rails.

Stand-offs are aluminum devices that clamp to the top of the ladder and hold it away from the building or roof. They enable one to work on, for example, the gutters or drip edge of the roof without the ladder having to lean against the work area. Both aluminum ladder jacks and stand-offs are available at most building supply stores.

Pipe scaffolding provides additional safety when working on high roofs.

Photo by author.

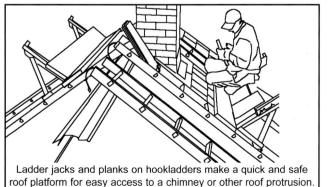

Ladder jacks and planks on hookladders make a quick and safe roof platform for easy access to a chimney or other roof protrusion.

Chapter Thirteen

INSTALLING SLATE ROOFS

A slate roof is a perfect roof. It's beautiful, natural, durable, recyclable, environmentally friendly, easy to maintain and costs less than just about any other serious roof when the life of the roof is taken into consideration. Although there are many myths and misconceptions about installing slate roofs, there's really nothing mysterious about building a roof of stone. The stone is dug or quarried from the earth, brought to the surface, then hand split with a hammer and chisel into thin sheets about a quarter of an inch thick (more or less). These slate shingles are then trimmed into particular shapes, usually punched for nail holes, and are then ready to be installed on a roof. They typically come from American quarries already trimmed and punched, so the roof installer only needs to know how to put them on the roof.

There is, however, no "one right way" to install roof slates. The covering of roofs with stone is an age-old tradition that has been practiced around the world for centuries. Different regions of the world have their own traditional slate roof installation techniques. We will look at some of the various types of international installation methods in this chapter, but we will focus primarily on the traditional American styles of slating.

Roof slates in the United States are tradi-

tionally nailed onto solid wood roof decks built of natural boards. This traditional style of construction has been tried and proven over centuries and can easily be duplicated when installing slate roofs on new or restored construction today. The primary characteristics of traditional American slate roof construction are as follows:

1) The slates are nailed, not hooked, onto a completely sheathed roof deck rather than a lath roof. Lath roofs are not uncommon in the U.S. but they are not preferable to solid board decks, as we will discuss later.

2) Full one-inch-thick, local, green or air-dried lumber boards are used for roof sheathing, or other comparable roof decking material such as 3/4" kiln dried boards, tongue-in-groove boards, or 1½" t-i-g lumber (more suitable for large institutional buildings). Laminated woods, plywoods, particle boards, and chip board decks are avoided.

3) The roof deck is covered with 30 lb. felt paper — self-adhesive underlayments are not needed. Even the felt paper is optional.

4) The slates are nailed onto the wooden sheathing using two nails per standard slate (no adhesives are used). Four nails per slate may be used in hurricane prone areas or when installing excessively large, heavy slates. The nails recommended today are copper or stainless steel, although most old

▲ Traditional British stone roofing style (above) involves the use of lath strips in the roof construction. This abandoned farm building is located on the Yorkshire-Lancashire border in the UK. Close-up on following page.   Photo by Dave Starkie.

▲ Close-up of lath and peg roof construction on old abandoned farm building in England showing the minimal amount of wood that can support a stone roof. The slates were attached with wooden pegs, not nails.

▼ Solid wood sheathing (decking) made from local mixed hardwoods only one week "off the stump." This Pennsylvania outbuilding roof will be slated with Vermont "sea green" slates. The finished roof can be expected to last 150 years or longer.

Top photo by Dave Starkie, bottom photo by Bob Sayre.

roofs used iron or galvanized steel.

5) A headlap measurement of three inches is typically used (more for lower slopes or ice-dam conditions).

6). Sidelaps, or the lateral spacing of the slates in relation to the courses above and below, should incorporate a minimum 3" overlap, if possible.

7) Slate roofs are installed on slopes greater than four inches of run in twelve inches of rise, although the most long-lasting roofs are installed on slopes that are too steep to walk on.

1) SOLID WOOD ROOF SHEATHING

The slates are nailed onto a completely sheathed roof deck — the roof deck is constructed of full boards abutting each other so as to cover the entire roof surface, leaving no, or negligible, gaps between the boards, other than perhaps "toe-holds" or spaces left between the boards for footholds during construction. This is in contrast to the practice of nailing the slates onto "slater's lath," which are narrow strips of wood spaced evenly on top of, and perpendicular to, the roof rafters so as to provide a place to nail or hook the slate (as shown at left). Roof lath construction only requires the bare minimum of wood — just enough to provide an anchor for the nails or hooks, and is therefore more conservative in the use of lumber. This is the preferred method of construction in parts of Europe where many of the traditional buildings are constructed of stone or masonry and wood is minimally used. American barns often have lath-type roofs, no doubt because of the lesser cost of the materials. Many homes in specific regions of the U.S. also have lath roofs, probably reflecting the ethnicity of the people who built them.

The practice of using roof lath to support slate may have developed as a response to the scarcity of forest resources in Wales and Europe in the 18th and 19th centuries. That scarcity did not exist in the United States during that time, when most roof decks were completely covered with wood as opposed to just strips of lath. Perhaps this technique can be traced back to Scottish ancestry, as the Scots choose not to make their roofs of lath, but to use a solid wood deck instead, for reasons we will discuss later in this chapter. This turned out to be a good idea for American roofs — it created a stronger, better insulated roof that could withstand the snow weights typical of the American northeast. A stronger roof makes for a stronger building, and

America's wooden building styles benefited from a roof of solid sheathing. Solid sheathing also provides much more nailing surface, making it easier to repair slate roofs, and therefore facilitating their later restoration.

It's interesting to note that roof slates in Wales were originally "hung" on roof lath using only wooden pegs driven through the slates (as shown at right). The thin lath were stripped, not sawn, from local logs. The hardwood pegs were split down to the size of a pencil stub from a block of wood, and a single peg was driven through the top center of the slate so that the peg was flush with the front of the slate, which was then hung on the lath. The heavy weight of the slates overlapping each other held the roof together.

Today, the Welsh use a modified version of the lath and peg system, which we will look at later in this chapter.

2) USE SOLID LUMBER

Full one-inch-thick, local lumber boards have been traditionally used for roof sheathing, unplaned, not tongue-in-groove, and not kiln dried. The author stayed in a house in Scotland, built in 1785 and still covered by its original slate roof (restored). The existing 215-year-old roof sheathing was made of 1" rough-sawn lumber. Kiln-dried and planed lumber, although not as commonly used a century ago probably due to the extra and unnecessary expense of drying and planing, has also been proven to work well as a roof deck under slate.

Green lumber is not kiln dried, and may not even be air dried, but it still makes a great roof deck. The boards simply dry in place after being nailed to the rafters. When fully dry, they'll have a gap between them of about 1/2" (wood shrinks more width-wise than lengthwise by a factor of seven). Some contractors may insist that the boards will warp, crack, and twist when they dry, but when nailed properly, the boards will remain flat.

In fact, many hardwoods such as oak *must* be used green. When completely dry they're too hard — nails won't penetrate them without bending. It's worth mentioning that a tree felled in the winter when the sap is down will dry more quickly and shrink less than one felled when the sap is up.

A one-inch-thick, solid wood roof deck is a strong, durable one which will last for centuries. And if a leak damages part of the roof sheathing, it can simply be cut out and replaced with the same material — local lumber.

▲ A "stone slate" from a 16th century abbey in Wales showing the actual peg and lath used in the construction of the roof.

▼ The author replacing a valley on a century-old, American barn with white oak lath construction. Note the deplorable condition of the disintegrated terne metal valley, 34 feet long, draping between the lath strips like paper. The oak lath was still solid, however, making the valley replacement a routine job. Imagine the nightmare if the barn had been sheathed in a shorter-lived laminated wood, which would have delaminated, rotted, and warped along the entire length of the valley!

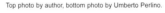

Top photo by author, bottom photo by Umberto Perlino.

Local lumber can be air dried, which simply means that the lumber is "stickered" for a few weeks or months before using. Stickering lumber means stacking it with dry sticks of uniform thickness (about an inch) between the layers of boards so air can circulate around the lumber, thereby drying it out. However, when building, framing or sheathing with green lumber, stickering is not necessary if the lumber will be nailed in place within two weeks of cutting. If the lumber is going to sit around longer than that, it should be stickered and protected from rain until it's time to use it; otherwise it may develop surface mold.

Local lumber is typically only sold by the sawyers who saw it; it usually needs to be ordered in advance and the buyer may wait for weeks when large quantities of specific species are ordered. It's also "green" (undried) when you buy it. Furthermore, it's heavier than the kiln-dried softwoods typically sold by building material centers, and therefore it requires more effort to work with. Because of the extra effort in locating a sawyer, ordering the lumber, waiting for it, hauling it, stickering it (if necessary), then lugging it up on the roof, only the most conscientious of builders use it nowadays. It is well worth finding those builders when installing a slate roof if you want the roof to last. If a contractor tells you he can't use local lumber because it's green and it will twist and warp, find another contractor. And it's well worth locating a local sawyer and ordering your roof decking directly from him if you're going to build your own roof.

You can locate local sawyers by looking under "lumber-wholesale" in the yellow pages of your phone book. Rough-sawn lumber is full size — a 1x8 is a full one inch by eight inches — therefore, the "board feet" required for the job is actually the square feet of roof area including overhang. When ordering local lumber, always allow for *at least* 15% waste when figuring your total area, and more when valleys and hips are being built. Don't deduct from the total area for skylights or chimneys.

Use #2 grade, one-inch-thick boards at least 6" wide. Preferred species are the softwoods such as white pine, yellow pine and hemlock, or the hardwood yellow poplar (tulip poplar) because they're lighter and therefore easier to work with, although almost any tree species suitable for lumber will do, including oak, cherry, maple, ash, chestnut, birch and beech. The boards do not need to be planed or joined, or milled in any way, although *it is important to get boards from a sawyer who does a good job of cutting his lumber.* Some sawyers are sloppy and their

boards will vary as much as 1/2" (or more) in thickness or width from one end to another. If you do run across a bad board in your lumber, set it aside and use it for something else — not for the roof. If you find a board with an excessively large knot or other defect, cut the defect out before using the board, or don't use it at all. This is one of the simple tricks to using local, green lumber (or any lumber) — sorting out the bad boards, and you probably will have some. It's the boards with the bad knots that are likely to warp when drying. When nailing the boards to the rafters, butt them firmly against each other, both at their sides and at their ends, and nail them down tightly. The slates are nailed to the roof sheathing while green; you need not wait for the wood to dry. The wood is felted while green as well — it will dry in place (providing you're not using self-adhesive underlayments, which are not recommended).

A suitable alternative to local lumber for roof decking is 3/4-inch-thick, planed, kiln-dried roof sheathing boards available from any lumber yard. These are usually made of spruce, fir, or pine and can be quite a bit more expensive than local green lumber, but are certainly preferable to plywood. You can also use tongue-in-groove roof decking under slate. One of the longest lasting roof sheathing lumbers is white oak, although it is heavy and somewhat difficult to work with, and hard to nail into when dry.

Nail 3/4- or 1-inch-thick roof sheathing to the rafters with eight penny common nails (2 1/2 inches long). Thicker decking will require longer nails.

THE PLYWOOD ROOF DECK CONTROVERSY

Unfortunately, a new generation of slate roofs is now being constructed in the United States using plywood roof decks, some as flimsy as 1/2" thick — even on multi-million dollar homes! These roofs can present many headaches for the slate roof restoration professionals of the future when the plywood begins to delaminate, sag, or rot.

Plywood is a material that is designed for the convenience of the contractor, not for longevity. Although it is a convenient material for the contractor and the architect, it should not be used on roofs expected to last a century or two. A slate roof will not be routinely ripped off and replaced every twenty years like its asphalt cousin. Plywood has not yet proven to have any longevity approaching the phenomenal life span of slate. Some argue that this is only because plywood hasn't been on the market long enough to prove itself. On the contrary, plywood has been used long enough on roofs to display an uncomfortable failure rate. For example, here's a quote from a roofing magazine:

> *"Though the building was only twelve years old, the roof was leaking at every conceivable point. The original shingles were guaranteed for twenty years, but poor ventilation had caused the plywood deck to delaminate in many places, and shingles to blow off during high winds. The roof was repaired on a consistent basis, but the efforts were in vain due to the delamination of the plywood."*
>
> Quoted from Roofer Magazine, November, 1997: The Roof Doctor's Prescription for Success, by Melinda North (page 27).

Today it's often recommended that all plywood roof decks be covered with a self-adhesive underlayment along the eaves, under valleys, and almost everywhere else on the roof deck in order to preserve the plywood. Not only is this more expensive than the traditional, tried-and-proven, wood boards and 30 pound felt roofing system, it is also more time consuming and less ecologically sound. It creates a roof deck that cannot breathe, adding new roof problems related to condensation and ventilation.

Wood decks made of boards are composed of pieces of wood perhaps six to twelve inches wide. In between each of these boards is a small gap. This gap not only allows for the roof deck to breathe, but it also allows for any water penetration to drip into the building and be quickly discovered (and repaired). Neglected leaks on old roofs can rot the board underneath, but the rot will be localized and easily repairable — it is not inclined to spread. Plywood, on the other hand, is made up of four foot by eight foot sheets of thin layers of wood glued together. If water penetrates the roof, the moisture wicks through the sheet causing delamination and rot to spread over a wider area. The plywood becomes punky and soft — any roofer who has put his foot through rotten plywood knows exactly what I'm talking about.

When slate roofs get to be a century old, they may have suffered a lot of neglect. Although slate roofs are fantastically successful roof systems, a lot of bad things can happen to them in a century or two. But because of the traditional methods of installation and the *traditional materials* used, there is almost no problem that can afflict an old slate roof that cannot be remedied.

3) NAILING THE SLATES

Roof slates are nailed onto the wooden sheathing using two nails per standard thickness slate, the nails being approximately 1.25" to 1.5" long and typically (on older roofs) made of hot-dipped galvanized (zinc-plated) steel. Alternatively, hardened copper nails of the proper length are used, which are the preferred nail on more expensive homes and buildings with slate roofs and on new roof installations today. Stainless steel nails may be preferable where older roof boards have hardened with age, which is typical of southern yellow pine and northern hardwoods. Roofs with thicker slates require longer nails. The rule of thumb is that the nail length should be twice the thickness of the slate plus one inch.

American roof slates are supplied by the manufacturer already pre-punched with nail holes, in contrast to European roof slates, many of which come from the quarry or mine without holes. American roofs tend to have solid sheathing, whereas many European roofs tend to use slating lath, so American slates can be pre-punched without worry as to whether the nail holes will line up with the lath. Many European roofers must "hole" their slates at the job site.

Nail holes should just clear the top of the underlying slate. This allows the slate to lay best. If the holes are too low you'll nail through the top of the underlying slate, which should be avoided. It's much more difficult to remove and replace "double-nailed" slates due to the inaccessibility of the upper pair of nails to the slate ripper. On the other hand, if the regular pair of nails are nailed too far up on the slate, the slates may not lie flat on the roof.

The slates should not be nailed down too tightly, but are actually "hung" on the roof nails, so they can "float" on the roof, although nailing snugly is OK. This prevents the slates from being subjected to damaging pressures which can crack the slate over time, referred to as "over-nailing." It takes some practice to nail roof slates properly, but it isn't difficult. The nail holes are pre-punched so the beveled side of the hole faces outward, allowing for the nail head to sit down into the beveled hole, thereby preventing the nail from rubbing against the slate over top of it and wearing a hole in it (referred to as "under-nailing"). When punching a hole in a slate, the strike against the slate should be against the back so the bevel appears on the side where the nail head is to rest.

Traditional Scottish slating, as well as other

older UK and European slating styles, require a single nail or peg at the top center of the slate or stone, known as "head-nailing." Standard American nailing is known in Europe as "center-nailing." This will be discussed in greater detail later in this chapter.

A BRIEF WORD ABOUT VENTILATING SLATE

Roof slate rock does not need to be ventilated. Exposure to air, sunlight, and heat all slowly cause slate rock to deteriorate. The longest lasting slate is the rock that's left in the ground unexposed to air. When you pry a century old slate off a roof and look at the back of it, you will see that those parts of the slate that had been in direct contact with the felt paper underneath with no exposure to air whatsoever are in the best condition. The greater exposure to air, the more deterioration. This is especially evident on old lath roofs where the part of the back of the slate contacting the lath remains sound, but the part between the lath exposed to air is flaking and delaminating. Is it wrong to ventilate the slates? No, but it's definitely not necessary.

This is not to be confused with the ventilation of the roof itself. Roofs should be able to breathe. Modern American roofing methods that utilize plywood, self-adhesive underlayments, and asphalt shingles create roofs that can't "breathe" and instead have ventilation problems. These roofs must have a ventilation system installed, usually through the ridge, but also out the gable ends or through other roof vents.

Interior roof spaces and attics on slate roofs should be ventilated if unheated. The airflow can be out gable end louvers, through roof vents, or via ridge vents designed specifically for slate roofs. Gable vents are preferable as they keep obstructions off the roof.

Any warm air that comes into contact with a cold surface will cause condensation. In cold climates, if warm household air is entering the roof space and coming in contact with the underside of the cold roof sheathing, then water droplets will form on the inside of the sheathing, thereby creating a condensation problem. Adequate insulation and proper ventilating techniques will keep the underside of the roof sheathing at the same temperature as the ambient outdoor temperature, preventing condensation. When insulating between rafters, leave an air space between the insulation and the roof sheathing boards, if possible. Always install a vapor barrier, such as a sheet of plastic, *interior to the insulation,* in order to prevent warm, moist air from exiting the living space and infiltrating the roof space.

FRAMING THE ROOF

Slate is heavy — standard 3/16" thick slate weighs about 700 pounds per square (100 square feet of finished roof) when new. This is about three times as much weight as standard asphalt shingle roof material, and it raises concerns among people who are afraid their structure won't stand up to the heavy load of slate. How should a slate roof be framed? What are safe timber sizes and spans?

Many older homes have 2x10 rafters on two foot centers and this has proven to work well. However, rarely is a 16' span on a rafter left unsupported without an inside wall or collar brace. It's recommended that a 2x10 rafter on two foot centers not exceed 10' in unsupported span (see chart). Many older slate roofs have only 2x5 or 3x5 rafters on two foot centers and have easily withstood a century of time, although the rafters tend to be oak or another strong hardwood, and are usually braced with collar ties and/or inside walls. And of course, the solid sheathing adds to the strength of the roof.

The author has designed and/or built many roofs, all of local (green) lumber, as was the style when most slate roofs were originally constructed at the turn of the last century. Building with full-sized, unplaned, rough sawn, green local lumber requires no special tools or equipment, and it allows for the construction of a solid, long-lasting roof. Again, the lumber shrinks and dries in place, causing no detriment to the structure or the roof. Instead, the structure becomes stronger as the lumber dries. This style of construction has been achieved on millions of buildings across the USA. Most of the men who built local-lumber buildings are dead, however, as are the slaters who roofed the buildings. Many of today's modern builders and architects have discarded the traditional building methods for ones that focus on contractor convenience. And now, unfortunately, it's difficult to find people in the building trades with knowledge of traditional building methods and materials.

As we are beginning to see, there are a number of factors involved in determining the strength of the roof. First, the weight of the roofing must be taken into consideration. Secondly, the type of lumber must be noted, as some lumber species are much stronger than others. Thirdly, the size and spans of the framing members are important, as unsupported spans that are too long will sag over time. And lastly, the slope of the roof is important, as the lower the slope of the roof, the more snow weight it must bear, while the steeper the roof, the more wind-force it must withstand (see chart).

The rule of thumb on slate roof slope is never to go below a 4:12 pitch. The lower the pitch, the longer the headlap, so that a 4:12 roof requires a 4 inch headlap. The steeper the roof, the shorter the headlap, so that a 12:12 roof can get by with only a two inch headlap, although a 3" headlap is recommended. Most slate roofs have a slope between 8:12 and 12:12, while 10:12 is perhaps the most common.

INSTALLING SLATE

Slates have a top and bottom, front and back, and they must be oriented properly on the roof. In general, always lay the slate face out and top up. Furthermore, butt the slates against each other on the sides. There is no advantage to leaving a space between the slates. The slates will not expand and contract and push against each other as wood does.

Most older slate roofs in America are laid in a standard pattern consisting of slates that are, on each roof, virtually all the same width, length, and thickness (3/16" to 1/4" is standard thickness). It is not uncommon for slate roofs to be made of slates of random widths, or with staggered butts, or ragged butts, or with slates that vary in both length and width (graduated roofs — to be discussed later), or that vary in thickness, or all of the above. All of these styles of slating must always follow two fundamental rules: sufficient headlap and sufficient sidelap must always be incorporated into the laying of the slate. Headlap is the overlap each slate has in relation to the second course below it. Sidelap is the lateral spacing of the slots (where the sides of the slate abut one another) in relation to the slots on the course above and below.

A rule of thumb is to use larger slates on larger roofs, and smaller slates on smaller ones (this rule is frequently broken). Larger slates go on faster, as fewer are required to cover a given area. The largest standard size is 14" wide by 24" long, and only 98 of these will cover one hundred square feet of roof surface with a three inch headlap, using about 200 nails. An 8x10 slate, on the other hand, requires 514 slates to cover the same area, and about 1,028 nails. Obviously then, the larger slates require less time and labor to install.

Some styles of laying slate were developed in order to conserve materials, such as the Dutch Lap (or "side-lap"), Open Slating, and "French" methods, which require far fewer slates to cover a roof than the standard overlap pattern does. These styles tend to be found on barns and outbuildings where the owner probably didn't want to spend extra money on materials. The main problem with these conservative slating styles is that the roofs are difficult to restore properly when they get old. The standard lap style is ingenious in that it allows for easy replacement of any broken slates in such a manner that the nail holding the replacement slate is covered by flashing and therefore rendered leakproof, or else a slate hook is used and no flashing at all is required. Side-lapped slates do not allow the nail and bib option for replacement slates and one is instead limited to the use of slate hooks or face nails (see Chapter 17). Also, these alternative slating methods leave much of the roof area covered by only a single layer of slate — a situation more vulnerable to such threats as hail damage. Therefore, when installing a slate roof, it's recommended that the slates be laid in a standard lap fashion. Of course, some traditional European styles of slating, notably the Old German style, require that *all* slates be side-lapped. However, of the traditional American slating styles, standard overlapping has proven to be the most durable, waterproof, strongest, and easiest to

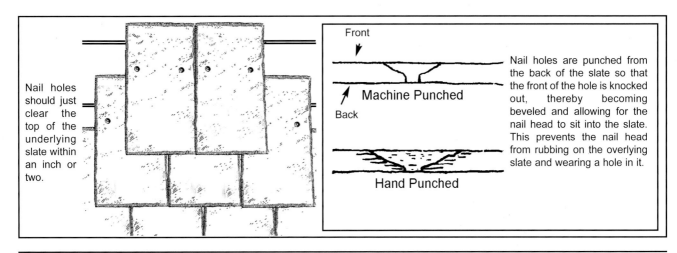

Nail holes should just clear the top of the underlying slate within an inch or two.

Front

Machine Punched

Back

Hand Punched

Nail holes are punched from the back of the slate so that the front of the hole is knocked out, thereby becoming beveled and allowing for the nail head to sit into the slate. This prevents the nail head from rubbing on the overlying slate and wearing a hole in it.

NAILS

Correct nail length is twice the thickness of the slates plus one inch (nail length=slate thicknessX2+1").

Recommended nails for new slate installations include solid copper or stainless steel. Type 304 stainless nails are made of nickel/chromium stainless steel. Type 316 contains slightly more nickel than Type 304 as well as 2-3% molybdenum, making it more resistant to corrosion caused by sea water or sulphuric acid. Both types of stainless show negligible weight loss and no corrosion when tested under one year of accelerated exposure. Under the same exposure, copper nails show a 26-28% weight loss and visible corrosion, while aluminum nails show a 32% weight loss and pronounced corrosion, and electrogalvanized nails show a 100% weight loss and complete corrosion.

(Source: Swan Secure Products, Fasten Data 105)

LENGTH (INCHES)	GAUGE	SHANK DIA. (INCHES)	COPPER HEAD DIA.	COPPER COUNT/LB	STAINLESS HEAD DIA.	STAINLESS COUNT/LB
3/4	12	109	21/64	360	5/16	380
1	12	109	21/64	288	5/16	348
1.25	12	109	21/64	236	5/16	255
1.5	12	109	21/64	202	5/16	218
1.75	12	109	21/64	176	5/16	189
2	12	109	21/64	156	5/16	170
3/4	11	120	3/8	302	11/32	323
1	11	120	3/8	229	11/32	244
1.25	11	120	3/8	187	11/32	199
1.5	11	120	3/8	155	11/32	170
1.75	11	120	3/8	139	11/32	149
2	11	120	3/8	124	11/32	136
1	10	134	7/16	187	3/8	212
1.25	10	134	7/16	147	3/8	166
1.5	10	134	7/16	123	3/8	139
1.75	10	134	7/16	112	3/8	126
2	10	134	7/16	93	3/8	105
2.5	10	134	7/16	81	3/8	91
3	10	134	7/16	67	3/8	78

NUMBER OF SLATES AND NAILS FOR 100 SQUARE FEET OF ROOF

FOR STANDARD THICKNESS (3/16" TO 1/4") SLATES USING 1.5" NAILS

SIZE OF SLATE (INCHES)	EXPOSED LENGTH (INCHES)	# SLATES 3" HEADLAP	# SLATES 4" HEADLAP	APPROXIMATE WEIGHT OF NAILS (POUNDS) GALV.	COPPER 11 GA.	STAINLESS 11 GA	SPACING OF LATH
6x10	3 1/2	686		7 7/8	8.85	8.07	3 1/2
7x10	3 1/2	588		6 3/4	7.58	6.91	3 1/2
8x10	3 1/2	514		5 7/8	6.63	6.04	3 1/2
6x12	4 1/2	533		6	6.87	6.27	4 1/2
7x12	4 1/2	457		5 1/4	5.89	5.37	4 1/2
8x12	4 1/2	400		4 5/8	5.16	4.70	4 1/2
9x12	4 1/2	355		4 1/8	4.58	4.18	4 1/2
10x12	4 1/2	320		3 5/8	4.13	3.76	4 1/2
7x14	5 1/2	374		4 1/4	4.82	4.40	5 1/2
8x14	5 1/2	327		3 3/4	4.22	3.85	5 1/2
9x14	5 1/2	290		3 3/8	3.74	3.41	5 1/2
10x14	5 1/2	261	288	3	3.37	3.07	5 1/2
12x14	5 1/2	218	240	2 1/2	2.81	2.56	5 1/2
8x16	6 1/2	277	300	3 1/8	3.57	3.26	6 1/2
9x16	6 1/2	246	256	3	3.17	2.89	6 1/2
10x16	6 1/2	222	230.40	2 1/2	2.86	2.61	6 1/2
12x16	6 1/2	185	192	2 1/8	2.39	2.18	6 1/2
9x18	7 1/2	213	220.69	2 1/2	2.75	2.50	7 1/2
10x18	7 1/2	192	198.63	2 1/4	2.48	2.26	7 1/2
11x18	7 1/2	175		2	2.26	2.06	7 1/2
12x18	7 1/2	160	171.43	1 7/8	2.06	1.88	7 1/2
10x20	8 1/2	170	180	2 3/8	2.19	2.00	8 1/2
11x20	8 1/2	154		2 1/8	1.99	1.81	8 1/2
12x20	8 1/2	141	150	2	1.82	1.66	8 1/2
14x20	8 1/2	121		1 7/8	1.56	1.42	8 1/2
11x22	9 1/2	138	145.46	2	1.78	1.62	9 1/2
12x22	9 1/2	126	133.34	1 3/4	1.63	1.48	9 1/2
14x22	9 1/2	109		1 1/2	1.41	1.28	9 1/2
12x24	10 1/2	114	120	1 5/8	1.47	1.34	10 1/2
14x24	10 1/2	98	102.86	1 3/8	1.26	1.15	10 1/2

[Sources: Radford's Estimating and Contracting (1913), p. 252; and Slate Roofs (1926), p. 12; # slates for 4" headlap from Slating and Tiling by J. Millar, 1937, pp. 21-22]

1.5" (11 ga.) copper roofing nails

for nailing standard thickness slates

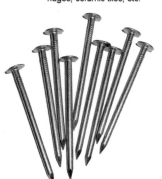

2.5" (12 ga.) copper roofing nails

for nailing thicker slates, some flashings, ridges, ceramic tiles, etc.

OLD SLATING NAILS

1919
82 years old

1890
111 years old

Copper or stainless steel nails are the preferred fastener for new slate roof installations. However, most old slate roofs were fastened with iron or steel nails such as those shown above right, taken from actual slate roofs of the age indicated. It is a very common claim that old slate roofs should be condemned because iron or steel nails were used to fasten the slates, a claim that does not bear up under scrutiny. The old galvanized nails on the left were installed in 1919 and were removed during the replacement of a valley at 82 years of age, in very good condition. The old cut steel nails above right, also removed during a valley replacement, were installed in 1890, and at 111 years of age are in excellent condition. Although nails are sometimes a cause of slate roof failure, they usually are not to blame.

maintain and restore.

Another style of slating that has been discredited and should be avoided at all costs is what has been known as the "economy method." This method eliminates the most critical element of a successful slate roof installation — head-lap, and instead utilizes a strip of felt paper between each row of slates to prevent water from penetrating in the slots between the slates. Although this looks like a normal slate roof after installation, as soon as the felt paper wears out, the roof leaks, cannot be repaired, and must be entirely re-slated in a proper manner, with headlap. This style of slating is only done by contractors who are totally ignorant about how a slate roof works, or totally unconcerned with the quality or longevity of their work.

It should also be mentioned that, no matter how much headlap is used, felt paper should *never* be laid between the rows of slate during installation as is done with wooden shake roofs. The felt paper on a slate roof is always completely *underneath* the slates and never between the courses. Remember that you are creating a roof that should last a century or two, and that it should be installed with future maintenance in mind. At some point in time some of the individual slates will need to be replaced for any number of reasons, including an accidental hit by a baseball, for example. When felt paper is layered between the courses of slates, it severely inhibits the use of a slate ripper, which slides under the slate in order to hook the nails and remove it. The felt bunches up and turns an easy job into a headache.

Today it's common to see new slate roofs laid with a random width pattern — the slates are all the same length, but the widths are random. Random width slates can be slightly less expensive than uniform width slates, and they work just as well, so they're preferred by some roofers. The joints, or slots between the slates should have at least 3" of lateral spacing where they overlap the course below, and this little bit of extra critical detail when laying the slate increases the overall labor time involved when installing a random width roof.

UNDERLAYMENT

The underlayment is the waterproofing layer that exists between the wood roof deck and the slates, typically 30 pound felt paper (weighing 30 per 100 square feet of roof). Its primary purpose is to provide temporary waterproofing while the finished slate roof is being installed. Once the slate is installed, the underlayment becomes obsolete as a

waterproof material — it is now full of holes from the slate nails and is no longer needed anyway. The slate, properly installed, will not leak. Instead, the felt will slowly dry up under the slate until it turns into a powder a century or so later.

Newly installed roof sheathing should be covered with one horizontally laid layer of felt paper overlapped about three inches at the top edges, six inches at the sides, and nailed to the roof with 1" hot-dipped or electro-galvanized roofing nails or other suitable nails such as plastic cap simplex nails. The nail spacing at all edges is approximately every six inches. In the center of the paper a diamond-shaped nail pattern is followed. If the felt paper must act as a temporary roof for a prolonged period (for example, over winter), then skim over each and every nail head with roof cement. This is not as big a job as it sounds, and it will keep the water out of the roof until the slate is installed. Vertical strips of wood may be necessary in order to hold the felt in place over a prolonged period of time, especially under windy conditions. Horizontal strips of wood will dam water on the roof and can leak like crazy.

Felt paper isn't absolutely necessary for the slate roof to function; many slate roofs, primarily barn roofs, don't have any felt paper at all and are still leak free after a century, with proper maintenance. Felt paper does, however, provide a temporary cover in the event of rain during installation. It also helps to insulate and waterproof the roof and cushion the slates. It also provides a suitable surface for chalk lines when installing a slate roof, so it is recommended to use 30 lb. felt in order to do the best job. For better protection, the 30 lb. felt can be installed in the "half-lap" fashion so two layers of the felt protect the entire roof. For larger institutional projects, a heavier felt such as a 55 pound smooth felt may be preferred.

SELF-ADHESIVE UNDERLAYMENTS

Most American slate roofs were installed about a century ago and the people who installed them are now long dead and gone. What they left behind was evidence of a simple, yet brilliant roofing system: stone on wood. Many thousands of these century-old homes still have roofs in good working order today consisting of nothing but slate over wood. The old roofers must have done something right.

In the interim, however, American roofers abandoned permanent roofing almost totally in favor of throw-away asphalt roofs. Asphalt roofs

GENERAL CHARACTERISTICS OF WOOD

1) Degree of workability with hand tools; 2) Tendency to warp; 3) Tendency to shrink or swell; 4) Relative hardness; 5) Comparative weight
H = High, I = Intermediate, L = Low

Species	1	2	3	4	5
Black Ash	L	I	H	H	I
White Ash	L	I	I	H	H
Basswood	H	I	H	L	L
Beech	L	H	H	H	H
Yellow Birch	L	I	H	H	H
Eastern Red Cedar	I	L	L	H	H
Western Red Cedar	H	L	L	L	L
Northern White Cedar	H	L	L	L	L
Southern White Cedar	H	L	L	L	L
Cherry	L	L	I	H	I
Chestnut	L	I	I	I	I
Cottonwood	I	H	H	L	L
Southern Cypress	I	I	I	I	I
Rock Elm	L	I	H	H	H
Soft Elm	L	H	H	H	H
Balsam Fir	I	I	I	L	L
Douglas Fir	L	I	I	I	I
White Fir	I	I	I	L	L
Red Gum	L	H	H	I	I
Eastern Hemlock	I	I	I	I	I
Western Hemlock	I	I	I	I	I
Pecan Hickory	L	I	H	H	H
True Hickory	L	I	H	H	H
Western Larch	L	I	I	H	H
Black Locust	L	I	I	H	H
Honey Locust	L	I	I	H	H
Mahogany	I	L	I	H	H
Hard Maple	L	I	H	H	H
Soft Maple	L	I	I	H	H
Red Oak	L	I	H	H	H
White Oak	L	I	H	H	H
Ponderosa Pine	H	I	I	L	I
Arkansas Soft Pine	H	L	L	L	L
Sugar Pine	H	L	L	L	L
N. White Pine	H	L	L	L	L
W. White Pine	H	L	I	L	I
S. Yellow Pine	L	I	I	H	H
Yellow Poplar	H	I	I	L	I
Redwood	I	L	L	I	I
E. Spruce	I	L	I	L	I
Sitka Spruce	I	L	I	L	I
Sycamore	L	H	H	H	H
Tupelo	L	H	H	H	H
Walnut	I	L	I	H	H

(Source: H. E. Brosius Co., Kittanning, PA; Home Handbook (1956), p. 4)

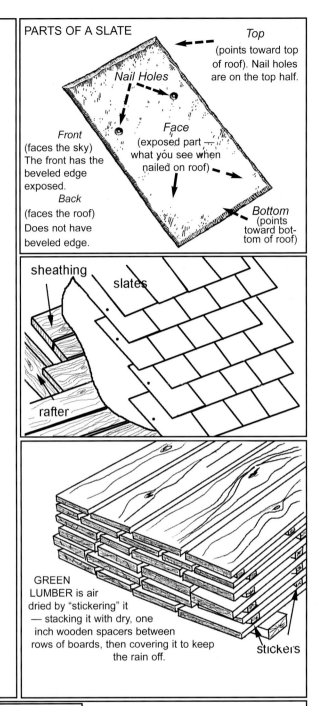

PARTS OF A SLATE

Top (points toward top of roof). Nail holes are on the top half.

Nail Holes

Front (faces the sky) The front has the beveled edge exposed.

Back (faces the roof) Does not have beveled edge.

Face (exposed part — what you see when nailed on roof)

Bottom (points toward bottom of roof)

sheathing

slates

rafter

GREEN LUMBER is air dried by "stickering" it — stacking it with dry, one inch wooden spacers between rows of boards, then covering it to keep the rain off.

stickers

LUMBER GRADES
Common or Board Lumber

No. 1: Best quality, most expensive, generally clear of knots, but may have small, tight knots.
No. 2: All around utility grade. Ideal for roof construction. May have occasional wood defects that need to be cut out.
No. 3: May have knot holes. Lower grade than #2.
No. 4: Lowest grade generally considered usable for construction of any kind.
No. 5: Not suitable for construction.

AVERAGE WEIGHT OF TIMBER

Species	Weight (lbs/cu. ft)
Ash	42
Chestnut	41
Hemlock	25
Hickory	53
Maple	49
Oak	32-48
Norway Pine	36
N. Yellow Pine	34
S. Yellow Pine	45
Spruce	25
Walnut	48

APPROX. WEIGHT OF DRY LUMBER PER 1000 BOARD FEET

Species	Weight (lbs)
Ash	3,500
Chestnut	3,400
Hemlock	2,100
Hickory	4,400
Maple	4,100
Oak	4,000
Norway Pine	3,000
White Pine	2,100
Yellow Pine	3,000
Spruce	2,100
Walnut	4,000

AVG. WEIGHT OF SLATE ROOF
(per 100 Square Feet Coverage)

Thickness	Weight (lbs)
3/16"	700-750
1/4"	1,000
3/8"	1,500
1/2"	2,000
3/4"	3,000
1"	4,000
1 1/4"	5,000
1 1/2"	6,000
1 3/4"	7,000
2"	8,000

[Source: *Slate Roofs* (author unknown), 1926, page 13]

WEIGHT OF SLATE ROOF PER 100 SQUARE FEET (SQUARE)

Slate Length (in.)	Weight in Pounds/Square for Thickness (in.) Shown							
	1/8	3/16	1/4	3/8	1/2	5/8	3/4	1
12	480	725	968	1450	1938	2420	2900	3870
14	460	688	920	1370	1845	2300	2760	3685
16	445	668	890	1336	1785	2230	2670	3568
18	435	650	870	1305	1740	2175	2608	3480
20	425	638	850	1276	1705	2130	2555	3408
22	418	625	836	1255	1675	2094	2508	3350
24	412	616	825	1238	1655	2066	2478	3307
26	408	610	815	1222	1630	2039	2445	3265

[Source: Radford, Wm. A. (1913), *Radford's Estimating and Contracting*; Radford Architectural Co., Chicago, p.253]

APPROXIMATE WEIGHT OF SLATE STONE (NOT ROOF) PER SQUARE FOOT

Thickness (inches)	1/8	3/16	1/4	3/8	1/2	5/8	3/4	1
Weight (pounds)/ft²	1.80	2.70	3.62	5.47	7.25	9.06	10.9	14.5

[Source: *Radford's Estimating and Contracting*, 1913, p. 253]

RECOMMENDED RAFTER SIZES FOR RIGID CONSTRUCTION

Rafter spacing, (inches on center) ⟶

Rafter spacing	12"	16"	24"

Unsupported Span ↓

[when 50 lbs/ft² live load (e.g. snow weight) anticipated]

Span	12"	16"	24"
6 ft.	2x6	2x6	2x6
8 ft.	2x6	2x6, 3x6	2x8, 3x6
10 ft.	2x8, 3x6	2x8, 3x6	3x8, 2x10
12 ft.	2x8, 3x8	2x10, 3x8	2x10, 3x10
14 ft.	2x10, 3x8	2x10, 2x12	3x10, 3x12
16 ft.	2x10, 3x10	3x10, 3x12	3x12, 2x14
18 ft.	3x10, 3x12	2x12, 3x12	2x14, 3x14
20 ft.	2x14, 3x12	3x12, 3x14	3x14
22 ft.	2x14, 3x14	3x14	
24 ft.	3x14	3x14	

[when 30 lbs/ft² live load anticipated]

Span	12"	16"	24"
6 ft.	2x4	2x4, 2x6	2x6
8 ft.	2x6	2x6	3x6, 2x6
10 ft.	2x6, 3x6	2x8, 3x6	2x8, 3x6
12 ft.	2x8, 3x6	3x8, 3x8	2x10, 3x8
14 ft.	2x8, 3x8	2x10, 3x8	2x10, 3x10
16 ft.	2x10, 3x10	2x10, 3x10	2x12, 3x12
18 ft.	2x10, 3x10	2x12, 3x12	3x12, 2x14
20 ft.	2x12, 3x12	2x12, 3x12	3x12, 3x14
22 ft.	2x14, 3x12	3x12, 3x14	3x14
24 ft.	3x12, 3x14	3x14	3x14

Rafter strength can vary greatly depending on species of wood.

[Source: *Slate Roofs* (1926), author unknown, pp. 36-39.]

Recommended Roof Slopes for Slate Roofs
(showing mimimum headlaps)

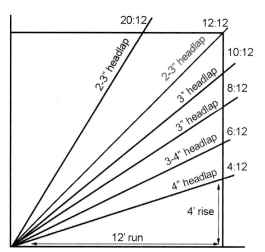

Most American slate roofs range in slope from 8:12 to 12:12, although many are steeper and shallower in pitch as well. Although a 3" headlap is standard and recommended, some older, steeper, slate roofs have only a 2" headlap.

Dead load = weight of roof rafter, sheathing and roof covering.
Live load = additional total weight roof may be subjected to, such as from snow or wind.

According to <u>Slate Roofs</u> (1926), roofs having a rise of 4 inches or less per foot of run, shall be assumed to have a vertical live load of 30 lbs. per square foot. A slope of more than 4:12, but less than 12:12 shall be assumed to have a live load of 20 lbs. per square foot. Slopes exceeding 12:12 shall be assumed as having no vertical live load, although provisions must be made to compensate for a wind force of 20 lbs per square foot. In localities where snow loads are an important consideration, the loadings shall be increased in accordance with local experience.

Roof Framing Terms and Diagrams
Courtesy of USDA, Forest Service, Agriculture Handbook #73, 1970

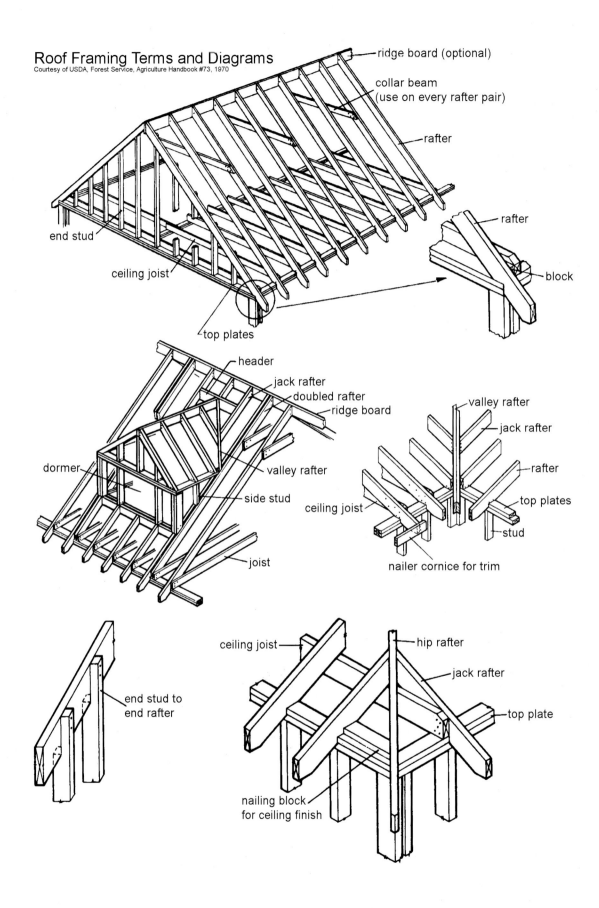

ridge board (optional)

collar beam
(use on every rafter pair)

rafter

end stud

ceiling joist

top plates

rafter

block

header
jack rafter
doubled rafter
ridge board

dormer

valley rafter

side stud

joist

valley rafter

jack rafter

rafter

top plates

ceiling joist

stud

nailer cornice for trim

end stud to
end rafter

ceiling joist

hip rafter

jack rafter

top plate

nailing block
for ceiling finish

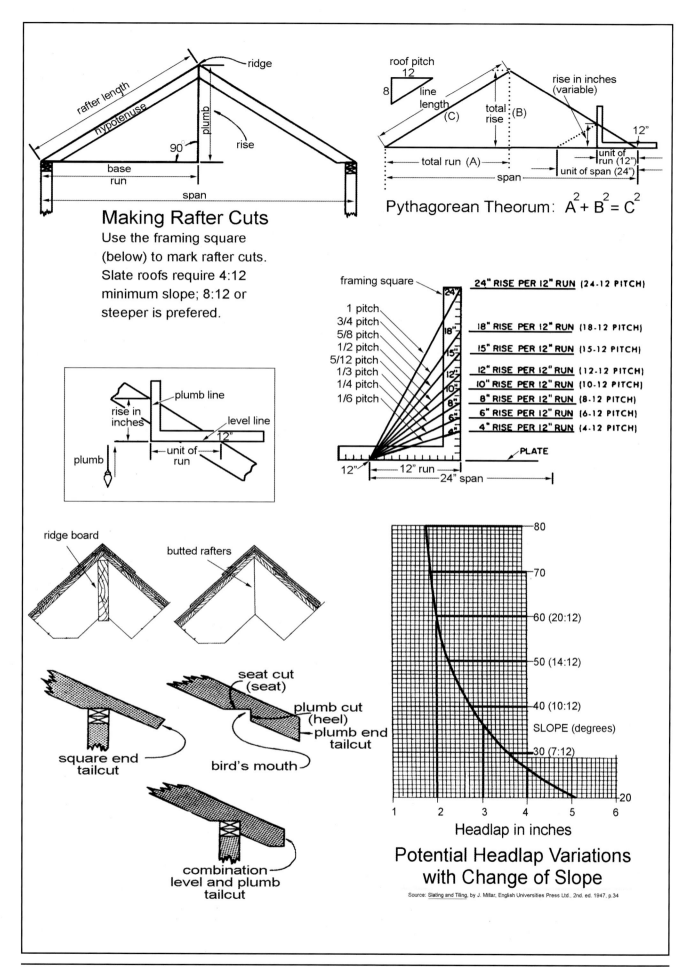

ridge

rafter length

hypotenuse

plumb

rise

90

base
run

span

Making Rafter Cuts

Use the framing square
(below) to mark rafter cuts.
Slate roofs require 4:12
minimum slope; 8:12 or
steeper is prefered.

roof pitch
12
8
line
length
(C)

rise in inches
(variable)

total
rise (B)

12"

total run (A)

unit of
run (12")

unit of span (24")

span

Pythagorean Theorum: $A^2 + B^2 = C^2$

plumb line

rise in
inches

level line

12"

plumb

unit of
run

framing square

24" RISE PER 12" RUN (24-12 PITCH)

1 pitch
3/4 pitch
5/8 pitch
1/2 pitch
5/12 pitch
1/3 pitch
1/4 pitch
1/6 pitch

24"

18"

15"

12"

10"

8"

6"

4"

18" RISE PER 12" RUN (18-12 PITCH)

15" RISE PER 12" RUN (15-12 PITCH)

12" RISE PER 12" RUN (12-12 PITCH)

10" RISE PER 12" RUN (10-12 PITCH)

8" RISE PER 12" RUN (8-12 PITCH)

6" RISE PER 12" RUN (6-12 PITCH)

4" RISE PER 12" RUN (4-12 PITCH)

PLATE

12"

12" run

24" span

ridge board

butted rafters

seat cut
(seat)

plumb cut
(heel)

plumb end
tailcut

square end
tailcut

bird's mouth

combination
level and plumb
tailcut

80

70

60 (20:12)

50 (14:12)

40 (10:12)

SLOPE (degrees)

30 (7:12)

20

1   2   3   4   5   6

Headlap in inches

Potential Headlap Variations
with Change of Slope

Source: Slating and Tiling, by J. Millar, English Universities Press Ltd., 2nd. ed. 1947. p.34

A PICTURE IS WORTH A THOUSAND WORDS. It has become a common and widespread myth that slate roofs are so heavy that they require extraordinary engineering and a massive roof frame underneath to support them. American architects and builders are constantly talking owners out of new slate roofs on the grounds that their structure cannot withstand the weight of the roof. In reality, traditional slate roofs such as this very old one (above and below) in Galicia, Spain, employing simple construction methods and using local materials sparingly, are a common sight throughout Europe. They last for many generations and possess an enduring beauty that cannot be imitated by artificial roofs.

These photos show how simple wooden poles gathered from the local environment created low-slope stone roofs that have lasted generations and still have long lives remaining. Despite all of the expert opinions about how a slate roof should be constructed today, an inspection of any old building with a sound slate roof will reveal the truth: natural materials and a variety of simple techniques will do the job quite nicely.

Slate roofs can be inadvertently under-built, which is a mistake that will show up in time as the rafters sag and the roof bows. However, standard traditional solid construction techniques are adequate for most slate roofs. On the opposite page, an ancient village spring in the Italian Alps is roofed with heavy stones on what appears to be a delicate framework (bottom). Roof slates are being removed and salvaged from a 120-year-old barn in Ohio, USA (top), which is in the process of being refurbished. The framework of 2"x6" rafters and 1"x4" slating lath is common on American barns.

Photos by author.

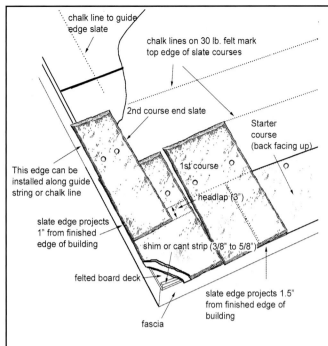

chalk line to guide edge slate

chalk lines on 30 lb. felt mark top edge of slate courses

2nd course end slate

Starter course (back facing up)

1st course

This edge can be installed along guide string or chalk line

headlap (3")

slate edge projects 1" from finished edge of building

shim or cant strip (3/8" to 5/8")

felted board deck

slate edge projects 1.5" from finished edge of building

fascia

1) Make sure that the fascia is completely installed before slating the roof and that the ends of the sheathing boards are firmly nailed.

2) Felt over the solid board decking (avoid plywood) with minimum 30 lb. roofing felt, lapped at least three inches at the top. Felt paper is optional, but recommended.

3) Nail or screw a wooden cant strip at the bottom edge of the lowest sheathing board — it should be 3/8" to 5/8" thick (standard slates only), and at least an inch wide (eight foot lengths are convenient). Cedar or redwood is ideal (cedar shim shingles will work), but the same lumber as the roof decking will do fine (not plywood). Cant strip will be thicker when installing thicker slates.

4) Chalk a horizontal line on the felt paper for the top edge of the starter slates, measuring the <u>width</u> of the slate up the bottom edge of the finished roof, deducting 1½ inches for the slate overhang. Then chalk a line for the first full row of slates, now measuring up the roof the <u>length</u> of the slate and deducting 1½ inches for the overhang. Lay starter slates sideways, back side out.

5) Now measure up the remainder of the roof equal distances equivalent to the *exposure* of the slate, and chalk lines accordingly. But first, make sure your second full row of slates will overlap the starter row by a minimum of three inches based on your measurements — if not, drop that second row down to where you need it to be, *then* chalk the rest of the roof with the exposure measurement. If laying the roof for a 4" headlap, adjust your chalk lines accordingly. If slating in an ice-dam-prone area, lay the bottom three feet of the roof with a 4" headlap and the remainder with a 3" headlap (assuming adequate slope — slate roofs should be too steep to walk on for greatest longevity).

Exposure is determined by subtracting the headlap from the total length of the slate, then dividing the remainder in half. For example, a 20" slate with a 3" headlap will have an 8.5" exposure (20 - 3 = 17, divided in half = 8.5"). Although 3" headlap is standard, 4" headlap is required for lower slope slate roofs and for ice-dam-prone sections of a roof.

6) Do not routinely bed the starter slates or any slates in roof cement or caulk. Adhesives make it very difficult to repair the roof in the future. Instead, a pair of 1½" copper or stainless steel nails per slate is a good rule of thumb which will ensure the secure attachment of all standard-thickness slates to the roof. Don't nail the slates *too* tightly; let them hang snugly on the roof. Do make sure the nail heads are set into the slate however, as nails that stick up will eventually wear a hole in the overlying slate and cause a leak.

7) Tap a couple of temporary nails into the side of the fascia on the gable end, one at the top and one at the bottom, and run a string up the edge of the roof positioned one inch out from the fascia. Use the string as a guide to align the outside edge of the slates as you nail them into place. Remove the string when you're done. Alternatively, chalk a vertical line on the felt up the edge of the roof, as illustrated above, as a guide for the end slates.

8) The butted sides of the slates on the first row must be staggered at least 3" laterally from the butted ends of the starter slates. All slates should have a minimum 3" lateral overlap in relation to the row of slates above or below.

9) You can work the first half dozen rows from a ground ladder or ground scaffold, then nail roof jacks and planks along the bottom of the roof and work up from there. Use more jacks and planks as needed. Leave a slate out periodically in order to have a place to nail the roof jack. The missing slates can be installed later with a stainless steel slate hook. See illustration this chapter.

10) Remember that all rules have exceptions in specific circumstances. Have fun! [Check jenkinsslate.com for instructional videos about slate installation.]

TOP ROW OF SLATES MAY NEED SHIMMED WHEN USING METAL RIDGE IN ORDER TO LAY FLAT.

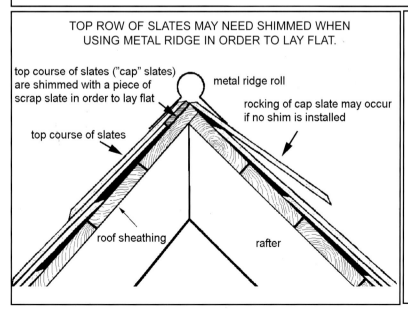

top course of slates ("cap" slates) are shimmed with a piece of scrap slate in order to lay flat

metal ridge roll

rocking of cap slate may occur if no shim is installed

top course of slates

roof sheathing

rafter

END SLATES CAN BE HALF SLATES OR 1 & 1/2 SLATES

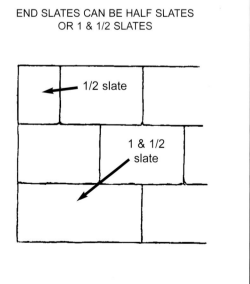

1/2 slate

1 & 1/2 slate

have to be replaced over and over again, which is great for roofers but bad for homeowners and the environment. Asphalt roofs can conveniently be laid over cheaper laminated roof decks — decks that have inherent weaknesses that are alleviated by such things as self-adhesive waterproof membranes.

After a couple generations of installing 99% asphalt roofs, plywood decks and membrane underlayments, many of today's roofing contractors and architects can't imagine anything else. Certainly a simple, natural roofing system such as stone on wood can't work! Nothing that simple will keep water out of a building. Or will it?

Slabs of stone worked into place with copper flashing can't possibly create a totally waterproof roof system, can it? In a word: yes. In three words: yes, *of course*. That's what this book is about — an ingenious roofing system made of natural materials that will not leak if properly installed and maintained. It's the slate and the flashings that keep the water out. *The underlayment is the least important part of a standard slate roof once the finished roof is installed.*

What about ice damming? Surely that's a good reason to use a beefy underlayment. However, ice build-up along eaves is usually an insulation problem. Uninsulated roofs thaw the ice and snow from the field of the roof and the water drains down to the eaves where it freezes again, at times building up dangerous levels of ice. You can largely prevent this by insulating the roof. You have leakage at the eaves anyway? Think of it this way: *If* the leakage is being caused by water penetration through the slates due to ice back-up, *then* the water penetration should occur *along the entire length of the eaves*. This is not usually the case at all. If, in fact, that is what happens, more headlap may likely solve the problem (unless the problem is due to insufficient slope or other design flaw, in which case you're beating a dead horse). In any case, water penetrating a slate roof anywhere, for any reason, means the roof was not installed or designed properly or the slates or flashing simply need repaired. When installing a slate roof in an area known for ice damming problems, install the bottom three or four feet of the roof along the eaves with 4 inches of headlap rather than the typical 3 inches and make sure the roof slope is 8:12 or greater. Architects must learn that slate roofs have certain design considerations that are different from other roofs — sufficient slope being one of them. The object is to create a waterproof roof, regardless of the underlayment. Underlayment material is only expected to have a relatively short career on the roof. To depend on it for the proper

functioning of a 150 or 200-year roof is only to fool oneself, or the home-owner.

Self adhesive membrane is a product designed to protect plywood from delamination — you will not need it on a slate roof. The author has never used a square inch of it in his over 30-year slate roofing career. Of course, this does not mean that you can install a low slope roof in a heavy snow load area and expect it to function well with a slate covering. If the slate roof system is poorly designed, it can leak. The solution is not to beef up the underlayment, but to design the roof properly in the first place.

LAY OUT THE ROOF

When laying out a roof in preparation for slating, chalk lines across the entire roof area marking the *tops* of each row of slate. *No metal drip edges are needed on slate roofs* such as the aluminum drip edges popular on asphalt shingle roofs. The purpose of these metal strips is to prevent the asphalt shingles from sagging at the edge of the roof. Slates don't sag. These small "drip edges" are not the same as the large, exposed, copper or stainless steel snow aprons installed at the eaves of roofs to eliminate ice damming. Snow aprons can function quite well on a slate roof.

When measuring for the starter slate and the first row, allow for the slate to hang beyond the drip edge of the fascia (or trim molding) one and a half inches. The starter slates are usually made of the same size slates as those on the main roof, turned sideways, and usually 1/4 of the length of the first one is trimmed off to allow the joints to be properly staggered in relation to the overlapping row. *All butt joints between slates should have a minimum of three inches of lateral clearance in relation to the butt joints of overlapping slates* (this is the sidelap already mentioned). On many old roofs, the starter slates are not laid sideways, but are simply the same slates as the rest of the roof — cut short — and again the joints are staggered. In any case, make sure the starter slates have sufficient headlap (3" minimum) under the course two rows above them.

In all cases, the starter slate must be laid over a cant strip about 3/8" to 1/2" thick (for standard thickness slates), which cocks the slate at an angle comparable to the angle of the slates on the rest of the roof. The starter row is typically laid back side out so that the bevel of the starter slate meshes with the bevel of the first row.

A thicker cant strip is needed when using

VARIOUS SLATING STYLES

Standard pattern (above), common in USA

Traditional German slating pattern (above)

Graduated pattern, shown on a Scottish roof (above)

"Triple Covering" shown with Italian slate. Note overlap.

"Open Slating" shown on a Spanish roof (above)

Random widths on an English roof (above)

Diamond pattern, showing hooked Italian slates (above left); Traditional Spanish slating style (above right)

THE VERSATILITY OF SLATE

Combinations of slating styles, textures, shapes, and colors can make a slate roof one of the most versatile, unique, and beautiful roofs in the world. Below are a few of many examples.

A collage of recycled slates creates a work of art (below) by Guido Lesser.

Proper color blends can make a roof a work of art (above). Artistic slate roof with rounded valley in Atlanta, Georgia (below).

A combination of thicknesses, colors, lengths, and widths creates a classic roof. This one is made of Vermont slate varieties.

Standard lap, uniform roof with bands of decoratively cut slates (below).

Another graduated Vermont roof with rounded valleys and dormers (below).

"RAGGED BUTT" SLATING — This roof in Lake Forest, IL, includes a mix of Vermont gray-black, Virginia gray, Vermont "sea green," and Vermont mottled purple slates (also shown below).

STAGGERED AND RAGGED BUTT SLATING STYLES

The roof slates are installed in the standard manner — the top edges of the slates are chalked on the roof before installation, but the slates used are of varying lengths. If the chalk lines are for 16" lengths, for example, 16" long slates are used as well as 18" and maybe 20" long slates. The extra length is left to hang down in order to create a "staggered butt" look (opposite page). The exposed butts can also be irregularly cut to produce a "ragged butt" style as shown above.

All photos by author.

heavier slates. The purpose of the cant strip is to tilt the first row of slates at the same angle as the rest of the roof, as no slate lays flat against the roof deck. When using thicker slates, you can figure out the thickness of the cant strip on-site by laying some slates on top of each other with the proper headlap and observing the gap beneath the bottom slates.

The slates that run up the side edge (rake) of the roof should extend beyond the gable end one full inch. Run a string up the edges of the roof to give yourself a straight line to follow when laying the slate (tie the string to temporary nails). Or, alternatively, chalk a vertical line up the roof near the edge to align the end slates.

When you reach the top of the roof, the top two rows of slates must be cut shorter in length to fit the roof. Sometimes the top row, the "cap" slates, must be shimmed under their top edge so they'll remain flat when the ridge metal or ridge slate is installed; otherwise they may cock crookedly and look bad. They can be shimmed with pieces of slate, usually the pieces that are cut off the top rows when the slate is laid, or a strip of wood can be used.

HIPS

Many hips on old American slate roofs are simply covered with metal ridge flashing, either galvanized steel (known as "ridge iron") or copper. These are simply nailed into place using nails of compatible metal (often 2.5" in length) spaced every foot or two, and the nail heads are caulked with a

Slate roofs can be multi-dimensional. They can include color variations, length variations, width variations, and thickness variations, all of which can be combined to create a roof of inimitable beauty. The roof shown here, at Lake St. Catherine in Vermont, exploits all four of these dimensions using a combination of Vermont slates of uncommon thickness.

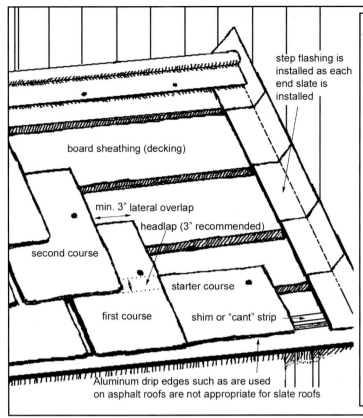

step flashing is installed as each end slate is installed

board sheathing (decking)

min. 3" lateral overlap

headlap (3" recommended)

second course

starter course

first course

shim or "cant" strip

Aluminum drip edges such as are used on asphalt roofs are not appropriate for slate roofs

material such as clear silicon of lifetime durability.

Hips made of slate are quite common and usually require less maintenance than metal hips. Common slate hips include the miter hip and the saddle hip. Mitered hips simply consist of the courses of roof slates butted against each other at the hip. Mitered hips sometimes utilize step flashings under the hip slates.

Saddle hips are made of slates positioned parallel to the hip rafter and laid over the standard field roof slates. Saddle hips can also have flashing underneath them, either a continuous piece or step flashing. Flashing under the hip slates reinforces the hips and is helpful in preventing leakage at the hip when the roof ages, since the hip slates tend to separate somewhat after eighty or ninety years. Nevertheless, flashing under hip slates is uncommon on older roofs and most were installed with no flashing at all.

Ridge metal can be installed over old, separated hips to cover any gaps or tarred joints.

RIDGES

Ridges are the horizontal peaks of the roof. In the United States they are often made of metal, especially galvanized steel or copper. When finishing slating along a ridge, it's important that the roof boards do not have any appreciable gap at the peak. If a gap exists (as is typically left when a carpenter sheaths a roof for ventilated ridge) the slates may not lay properly and the ridge metal will not have a sufficient base on which to nail.

Alternatively, ridge vents designed specifically for slate roofs are commercially available. These require that a gap be left at the peak of the roof between the sheathing boards. Follow the instructions from the manufacturer when installing these ridges (see back of book for sources).

The Welsh are quite fond of lead ridges, generally unavailable in the United States, but certainly superior to galvanized steel ridge. Copper ridge is readily available in the U.S., either custom made or prefabricated from roofing supply outlets.

Slate ridges are quite common in the U.S. Europeans often use ceramic tile ridges on their slate roofs, as can be seen throughout this book; these ridges are perhaps the most durable of all. They're more compatible with the masonry-walled construction of Great Britain, but they're rare in the United States.

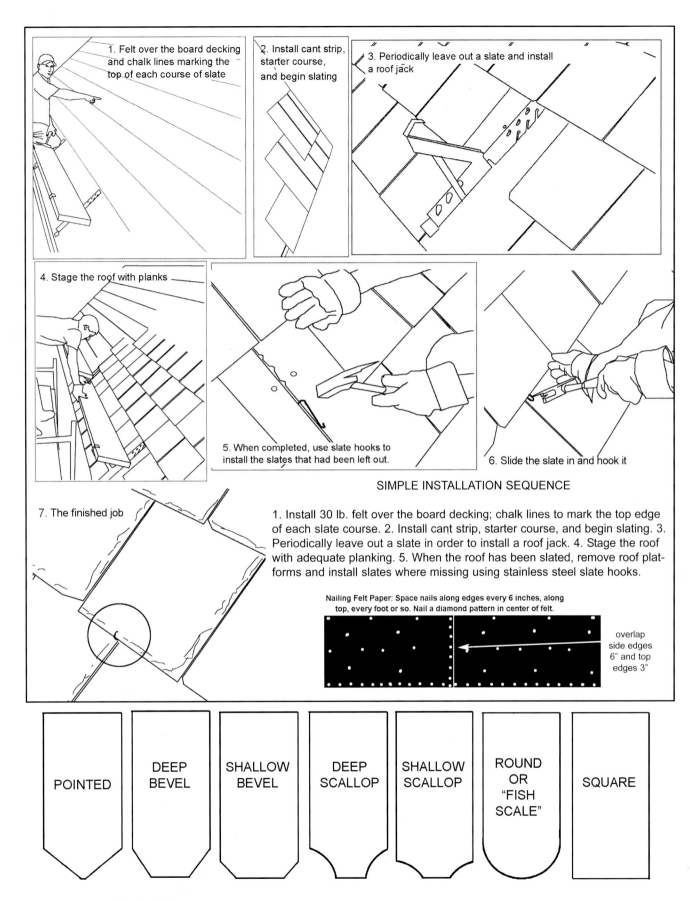

1. Felt over the board decking and chalk lines marking the top of each course of slate

2. Install cant strip, starter course, and begin slating

3. Periodically leave out a slate and install a roof jack

4. Stage the roof with planks

5. When completed, use slate hooks to install the slates that had been left out.

6. Slide the slate in and hook it

7. The finished job

SIMPLE INSTALLATION SEQUENCE

1. Install 30 lb. felt over the board decking; chalk lines to mark the top edge of each slate course. 2. Install cant strip, starter course, and begin slating. 3. Periodically leave out a slate in order to install a roof jack. 4. Stage the roof with adequate planking. 5. When the roof has been slated, remove roof platforms and install slates where missing using stainless steel slate hooks.

Nailing Felt Paper: Space nails along edges every 6 inches, along top, every foot or so. Nail a diamond pattern in center of felt.

overlap side edges 6" and top edges 3"

| POINTED | DEEP BEVEL | SHALLOW BEVEL | DEEP SCALLOP | SHALLOW SCALLOP | ROUND OR "FISH SCALE" | SQUARE |

SHAPES OF ROOF SLATES: Roof slates were once routinely manufactured in decorative shapes in order to add an elegant appearance to the roof. Today's roof slates are mostly manufactured in a rectangular shape (above right), however they can still be ordered from some suppliers with decorative shapes, or they can be cut to match the patterns of old roofs using a hand cutter.

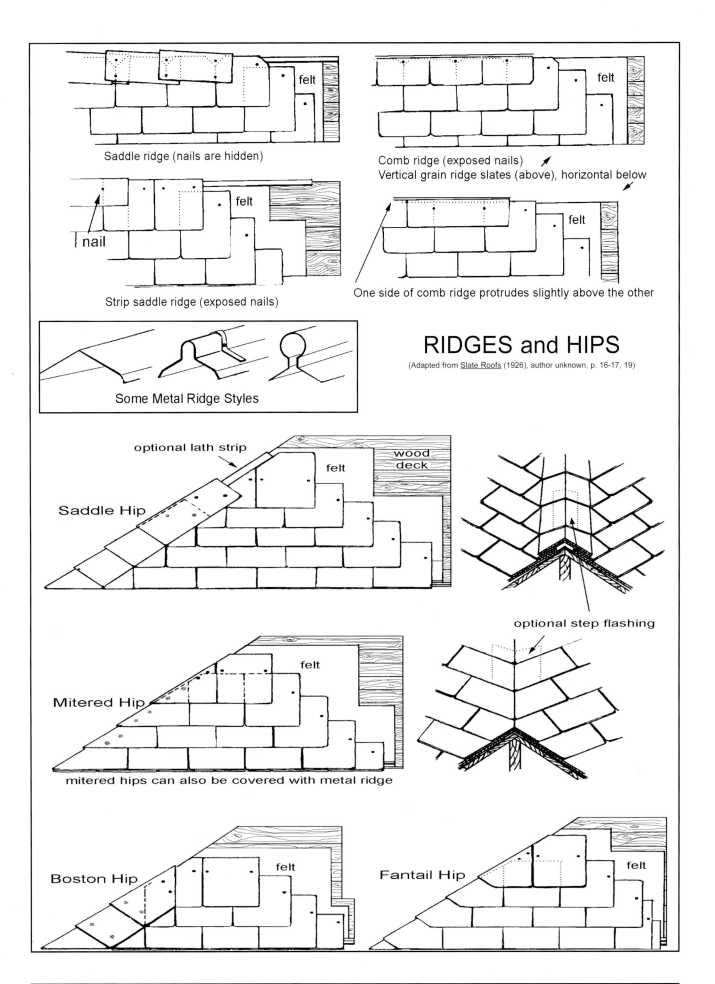

Saddle ridge (nails are hidden)

Comb ridge (exposed nails)
Vertical grain ridge slates (above), horizontal below

nail

felt

Strip saddle ridge (exposed nails)

One side of comb ridge protrudes slightly above the other

Some Metal Ridge Styles

RIDGES and HIPS

(Adapted from Slate Roofs (1926), author unknown, p. 16-17, 19)

optional lath strip

wood deck

felt

Saddle Hip

optional step flashing

Mitered Hip

mitered hips can also be covered with metal ridge

Boston Hip

Fantail Hip

Above left: Lead ridge common in Great Britain, here shown in Devon, England. Above right: Copper ridge in Massachusetts by Doug Raboin. Below left: Lead hips and valleys plus ornate ridges on this slate roof in Worcester, England, demonstrate the use of durable materials to finish off a roof. Below right: Smithsonian Institution Building with Buckingham slate ridge installed with exposed nails.

Photo credits: top left by Dave Starkie, top right by Doug Raboin, bottom left by Dave Starkie, bottom right by author.

WHAT NOT TO DO WHEN INSTALLING A SLATE ROOF

1. Do not use laminated wood roof decking — use solid lumber.

2. Do not use insufficient headlap — use three inches of headlap — more on lower slopes or ice dam prone areas. Use a minimum of 3" of lateral spacing (sidelaps).

3. Do not routinely walk on the slates or sit on them carelessly during installation — if possible, work from the side or from hook ladders or on planks. Walking on slate roofs during installation is one of the most common causes of breakage and subsequent failure with new slate roofs.

4. Do not use "electrogalvanized" nails for slate — use copper or stainless steel nails on new installations and hot dipped galvanized nails, or better, on recycled roofs. When doing restoration work, use nails similar to those already existing on the roof.

5. Do not rely on the underlayment to permanently waterproof the roof — you will puncture it profusely when you install the slate. A properly installed slate roof will not leak, underlayment or no underlayment.

6. Do not use aluminum drip edges — they're made for asphalt shingle roofs. These are not the same as copper or stainless steel aprons installed at the eaves of roofs to eliminate ice damming.

7. Do not use ventilated ridges unless they're specifically designed for slate roofs — if possible, ventilate through gable ends or through individual roof vents.

8. Never install slates with felt paper overlapping each course. Install the felt *under* the slate.

9. Do not select a slate based only on how it looks when new. Choose a good quality rock that has a history of successful use as a roofing material. The following quote from a misled home-owner reveals a situation that you want to avoid at all costs:

"We built a new home here [Florida] last year and put a slate roof on it. When we specified black slate our general contractor obtained samples from various suppliers and we chose the one that was the blackest. It was offered by [Company X] Slate, Inc., and was represented to be ASTM S1 and domestically mined. It was put on in June 1996, and by August it had large, red rust spots all over it. It got much worse very rapidly, and every time it rains, it leaks rust down on my white stone entrance, walks, etc. When we investigated, it turns out the slate was shipped from overseas and had large amounts of pyrite. Independent testing revealed a modulus of rupture and an absorption rate that were both so poor as to not even rate this slate as an S3! The slate company is now expected to replace the entire roof, and the threat of litigation is coming up repeatedly."

10. When ordering roof slates, make sure standard thickness (3/16" to 1/4") slates have nail holes that are punched rather than drilled, or if drilled, also counter-sunk. The nail head must be able to sit down into a counter-sunk depression in order to avoid rubbing on the overlying slate and working a hole in it. This is not such an issue with 1/2 inch or thicker slates.

GRADUATED SLATE ROOFS

Roof slate is available in many types with many different characteristics. In the old days, when roofing stone was wrestled from deep quarry holes and dark mines using hand tools and beasts of burden, the splitting of roofing shingles from rock was an arduous and exacting art. Some European slate veins contained very hard, rough-textured rock strata that could not easily be split into the uniform, thin sheets required for shingles. For the sake of efficiency, the stone was split into the largest slate shingles possible, creating a supply of coarse shingles that varied considerably in size — some larger, some smaller. In order to make good use of all of these slates, a certain style of roof was developed of necessity — *graduated* slate roofs.

In this style of roofing, the largest stones, sometimes massive, are installed at the bottom of the roof. This allows for the heavier weight of these large slates, perhaps 30" long and an inch thick, to be carried by the wall of the building. It also relieves the roofer of having to lug the flat stones, which may weigh 200 pounds, to the very top of the roof. Furthermore, the bottom of the roof is exposed to more water than any other part and heavier slates are more apt to withstand the excess erosion and weathering that occurs near the drip edge.

As the roof installer progresses up the roof, smaller and smaller slates are used, with the smallest slates, perhaps only 12" long, fastened near the top. The result is a roof that "graduates" in size from large at the bottom to small at the top, yielding an architectural style that is utterly unique and quite pleasing to the eye. Traditional graduated roofs also utilize random *width* slates. There are still many of these roofs in good condition scattered throughout the United States and Europe, yet the art of installing graduated slate roofs is a disappearing one.

A good place to look at the long history of graduated slate roofs is in Scotland, England and Wales. Scottish slate tends to be a rugged, coarse, and extremely durable material. Unable to split large, uniformly thin slates from the raw material available in Scotland, the Scots created a distinctive roofing style with a rough texture in keeping with the stone architecture so characteristic of Scotland's traditional buildings. This graduated slating style was also popular in England and Wales for the same

▼ A GOOD EXAMPLE OF BAD WORKMANSHIP

Poor sidelap technique ruined a slate installation (below) on a bad "economy method" re-installation of what started out to be an excellent slate roof.

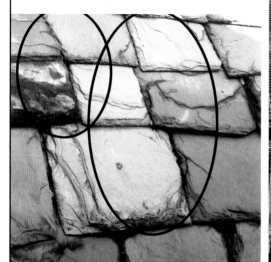

▼ Roof jacks and planks (below) make a handy and safe platform from which to work when installing a new slate roof.

Curved eaves may require shorter pieces of slate to be used at the eaves curve — the sharper the curve, the shorter the slate. Broken slates are repaired in the same manner as any other slate on the roof — with a nail and bib or a slate hook. Large gaps underneath the slates where the curvature occurs may benefit from horizontal battens, which help in nailing the slates to the roof by providing additional bite for nails. The battens also provide some backing support for the slates.

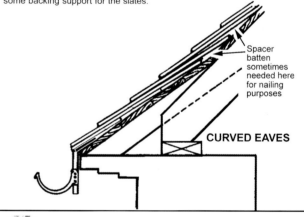

Spacer batten sometimes needed here for nailing purposes

CURVED EAVES

IN MEMORY OF
ALL SLAVIC IMMIGRANTS
AND FIRST GENERATION
SLAVIC AMERICANS WHO
LABORED IN THE SLATE
INDUSTRY IN THE LATE
19TH AND EARLY 20TH
CENTURIES

MITERED HIPS on a 90-year-old Vermont purple slate roof, in Grove City, PA, are being covered with new metal ridge. The hips had opened enough over the years to cause slight leaking, and therefore had to be covered. Copper ridge is excellent for this purpose.

Photos this page by author.

Beau and Liz Heath Residence; Grove City, PA; 30' rafter, 3" headlaps; VT slates. Unusual graduated slate roof with uniform standard thickness slate.	**Ketler House**; Grove City, PA; 3" headlaps; VT slates	**Ketler Garage**, 25' 8" from drip edge to ridge; 39 courses; 3" headlaps; VT slates

# courses	length of slate
[Top of Roof]	
13	. . . . .16"
11	. . . . .18"
9	. . . . .20"
6	. . . . .22"
5	. . . . .24"
[Bottom of roof]	

# courses	length of slate
[Top of Roof]	
etc.	
1	. . . . .27"
1	. . . . .28"
1	. . . . .29"
1	. . . . .30"
[Bottom of roof]	

# courses	length of slate
[Top of Roof]	
9	. . . . .13"
1	. . . . .14"
6	. . . . .18"
3	. . . . .17"
5	. . . . .16"
3	. . . . .19"
3	. . . . .20"
3	. . . . .21"
3	. . . . .22"
3	. . . . .24"
[Bottom of roof]	

This 90-year-old roof had five slate lengths which graduate in 2" increments according to an apparently random scheme. The slates on this roof are 1/4" in thickness.

This 90-year-old roof graduates one inch per course from bottom to top. The bottom slates are 1" thick, the top slates are 3/16".

This 90-year-old roof graduates randomly. The bottom slates are 1" thick, top slates are 3/16".

FIVE EXAMPLES OF GRADUATED SLATE ROOFS (ABOVE AND OPPOSITE PAGE)
It's obvious that the sizes of slates and number of graduations are a matter of style and/or personal taste.

Photos by author.

reason — it allowed for the efficient use of a stubborn material.

On one of the author's trips to Wales researching slate, he happened to meet a young slater who was installing a slate roof. Interestingly, there were several steps involved in the slate installation that roofers in the U.S. almost never encounter. For one, the roofer was obligated to "hole" each slate, as no nail holes are punched in the slate at the quarry as is typically done in the U.S. This is a carry-over of the days when graduated slate roofs were the norm and nail holes had to be punched in the slate on the job after the proper overlap had been determined for each diminishing course. The position of the holes in the slate was particularly critical when using lath, which left little room for error. Sawn lath strips developed from the practice of using hand split lath, as mentioned earlier, and continues to this day as much from tradition as from a lack of lumber resources in England and Wales.

But another surprising practice was the *sorting* of the slate prior to installation. This was a roof of uniform sized slates — *not* a graduated slate roof. Yet, the roofer, according to custom, sorted the slates according to thickness before carrying them up onto the roof, the thicknesses being termed "very heavies," "heavies," "mediums," and "lights." The very heavies were installed at the bottom of the roof, and so on until the lights finished off the top. I found the variance in thicknesses to be minimal, yet the roofer, with tradition in his blood, carried on a custom that began with the graduated roofs of old: sorting prior to installing.

Glenridge Hall; Sandy Springs, GA; 3" headlaps; mixed VT slates

# courses	length of slate
[Top of Roof]	
18	14"
19	16"
24	18"
8	20"
5	22"
4	24"
[Bottom of roof]	

This newly re-installed roof graduates randomly. The bottom slates are 1" thick, the top slates are 3/16" thick.

Harbison Chapel, Grove City College, Grove City, PA — 32' from drip edge to ridge; 55 courses; 3" headlaps; VT mixed slates

# courses	length of slate
[Top of Roof]	
16	12"
14	14"
4	16"
8	18"
4	20"
5	22"
4	24"
[Bottom of roof]	

This 75-year-old roof graduates randomly. The bottom slates are 1-1.5" thick, the top slates are 3/8-3/4" thick.

Today, the sorting of the slate prior to and during installation is critical to the successful creation of a graduated slate roof. It requires careful advance planning for the job to be well done. The number of courses required on the roof must be determined beforehand, and the number and degree of graduations, both in thickness and length, must also be part of the planning of the roof installation. There is no one correct formula for this. Diminishing lengths can occur with each course, or they can occur only with every several courses. In any case, once the particular formula for a particular roof job has been determined, then the correctly sized slates can be ordered from the quarry. To provide examples, the measurements of the graduations of five random graduated slate roofs are illustrated above.

Graduated slate roofs utilize slates of varying lengths, typically with varying thicknesses, as well as slates of random *widths*. The installation of slates with random widths is an art in itself, as adequate *sidelaps* must be carefully maintained. That is to say that the side-butts (where the slates butt against each other at their sides) of each pair of slates should be spaced three inches laterally from any side-butt above or below. If the side-butts are spaced too closely to each other, the roof could leak. A sloppy roofer will install random width slates with close side-butts; a master roofer will not.

Finally, a graduated slate roof can be made a work of art by blending together a variety of slate colors. A common color scheme involves a mix of Vermont slates, including purples, unfading greens, sea greens, grays, and perhaps Vermont black, and/or New York red. Sometimes Pennsylvania blacks, Virginia grays, or European, Chinese, or other imported slates are mixed in as well. The percentage of each color must be determined before the slate is ordered, and with a variety of lengths, widths, and thicknesses also to consider, careful pre-planning is a necessity in order to ensure a successful job when creating a graduated slate roof. Some suggested color combinations by Rising and Nelson Slate Co. include: 1) 70% semi-weathering gray green with 30% variegated purple; 2) 50% semi-weathering gray green and 50% variegated purple; 3) 60% unfading mottled green and gray with 40% unfading green; 4) 50% semi weathering gray green with 20% variegated purple, 20% unfading green and 10% Vermont gray black; 5) 70% unfading green and 30% unfading mottled green and purple.

Once completed, a new graduated slate roof can be expected to grace a building and charm a community for at least a century, and maybe two.

EYEBROW DORMERS

Perhaps one of the most elegant of slate roof installation techniques is the eyebrow dormer. In the illustration at the top of this page, the 120-year-old Peach Bottom slates (in perfect condition) on the field of this roof in St. Louis, Missouri, USA, are 10" wide, but the slates that curve over the dormer are 5" wide, allowing them to sweep across the dormer in a smooth arc.

Above, an eyebrow dormer on Glenridge Hall in Sandy Springs, Georgia, is stripped of its slates in preparation for reslating revealing the sheathing pattern underneath. Slated dormer is shown at right. 20 ounce overlapping copper strips, 16" deep, fortify the curved sections.

Opposite page, top right: There is a sheet of zinc flashing underneath each course of 5" slates shown in the photo where a slate has been pushed aside to reveal the hidden flashing.

Opposite page, top left: This eyebrow dormer in New Jersey on the Zimmerman Estate is made of 120-year-old Vermont sea green slates.

Opposite page, bottom: An eyebrow dormer in Grove City, Pennsylvania, USA, at the Ketler Estate, made of a mix of Vermont slates installed in a graduated pattern.

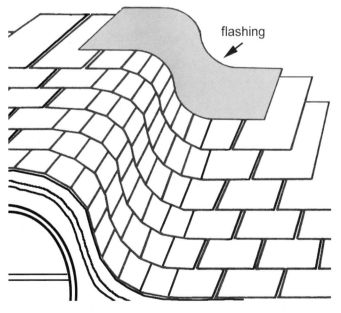

flashing

The two Glenridge Hall photos are by Ron Stokes. Remaining photos by author.

ROUNDED AND CONICAL TURRETS AND TOWERS

The photos on this page are of a 120-year-old Peach Bottom region slate roof located near St. Louis, Missouri, USA. The slates are nailed to the roof with iron nails. Note how the slates are installed in narrow pieces in order to conform to the curved shape of the turrets. Vary narrow slates are nailed on with only one nail. Flashing is not needed between the courses of slates due to the extreme slope of the roof. The photos on the opposite page showing the rafter framework are of the towers below.

Photos by author.

▲ Note the horizontal framing members between the rafters. The narrow sheathing strips are installed vertically (parallel to the rafters) and nailed or screwed to these curved framing members.

The tower at right is covered with 120-year-old Vermont unfading green slate in excellent condition.

The illustration below indicates that the slates' sides may need to be slightly tapered in order to lay well on the roof.

Surface area of a conical roof = 3.1416D X S/2
when D is the diameter of the base and S is the slant height.

Photos by author.

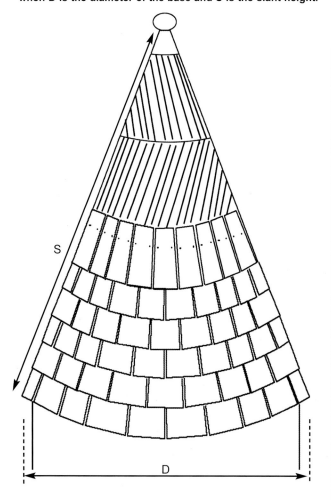

ALL IN A PENNSYLVANIA DAY'S WORK

1. A typical American low-slope particle board and asphalt roof on a newly purchased structure proves unacceptable to the new owner. 2. Hemlock 2x8 rough-sawn rafters are cut a day ahead to prepare for the construction of a 12:12 sloped recycled-slate roof. 3. A volunteer community group arrives, tears off the old roof, and begins installing the new rafters. Note the stickered rough-sawn hemlock roof sheathing in foreground. 4. The 1" hemlock sheathing is nailed over the rafters then covered by 30 lb. felt paper. Note the toe-holds (spaces) left between the sheathing boards, visible on the back side. The old roof was removed and the new roof was framed, sheathed, and papered in a day. 5. The slates are installed, one side at a time. Each side requires a day's hard work by two experienced slaters. 6&7. The finished roof, made of three different recycled Vermont slates and copper ridge, will last the life of the owner and many years more.
Photos by author.

Chapter Fourteen

TRADITIONAL EUROPEAN SLATING METHODS

This section is not meant to be a comprehensive explanation of international slating styles and methods. Instead, it is only an introduction, as a thorough look at such techniques would require another volume. For more information about the various countries mentioned below, please refer to Chapter 10 — International Slate.

BRITISH ISLES

As stated earlier in this book, the Welsh and British originally hung their slate on split wooden lath using wooden pegs as hangers. The pencil-stub-sized pegs were split out of a block of wood, then driven through a round hole that had been punched or drilled into the top center of the slate or stone. This force-fitting caused the peg to be firmly wedged in the slate. The peg hung over a thin lath while the weight of the stones held the roof together. This system worked very well and lasted quite a long time.

It was modified over the years, however, so that the slates were eventually nailed into the lath — the lath itself is no longer split, but sawed into approximately one inch by two inch strips. Felt paper is installed *under* the lath where it drapes between the roof rafters, waiting to catch a leak should one occur. The advantage to felting under the lath is that the roof nails do not penetrate the felt. When a roof of this style gets old enough to require replacement, the old slates are carefully removed, the lath and felt are replaced with new material, then the roof is reslated, either with new slates or old.

The slates are sorted by thickness before being nailed to the roof. The "very heavies" are placed at the bottom of the roof, the "heavies" above them, the "mediums" above the heavies, and the "lights" at the top. Unlike American slates, Welsh slates don't come pre-punched for nail holes from the quarry, but instead must be "holed" by the roofer prior to installation.

SCOTTISH SLATE ROOF
Photo by author.

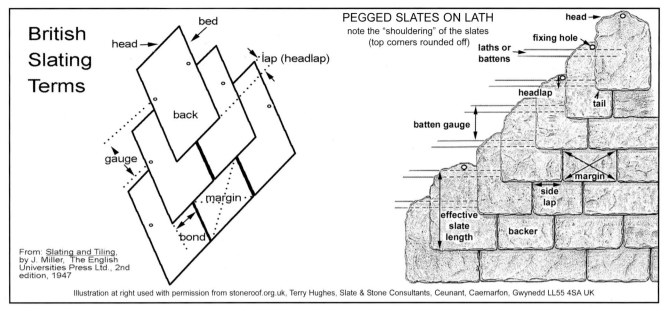

British Slating Terms

bed
head
lap (headlap)
back
gauge
margin
bond

From: Slating and Tiling, by J. Miller, The English Universities Press Ltd., 2nd edition, 1947

PEGGED SLATES ON LATH
note the "shouldering" of the slates
(top corners rounded off)

head
fixing hole
laths or battens
headlap
tail
batten gauge
margin
side lap
effective slate length
backer

Illustration at right used with permission from stoneroof.org.uk, Terry Hughes, Slate & Stone Consultants, Ceunant, Caernarfon, Gwynedd LL55 4SA UK

Another technique exhibited on European slate roofs is "weather-clipping" (illustrated in this chapter) — cutting the corner off the bottom of the slates at the gable ends of the roof (known as the "gable verge" in the UK). The purpose of this practice is to draw the rain water toward the center of the roof as it runs down the "verge," or the outer edge of the roof.

SCOTLAND

Scottish slates tend to have a more rugged character and texture than Welsh slates and Scottish slating techniques developed their own unique characteristics as a result. Traditional Scottish slating tends to utilize random width slates laid in a graduated pattern. The top corners of the slates are "shouldered," allowing the rough slates to be more easily installed and, if necessary, replaced. Scottish slates are head-nailed, utilizing a single nail located at the top center of the slate. Perhaps the most obvious difference between Scottish slating styles and British is the use of solid board roof sheathing by the Scots (referred to as "sarking" board), instead of the battens or lath preferred by the British and Welsh.

ITALY

Italian slates tend to be high in carbonate and although black when quarried, will turn almost white with exposure to the weather. Although high carbonates are usually considered to be a weakness in roofing slates, this problem is offset in Italy by splitting the slates to a thickness of about a half inch, which is fully twice as thick or more than a

standard American slate. Traditional Italian roofs are sometimes installed with a low slope, as is also common on old roofs in Spain and elsewhere. The reduction in drainage caused by a lower slope is offset by a slating method known as "triple-covering" in which the head lap extends completely down to the top of the third slate below, effectively creating a triple layer of slate over the entire roof. Today, Italian slates are hung on stainless steel slate hooks.

Modern slating methods in Italy include numerous styles, although the slate tends to be installed thick and hung on hooks, or occasionally nailed. Some of the styles include the "French" method, "triple covering," and standard lap slates of various shapes.

SPAIN

Traditional Spanish slating involves the use of un-trimmed slates laid in a random pattern. The slate is so haphazard in its appearance that it looks like a work of art. This is in stark contrast to modern slating methods being taught in Spain's roofing schools today, which insist upon installing all slates in a uniform manner with standard thicknesses using only slate hooks. The trend to install roof slates with slate hooks rather than nails is a dominant one in much of today's European roofing circles. Slates installed with hooks can be thinner, therefore yielding more shingles per block of slate quarried. Thinner slates, however, may not last as long, so those who are intent upon constructing roofs that will last a century or two tend to stick with traditional methods by using thicker slates and nailing them into place. Hooks work well for fastening

Various Italian slating styles, including standard lap (left two photos), triple covering (above) and "French" pattern (right).

Photos by author.

slates, especially stainless steel hooks, but tradition-alists find that the glint of exposed steel across the roof in the sunlight detracts from the beauty of the stone roof.

The Spanish are the world's greatest slate producers with a reputation for quality, integrity, and craftsmanship in the production of roofing slate. Spanish slate is prized in Europe and increasingly sought after in America. Roofing schools in Spain teach many methods and styles of slate roof installation. Although our focus here is on the older traditional methods, this is not to suggest that traditional slating is all that is done in Spain. Many modern slate roofs patterned after the UK style of slating with rectangular slates on battens, hung with either hooks or nails, are a common sight in Spain today.

GERMANY

Although trained German slaters are undoubtedly capable of any sort of slating style, their *traditional* methods are quite unique, which is what most sets the Germans apart from the rest of the world in the realm of slate roofing. "*Altdeutsche*" or Old German slating and its very similar modern cousin, *Shuppen* (fish scale) slating are unheard of in the United States, although such roofs can be seen throughout Europe outside of Germany, wherever German slaters display their incredible craft. Of course, traditional German slate roofs are seen throughout Germany, too.

The slating style is unique in that the slate is cut in the shape of a parallelogram rather than a rectangle, then laid on the roof with one side edge and top edge overlapped in such a manner that only one corner is exposed. This corner is cut into a curved shape, usually by hand using a slater's hammer and stake right on the job site as in the Altdeutsche method, where the slates will vary in size, or manufactured as such in the Shuppen method where the slates are uniform in size. Each piece of roofing slate is nailed in place with three nails onto a solid board roof deck rather than the lath roofs more common in

the UK and other parts of Europe. The result is a totally unique style of slating that enables the roofer to dispense with much of the exposed flashing metal common to American roofs and instead simply wrap the slates over the valleys and dormers in a smooth, sweeping style that is both artistic and functional. Needless to say, German slating techniques require specific training and time to master, which is perhaps why they are not more widespread throughout the world when compared to the much easier, simpler, but less exotic system of standard lap rectangular slating. With 1,700 years of tradition behind them dating back to the Romans, the Germans have kept alive and perfected a unique style of slating that is incredible and beautiful. A thorough discussion of German slating techniques would require a book unto itself. Nevertheless, this pictorial introduction should give the reader a good idea of the nature and uniqueness of Germany's slating traditions.

EUROPEAN ROOFING SCHOOLS

Roofing is a serious and respected trade in Europe, perhaps because the roofs there tend to be high quality and long-lasting. In contrast, average American roofs are often marveled at if they last more than twenty years, and American roofers are often placed in a category on the social totem pole slightly above "ex-convict." The people who learn the roofing trades in Germany are expected to undergo a three-year period of formal training followed by field experience, then perhaps an additional training for those who will be the "masters." European roofing schools cover slate roof installation methods — modern ones as well as traditional ones that may be specific to the country where the school is located (such as German traditional slating methods being taught in German roofing schools). The roofing schools may also cover ceramic tile roof installations, artificial slate, metal, and even low-slope, single-ply roof systems.

COMMON SLATING METHOD IN ENGLAND AND WALES

At left are exposed rafters on an old Welsh slate roof. The old slates and lath have been removed by Robert Jones and David Hussey in preparation for re-roofing.

The roofing felt is installed directly over the rafters.

The slating lath is nailed to the rafters directly over the felt paper. The slates are then nailed to the lath using two nails positioned about a third of the way from the top of each slate and about an inch or inch and a half in from the edges.

Photos by author.

INTERNATIONAL FEDERATION FOR THE ROOFING TRADES (IFD):

Internationale Foderation des Dachdeckerhandwerks e.V., Fritz-Reuter-Str.1, D-50968 Koln, Germany; Tel: +0049-221 3980380; Fax: ++49-221 39803899; email:info@ifd-cologne.de www.ifd-cologne.de

▲ Twenty-seven-year-old Neil Berridge, with ten years of slating experience under his belt, prepares this new life-boat shed in Barmouth, Wales, for slate. The roof is papered directly over the rafters and the slating lath is installed over the felt paper, as is common in England and Wales. The slates have been sorted into "very heavies," "heavies," "mediums," and "lights," depending on thickness — the heavier ones are nailed to the bottom of the roof. The slates have also been "holed" by Neil (nail holes punched).

▼ The finished roof is shown below.

Photos by author.

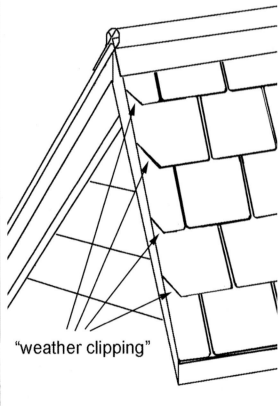

"weather clipping"

Two very old, low-slope Italian slate roofs during a rain.

▲ The underside of the roof on the opposite page, top.

▼ "Triple-covering" on the same Italian roof: The bottom of each slate extends down to the top of the third slate beneath it.

Photos by author.

Old slate roofs in the mountainous Orense region of Spain where much of Spain's roof slate is quarried.

▲ Traditional Spanish slate roofs, such as on the slate quarry building above, are as much art as they are functional.

▼ A traditional Spanish slate roof being prepared for installation is inspected by Mr. Vime and Ms. Pilar Cubelos of the Franvisa slate company in northwest Spain.

Photos by author.

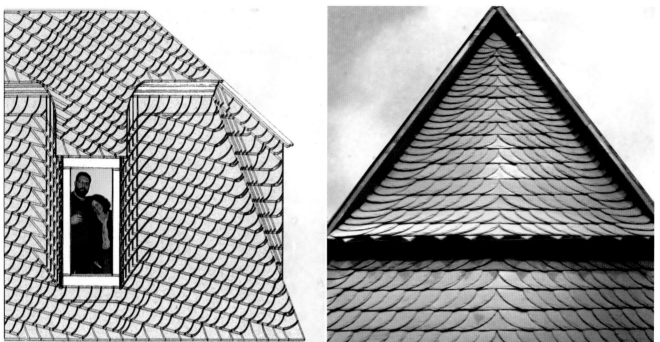

Traditional German Slating Styles

Altdeutsche (old german) Slating

Schuppen (fish scale) Slating

slate shingles
are cut from a
parallelogram

overlap

Photos both pages by author.

More German Slating Styles

schuppen

altedeutsch

rechteckdoppel

bogenschnitt

waben

spitzwinkel

The roofing school in Mayen, Germany (opposite page), allows students to learn every aspect of German slating techniques, as well as metal work, tile roofing, and even wall cladding with slate. A student in the same school (below) works out the details of Altdeutsche slating. Another German roofing school in the Fredeburg area (above) is also fully populated with highly motivated young men. Photos by author.

Slate wall cladding at the Rathscheck Company, Germany's largest slate merchant.

Chapter Fifteen

ROOF INSCRIPTIONS AND DESIGNS

One of the interesting characteristics of slate roofs is that they can be installed in decorative styles using slates of various colors. These decorative roofs can be inscribed with installation dates or with words, names, or abstract designs. In any case, the easiest procedure to use when installing an inscription or design on a slate roof is to draw a schematic of the roof beforehand. Make sure the schematic shows every slate in its proper proportion. The dimensions of the roof as well as the size of the slates must be known in order to do this — a job made easier by the use of a computer, although a computer was obviously not available in the 1800s when many dates were installed on slate roofs. Once the schematic has been created and a number of copies made, the artist can play around with various designs until he or she has settled on one that looks good. Then it is only a

matter of referring closely to the schematic during installation while using slates of contrasting colors in order to make the design appear on the roof. This does not add a lot of time to the job, but does create a lifetime landmark roof.

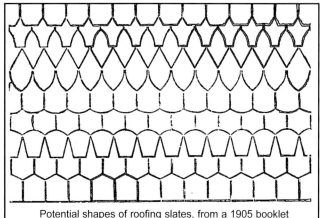

Potential shapes of roofing slates, from a 1905 booklet called "The Slate Roofer."

▲ Hathaway barn, near Rutland Town, Vermont.

Opposite page: Intricate slate wall cladding at Rathscheck Slate Company in Mayen, Germany.

Photos by author.

▲ NY Red slate against a Vermont sea green background on a barn in northwestern Pennsylvania reads
"P. E. Wood Fountain Farm 1900."

▼ Vermont unfading green slate accents a Cwt-y-Bugail Welsh black slate background on the Cathedral of St. Andrew in Little Rock, Arkansas, USA (below). See additional photos on pages 195, 228 and 229.

Photos by author.

Decorative artwork blends with skilled craftsmanship as indicated by the figure emerging from the wall (right), created in a German roofing school. The snail-like figure (bottom) was worked into a slate roof being installed at the World Slating and Tiling Championships, 2002, in Dublin, Ireland. The Belgium team created this artistic roof using a sheet metal pattern that swiveled on a nail hinge (below).

Photos by author.

The Slate Roof Bible — Chapter 15 — Inscriptions and Designs

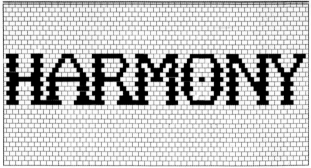

The schematic at left was used for the roof inscription above. Slates are 12x22 sea green background (salvaged) with new 12x22 Pennsylvania black slates, donated by Williams and Sons Slate and Tile in Wind Gap, PA. The building is a student-designed barn at the Macoskey Center for Sustainable Systems Research and Education at Slippery Rock University, Slippery Rock, PA, also known as the Harmony Homestead, located on Harmony Road.

Photo above by author.

Right: Aerial view of the Cathedral of St. Andrew in Little Rock, Arkansas, USA, re-slated in 2002-2003 by Midland Engineering Co. of South Bend, Indiana. Slate is Welsh black and Vermont unfading green supplied by Hilltop Slate Co. of Middle Granville, NY. The original roof was approximately 125 years old at the time of replacement. Flashings used on this project were primarily terne coated stainless steel.

Photo at right by Lyle Bandurski.

Opposite page: Amazing slate work done at the German roofing school in Mayen, Germany.

Photos on opposite page by author.

▲ Apartment house near Monkton, Vermont reads "Oct 1884 N. J. Allen." Bottom, this page: Date inscribed on a barn near Charlotte, Vermont, probably reads "B" for the initial of the family name, and "1893."

Opposite Page — Top: Purple slated barn near Brandon, Vermont with green slate date. Bottom: Two types of recycled roofing slates allowed the owners in Plain Grove, PA, to create an interesting design on their garage roof. A blank schematic was drawn on a piece of paper beforehand, then copied to allow for several designs to be considered. The final pattern, a Vermont purple slate on a Vermont sea green background, has created a landmark roof. Middle right: Front side of a companion garage built beside the other garage in Plain Grove, PA and roofed in a matching pattern. Middle left: Back side of the companion garage with date.

Photos both pages by author.

▲ New house near Castleton, Vermont.
▼ New cottage near Charlotte, Vermont. Harriet reserves the right to change the "S" to "D" if Richard doesn't behave.
Photos by author.

▲ Dean brothers farm near Brandon, Vermont. Dean family members had their names memorialized in the slate roofs of the farm.
▼: What's left of the inscription that read "Otter Creek Stock Farm 1911" near Whiting, Vermont.

Photos by author.

▲ Smid building near Brandon, Vermont, built and roofed by Chuck Smid, owner of The New England Slate Company.

▼ Old barn in Granville, NY is a perfect example of how any business owner can create a beautiful and permanent sign by using roofing slate on a wall or a roof!

Photos by author.

Sample lettering schematics published in a 1905 booklet called "The Slate Roofer" by Auld and Conger Co., Cleveland, Ohio. Each letter requires the following number of slates: A-36; B-38; C-29; D-41; E-37; F-31; G-32; H-44; I-21; J-26; K-38; L-27; M-51; N-43; O-40; P-32; Q-43; R-39; S-33; T-31; U-34; V-31; W-50; X-39; Y-33; Z-36; "1"-20; "2"-27; "3"-28; "4"-29; "5"-27; "6"-30; "7"-23; "8"-32; "9"-30. (Special thanks to the folks at Vermont Structural Slate Company). Below left: Slate designs on the Slate Valley Museum roof in Granville, NY. Below right: roof design on a residence in Whiting, New York.

Photos by author.

Chapter Sixteen

RECYCLING SLATE ROOFS

One of the unfortunate consequences of America's throw-away mentality is the loss of thousands of perfectly good slate roofs, which are ripped off and destroyed by uncaring roofers who can't be bothered with recycling anything. The slates are dumped in landfills and the roof replaced with disposable petro-chemical roofing which is also destined to soon clog a landfill. Most of the harder slate roofs are quite recyclable, however, and care should be taken when removing them, to salvage the slates either for slate roof repairs or for a completely new roof.

The slates should be pried loose with a flat pry bar ("wonderbar") starting at the top of the roof, and then allowed to slide down the roof and collect on planks that run across the bottom of the roof on roof jacks. While someone is prying the slates loose, someone else can be gathering them up and either carrying them down a ladder, dropping them down over the eaves or through a hole in the roof in a rope and harness, or sliding them down a chute to someone else. Ideally, they would be slid directly into the back of a waiting truck where they'd be carefully stacked on edge before transit.

Salvaging roof slate can be a dangerous job not only because of the heights involved, but also because the old buildings can be in poor repair and may have rotten spots in the sheathing that can collapse under the weight of a person. Furthermore, when the slates are slid down the roof, great care must be taken to ensure that they do not slide underneath the plank that is supposed to catch them. Otherwise, the falling slates can pose a grave hazard to anyone on the ground. It is advisable to lay 2x10 planks on roof jacks, then catch the first slates that come sliding down the roof and carefully position them flat on the planks so they will prevent the other slates from sliding under the planks. Finally, slate edges can be as sharp as razor blades, especially if broken. A slate sliding down a roof can therefore pose a serious hazard for anyone working below without gloves.

Some of the slates will break during the process of removal, but it's better that they break there than after they've been nailed onto another roof later. Some slates will develop hairline cracks or other flaws and must be discarded. A good slater can tell a bad slate by simply holding it in his hand and tapping it. A bad slate will give a dull thud, a good slate will ring. When removing slates from an old roof, it's better to pry each nail loose than it is to pull the nails through the slates. Prying the nails out preserves the old nail holes, while pulling the nails through the slates ruins the holes which must then be re-punched with a slate hammer before the slate can be nailed to another roof, unless the slate is used for repairs where the original nail holes may not be used at all.

Recycling slate roofs is a good way to get an excellent roof for a new building with a unique antique look that could last a century. Often the recycled slates from one building don't yield enough quantity to cover another entire roof, and recycled slates from two or more roofs must then be collected before a sufficient quantity is obtained for the job. When this situation occurs, it is imperative to "shuffle" the slates together before nailing them onto the new roof, as each old roof over time has developed its own weathered appearance, and in order for the recycled slate roof to look right the slates must be randomly mixed. For example, if you remove a thousand slates from old roof "A," and five hundred from old roof "B," and you need fifteen hundred for the new roof you're going to install, then for every two "A" slates you carry up onto the new roof, you must carry up one "B" slate. If you can carry twelve slates up a ladder at a time by hand, then eight of them should be "A" slates and four of them "B" slates. This is how slates can be easily "shuffled" in order to make the job look right.

Rather than shuffle two different batches of slates together, one batch can be used on one side of the new roof and the other batch used on the other side, as both sides are not visible at the same time from the ground.

Otherwise, differing batches can be creatively combined as shown on page 209. If you have collected slates from two dramatically contrasting

Opposite page: Spiral roof at the roofing school in Mayen, Germany. More German slate work is illustrated in Chapters 14, 15 and 20.

Photo by author.

▲ Chuck Smid, owner of The New England Slate Co., Inc. in Pittsford, Vermont, relaxes in a room he has beautifully decorated with recycled roof slate flooring and chimney cladding.

▼ A blend of recycled roof slates creates a unique look on a vertical wall surface in a Pennsylvania kitchen. The slates are attached to cement board with thin-set epoxy mortar, then grouted. The cement board is screwed to the existing wall boards.

Photos by author.

This huge barn near West Middlesex, PA, shown above left, became scheduled for demolition, so its roof slates were salvaged.

Architect Chris Leininger, left, pries off the roof slates, which slide down to a waiting plank.

They are then slid down a chute into the barn, landing on an old mattress (above).

Finally, the slates are taken to Pennsylvania's Slippery Rock University Harmony Homestead and installed on a new building, shown at left with some of the student construction workers (all women!).

Photo at left shows the slates on an American lath roof being salvaged by Umberto Perlino using the same method as above — roof jacks and planks along the bottom of the roof collect the slates as they are pried off and slid down. The slates are then collected from the plank and either carried down a ladder, slid down a chute, or dropped to the ground using a rope and crate.

Photos by author.

A tornado flattened this Mercer, PA, garage (top), but the slate roof was salvaged by astute owners Mike and Diane Sharr. When the new garage was built (center), the original slates went back on.

This Hooker, PA, roof (bottom), designed by Guido Lesser, displays a clever mix of sizes, shapes and colors, yielding an aesthetically unique recycled slate roof.

Top photo by Mike Sharr, center photo by the author, bottom photo by Guido Lesser.

All of these roofs are made of recycled slate and should last several human generations, if not a century or more. They include a writer's residence (above), a professor's home (right), a beach house/sauna (below left), and a poet's retreat (below right).

Photos by author.

roofs, such as from a green roof and from a purple roof, then you may want to consider designing a pattern into the roof to take advantage of the color contrast (see also Chapter 15).

It's often a good idea to increase the headlap on recycled slates when nailing them to a new roof in order to cover up the weather marks that remain on the old slates. If the roof that the slates were removed from had a two-inch headlap, the new roof should have a two-and-a-half inch headlap, or even a three-inch headlap. This will give the finished recycled-slate roof a cleaner appearance. This extra lapping sometimes poses a problem, however, as the nail holes may then become too low and the nails will penetrate the top of the slates in the underlying row, which should be avoided if possible. If that happens you should punch new holes in the slate, higher up, before nailing.

OTHER USES FOR RECYCLED ROOF SLATES

Slates can also be recycled for purposes other than roofing. Painting and decoupage on old roof slate is popular among craftspeople, for example. Slates can also be cut up and stacked to make sculptures, or assembled into such things as doll houses, candle stick holders, or vases, with the proper adhesives. Roof slates can be engraved with stone engraving chisels, or lettering and designs can be sandblasted into them using a sandblast stencil.

Old roof slates that are good and solid make a good floor covering too, especially on concrete surfaces. They can be laid directly onto the wet concrete, perhaps with a bonding agent painted to their underside, or glued to cured concrete or plywood with an epoxy (thin-set) tile adhesive, or even with trowel grade roof cement (allow a few weeks drying time when using roof cement as an adhesive). Roof slates can be walked on when laid flat over an unyielding surface, but should not be walked on when on a roof surface where they're overlapping each other and likely to break. It's imperative that the floor have no "give" to it when using old roof slate as a floor covering. Slates can also be glued to drywall or plywood walls to make a very interesting and beautiful wall surface simulating cut stone in appearance. Again, epoxy tile cement is best, and with a cutter, a variety of slates, and an imagination, the design possibilities are endless.

Finally, roof slates can be epoxied to brick surfaces, such as old chimneys, to make them appear to be stone. In all cases, when recycling roof slate for decorative purposes, make sure the slate is not soft and flaking (use only good, hard slate), and thoroughly clean each slate with soap and water before use. And remember, if you cut the slate with a hand-operated slate cutter, the edges will be beveled. If you want square edges, you'll have to cut the slate with a masonry blade or a diamond blade on a circular saw or a grinding tool.

▼ Several companies in the United States specialize in the recycling of roofing slates and tiles including Durable Slate Company, Inc. of Columbus, Ohio, whose stockyard is shown below with co-owner Gary Howes. Lists of both new slate and salvaged slate suppliers can be found at jenkinsslate.com and at the back of this book.

Photo by author.

▲ Two garages roofed with recycled slates in Plain Grove, PA, blend Vermont purple slates with Vermont sea green slates to create a unique artistic look.

▼ The residence of Architect N. Lee Ligo AIA and wife Linda MacWilliams Ligo, *Tall Chimneys*, was built in 1994. Among numerous architectural antiques incorporated into the residence were the sixty squares of recycled 19th century slate originally installed on Pennsylvania barns. Because of the random colors, shapes and sizes salvaged from multiple demolition sites by Amish crews, the architects Brett Ligo AIA and Lee Ligo AIA made multiple daily trips to make some sense of organization to the visual integrity of the roof. The slate was installed over 1x4 hemlock sheathing spaced 4 inches apart. The Amish crews traversed the roof during construction by inserting the toes of their boots into the four-inch slots between the slats. In most cases the copper nails were driven through the original punched holes in the recycled slate.

Top photo by author, bottom photo by Ligo Architects.

Chapter Seventeen

REPAIRS AND RESTORATION

Now I lay me down to sleep; I pray the Lord my roof will keep.
If I should die before it breaks; I thank the Lord I made it slate.

Pam Sykola - slate roof owner, Pennsylvania

The things people do to slate roofs to ruin them could fill another book. Much of this abuse is, ironically, at the hands of professional roofers, which is one reason the author came up with the clever theory that Neanderthals never became extinct — they simply evolved into roofing contractors. The greatest threat to American slate roofs today is the roofing profession, both due to the improper installations of new roofs and to improper maintenance and repair of old ones. The vast majority of roofers today in the United States make their living by *re-roofing* buildings. They have a vested interest in destroying older slate roofs because once the slate roof is gone, the brand new "premium" asphalt replacement shingles will be installed, then replaced regularly and endlessly. Smart people replace their old slate roofs with new slate roofs, but they have to be careful that the roofers installing their slate know what they're doing. But let's not be repetitive here — slate *installation* has already been covered in Chapter 13.

There are a variety of routine repair and maintenance jobs that, if done properly, will keep a slate roof in good condition for generations. Our focus will be on repairs and maintenance above the *drip edge*, which is the very bottom edge of the slate where the water drips off during a rain. We will not include anything below the drip edge (such as rain spouting) in any detail, although rain spouting must be briefly discussed here because many American spouting contractors nail rain gutters on top of slate roofs, which is a mistake that damages them. Flashing and chimneys will be covered more extensively in Chapters 18 and 19.

In order to best understand how to do routine repair jobs on slate roofs, one should first become familiar with the *parts* of the roof, which are the drip edge, the sheathing or roof deck underneath the slate, the slate itself, the valleys, ridges, flashings and chimneys. Let's start with the drip edge.

DRIP EDGE

The drip edge of a slate roof often needs repaired because contractors looking for a quick and cheap way to fasten gutters to a house simply nail them through the slate using strap hangers. This damages the slates, which eventually have to be repaired or replaced.

It's important *not* to allow strap hangers to be used on slate roofs *unless* the hangers are fastened *underneath* the slate. Originally, many slate roofs did use strap hangers nailed to the sheathing under the slate because the fascia boards on the older houses were not plumb (straight up and down), making it difficult if not impossible to fasten the gutters to the fascia. When the old gutters wore out and the old strap hangers rusted away, the contractors' solution was to nail new strap hangers on top of the roof right through the slate, thereby damaging the roof.

The proper solution, however, is to rebuild the fascia so that it is made plumb, and then to hang the gutters on the fascia using fascia hangers. An alternative is to remove the slate where the hangers are to be installed, nail the gutter hangers to the wood sheathing, then replace the slate over the hangers.

Often when repairing the slates along the drip edge of a roof, the gutter hangers must be removed one at a time in order to replace the broken slates underneath them. When the broken slate is off, the hanger can be nailed back directly onto the wood sheathing underneath the slate and the replacement slate can then be installed *over* the gutter hanger. The slate along the drip edge can be repaired this way without the need to take down the entire rain gutter.

A larger problem resulting from the unfortunate practice of nailing strap hangers through slate is the damage to the wood sheathing underneath the slate caused by the leaking roof. Because more water runs over the drip edge than any other part of the roof, this is the worst place for a leak to occur, and the wood underneath will eventually rot. In some cases, not only do the slates have to be removed and replaced along the drip edge, but so do the boards. In severe cases even the rafter ends will be rotted and must be rebuilt. This is why roof owners must be especially vigilant in preventing unscrupulous contractors from ruining the drip edge of their slate roof by nailing strap hangers through their slate. The spouting people who do this common but shoddy work never have to repair the roofs later — that's left to the slaters, and the roof owners must pay twice — once to have their roof damaged and then again to have it repaired!

Much of the older American spouting was the galvanized "half-round" type, hung on cast iron strap hangers fastened directly to the roof boards before the slate was laid. Today, however, aluminum or copper spouting are preferable to galvanized metal as they are much more durable, don't rust, and don't need to be painted, although one must be careful to use the heavier gauge (.032") aluminum and not the thinner stock, which is not worth putting up. An alternative to aluminum or copper spouting is stainless steel, which is harder to get *and* expensive, but exceptionally strong and durable. In England, heavy cast iron spouting is popular.

OK, so your drip edge is bad and you have to repair it. It's not hard if you know how a drip edge is put together. The bottom course of slate is called the *starter course*, and it is usually laid sideways. It is an invisible row of slate as it lies *under* the first course. Under the starter course is a *shim or cant strip*, which is usually a narrow strip of wood approximately 3/8" to 5/8" thick, nailed or screwed to the sheathing and running horizontally along the bottom edge of the

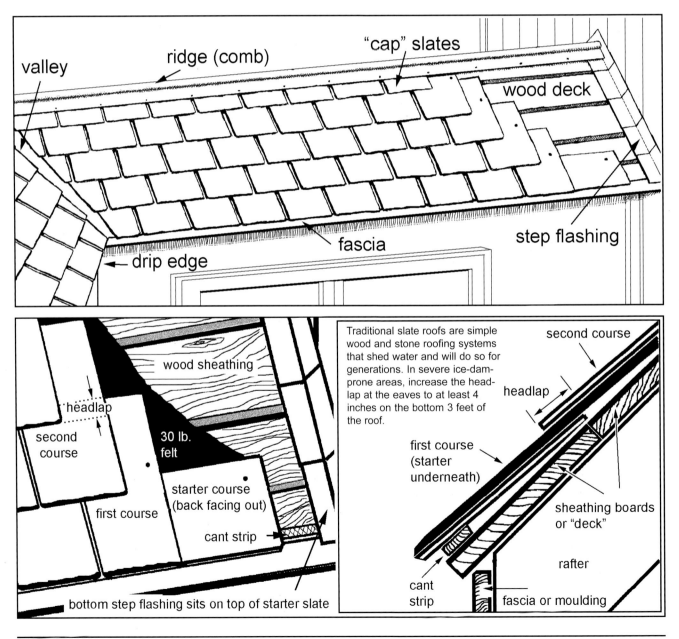

valley

ridge (comb)

"cap" slates

wood deck

step flashing

fascia

drip edge

headlap

second course

wood sheathing

30 lb. felt

starter course (back facing out)

first course

cant strip

bottom step flashing sits on top of starter slate

Traditional slate roofs are simple wood and stone roofing systems that shed water and will do so for generations. In severe ice-dam-prone areas, increase the headlap at the eaves to at least 4 inches on the bottom 3 feet of the roof.

second course

headlap

first course (starter underneath)

sheathing boards or "deck"

rafter

cant strip

fascia or moulding

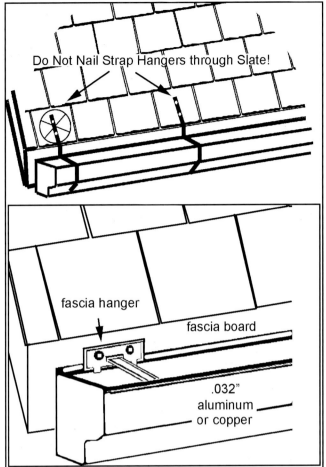

Do Not Nail Strap Hangers through Slate!

fascia hanger

fascia board

.032" aluminum or copper

Rain gutters should be attached to the fascia board using a fascia hanger. If the fascia is not plumb, it must be replaced or modified to allow for fascia hangers. Alternatively, gutter straps can be nailed or screwed directly to the roof deck *underneath* the slate.

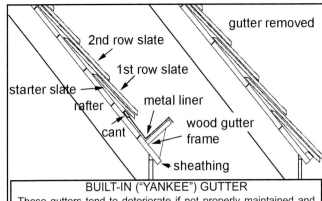

2nd row slate

1st row slate

starter slate

rafter

metal liner

cant

wood gutter frame

sheathing

gutter removed

BUILT-IN ("YANKEE") GUTTER
These gutters tend to deteriorate if not properly maintained and must eventually be removed from the roof. Gutter, liner, old wooden cant strip, and old starter slate are completely removed. New cant strip is installed at bottom of sheathing, new starter slate is installed over cant, and new 1st (and maybe 2nd) row of slates are installed. Note that a 3" headlap is maintained for all slates. Upper row is attached using nails and bibs or stainless steel or copper slate hooks.

roof. The purpose of the shim is to cock the starter slate at an angle so it matches the angle of all the rest of the slate on the roof. A good shim would be made of cedar or other rot-resistant wood, although almost any solid wood strip will work, and it's convenient to make the shim out of the same wood as the roof decking itself (not plywood, though).

After the starter course, the remaining slates are nailed in the standard way allowing for a standard 3" headlap, increased to 4" on lower sloped (4:12 or 4" of rise in 12" of run) roofs. *The headlap is the amount of overlap each slate has in relation to the <u>second</u> course of slate above or below it.* Every slate is overlapped by the course above it, but it's the overlap on the *second* course that's critical (Roof slope and slate installation are discussed in greater detail in Chapter 13).

As with any repairs, repairing drip edges requires removal and replacement of the bad slates (which will be cracked, broken, or tarred), and removal and replacement of rotten boards. The

wood sheathing traditionally used on slate roofs is one-inch-thick, rough sawn (unplaned), solid lumber usually from a local source, and usually installed "green" (not kiln dried). Local, green lumber can't be bought at standard lumber yards, but is available at sawmills, which are abundant in any forested area such as in most of the northeast U.S. Plywoods, laminated woods, and particle boards are *not* recommended for slate roofs as these materials can, and do, de-laminate over time. Local lumber is easy to get, costs less, and does *not* need to be dried before using. Alternatively, 3/4" planed, kiln dried planking will work fine. This issue is discussed in greater detail in Chapter 13.

It should be added that many older slate roofs have or had built-in gutter systems on the roofs. Many of the old built-in gutters have been removed because they weren't properly maintained (the metal wasn't kept painted), and they rusted and leaked. You will sometimes see old slate roofs with a couple of layers of asphalt shingles along the bottom edge because the people who removed the built-in gutters covered the exposed sheathing with asphalt shingles instead of slate as they were supposed to. These roofs can be restored by removing the shingles and reslating the drip edge.

THE BASIC REPAIR

The basic repair job on American slate roofs involves the removal and replacement of individual slates, which must be removed because they're broken or tarred, but also when they're covering flashing or sheathing that must be replaced. In almost any slate roof repair situation, slates must be taken off the roof and then put back on.

At the bottom of the roof are the *starter slates*. Above the starter slates are the *standard* slates, which are full size, uncut slates. At the top of the roof along the ridge are *cap* slates under metal ridges, which can be relatively small. At the gable ends of the roof are *end* slates or half slates, which are cut approximately in half lengthwise. Then there are the *valley* slates, or *flashing* slates, which are cut to any size or shape to fit against or on top of flashing, and finally we have *ridge* and *hip* slates, which are used as ridges and hips (also known as *saddle* or *comb* ridges or hips) in place of metal ridge.

The basic slate repair involves the standard slates, which are the full size, uncut slates that make up the bulk of the field of the roof. If one of these slates breaks, it must be removed and replaced. This is simply done by using a slate ripper to pull the two nails out of the roof which are holding the slate in place, then sliding a new slate in place and nailing it with one 1 1/2" hot dipped galvanized or copper nail, through the slot overlying the slate. Sometimes two nails must be used to get a solid replacement job (when replacing a row of slates, every few slates should be double nailed to prevent the slates from becoming crooked over time). The nail head is then covered with a piece of metal flashing called a *bib* flashing, which is slid under the overlying slates, but over the nail head to make the repair leak-proof.

A common alternative to this method involves the use of the *slate hook* to hold the replacement slate in place. A slate hook is a hook that nails into the roof decking between the two slates underlying the replacement slate (see illustration, previous page). The hook is made of copper, galvanized, or stainless steel, although the stainless steel hooks are recommended because they're stronger and longer lasting. After the slate hook is nailed in place, the replacement slate is slid into place and the bottom of the slate is nestled into the hook. The advantage of the slate hook is that no holes puncture the roof when a slate is replaced, thereby yielding a permanent repair that is virtually invisible from the ground. They are also very handy when repairing asbestos roofs, side-lapped slates, and other unusual roofing situations. The disadvantage of slate hooks is their limited applicability — they can't be used on valley slates, drip-edge slates, and other areas where there is no where to nail them.

It's common for some roofers to just "face-nail" replacement slates, which means they drive a nail or two through the face of the slate and leave them there, exposed to the weather. Sometimes the nail is gasketed, sometimes it is simply covered with roof cement or caulk, and sometimes it is neither gasketed nor covered. This is *not* the right way to replace a slate (except in unusual situations) because face nails, including gasketed ones, will eventually leak. An exception is made in the case of dilapidated *soft-slate* roofs when the roof is scheduled for replacement and the caulked face nails may temporarily help hold the roof together until it can be reslated. Contractors face-nail slates because it's easier and cheaper than hiding the nail in a slot and flashing over it, or using slate hooks, and most homeowners don't know whether their contractor is doing the job properly or not.

Slates such as cap slates and end slates don't have slots to nail through, and are therefore the only acceptable slates for face nailing on hard slate roofs, and even they don't *have* to be face-nailed, as we will see. Let's take a look at the various slates and how to fasten them in place when replacing them.

In order to replace a *starter* slate, the overlapping slates usually need not be removed. The starter slate can be pulled out with the ripper by hooking its nails and pulling them out, thereby removing the old slate. Then a new starter can be inserted underneath the first row of slate and nailed in place through an exposed slot (or two) with a nail and bib. Sometimes the overlying slates must be removed in order to get a starter slate out. Once the overlying slates are removed, the replacement starter slate can easily be nailed in place, then the overlying slates replaced, with *their* nails hidden in the slots (or with slate hooks). Starter slates lay on a shim strip, and that shim may need to be replaced (at least in part) in order for the starter slate to lie at the proper angle.

Cap slates do not have exposed slots allowing for replacement nailing in the standard fashion, so they must be nailed under the ridge metal by prying the ridge metal loose and lifting it enough to nail under it. The ridge metal has to be pried loose anyway to get the old cap slates out. In some cases the cap slates can be fastened by nailing right through the ridge metal with an 8 penny nail, then caulking the nail head with lifetime silicon caulk. The disadvantage to this technique is that when the ridge iron is removed for replacement, those cap slates will fall out.

When a ridge or hip is made of slate, the slate must be removed in order to replace underlying slate .

End slates are only half as wide as the standard slates and don't have a slot over them to nail

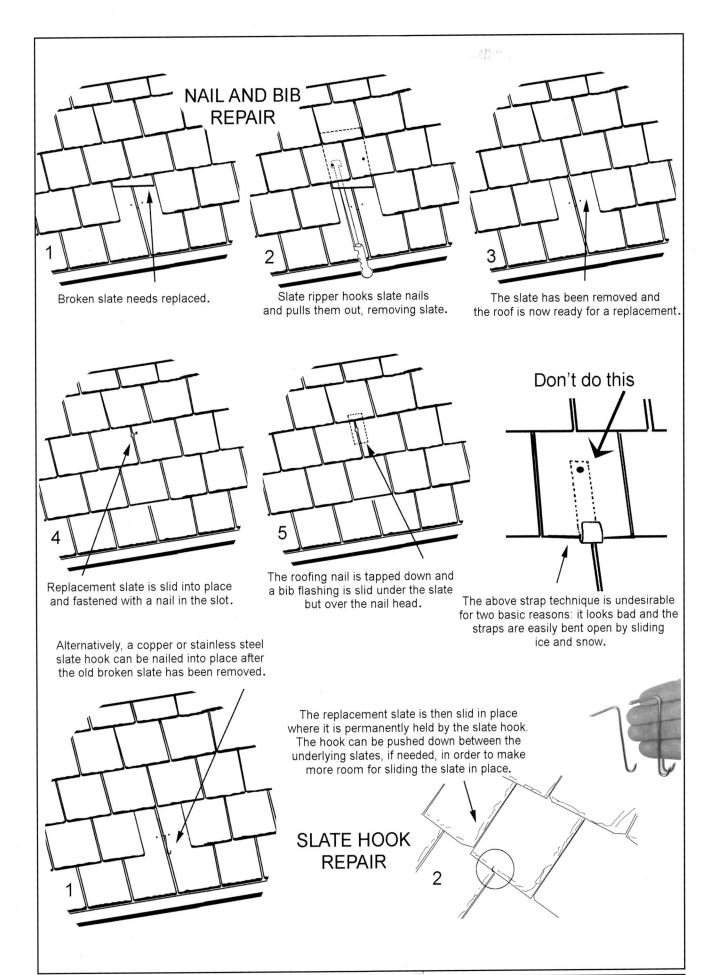

NAIL AND BIB REPAIR

1 Broken slate needs replaced.

2 Slate ripper hooks slate nails and pulls them out, removing slate.

3 The slate has been removed and the roof is now ready for a replacement.

4 Replacement slate is slid into place and fastened with a nail in the slot.

5 The roofing nail is tapped down and a bib flashing is slid under the slate but over the nail head.

Don't do this

The above strap technique is undesirable for two basic reasons: it looks bad and the straps are easily bent open by sliding ice and snow.

Alternatively, a copper or stainless steel slate hook can be nailed into place after the old broken slate has been removed.

The replacement slate is then slid in place where it is permanently held by the slate hook. The hook can be pushed down between the underlying slates, if needed, in order to make more room for sliding the slate in place.

SLATE HOOK REPAIR

1

2

through. Here is where a face-nail is often used, although the best method of replacing an end slate is to simply remove the overlying slate, nail the new end slate in properly with two nails, then reattach the overlying slate with a slate hook or nail and bib. End slates (half slates) often benefit from a dab of caulk or roof cement underneath them when being replaced (this is not needed or recommended during routine installations).

Gable-end slates have the unfortunate reputation of easily blowing off some roofs (such as old, wind-exposed barns). When gable-end slates are being replaced on a roof that suffers from chronic wind damage, the slates should be bedded in roof cement. This is done by applying roof cement *under* the slate before the slate is nailed in place, so the slates are glued together like a series of peanut butter sandwiches, and no roof cement is visible. This little trick will keep those gable-end slates from blowing off again.

Most *flashing* slates, such as slates along a valley, have a slot overlapping them, allowing a place for a standard replacement nail and bib. Otherwise, the overlying slates must be removed, the flashing slate replaced, then the overlying slates replaced in the standard fashion. When very small slates are replaced, it's often a good idea to bed them in a little roof cement or silicon before nailing. Punch a hole in the small slate with a slate cutter pin or other sharp object, put a dab of roof cement or caulk *under* the slate where the hole is, position the slate on the roof, then nail in place. The cement helps hold the small piece of slate in its proper position.

TEMPORARILY SEALING LEAKING VALLEYS

Valleys are a common source of leaks. Valleys carry more water than any other part of the roof because they act as a channel collecting water from two roof planes. When the valley wears out, holes develop which can leak a large amount of water into a building, and a pinhole in a valley can leak buckets of water. The solution to leaking valleys is simply to remove the old valley metal and replace it with new metal. This is a routine job, discussed in Chapter 20, and is a sure and permanent cure for any leaky valley, especially if the replacement metal is copper, stainless steel, or heavy gauge aluminum.

Sometimes the roof is not worth the cost of replacing entire valleys because the slate is soft and nearing the end of its life. In this case, simply spread trowel-grade roof cement on the valley metal, roll out some fiberglass roofing mesh onto the cement,

trowel the mesh in, then coat it with another layer of roof cement, being careful to make sure *no cement overlaps the slate*. Repeat, the cement should be worked *under* the slate with a trowel, *never* over it.

There are many roofs with leaking valleys where a homeowner or roofing contractor has tarred both the valley *and* the slate on both sides of the valley. This defaces the slate and ruins the appearance of the roof, while adding nothing to the effectiveness of the valley repair. If anything, it will make the valley leak worse over the long run, because the water will run down the roof and seep under the roof cement, which will act like a dam drawing the water into the roof. It is imperative to work the cement *under* the slate, then when the valley is eventually removed and replaced, the slate will still be good, and the valley replacement job will be much easier.

This three-step cement and fiberglass repair method is an inexpensive temporary measure to be used when the roof is not expected to last much longer, or in situations where the roof owner doesn't have the money to pay for a permanent roof repair, or in an emergency situation. A roof cement and fiberglass repair may last 10 years, and can also be used to repair almost any leaking surface (flat roofs, metal roofs, chimney flashings, and built in gutters, for example) *except* slate surfaces. Eventually, the roof cement and fiberglass repair will wear thin, but it can easily be redone and will last quite some time this way.

REPAIRING HOLES AND "HIDDEN LEAKS"

DON'T USE TAR

When valleys leak, the solution is simple — replace them as described in Chapter 20. If you can't afford to replace them, seal them temporarily. Yet there are often leaks on slate roofs that aren't in the obvious places like valleys, chimneys, flashings, and missing slates, and these are hard to pinpoint. This is where a bit of experience comes in handy. Being able to find the source of leaks (and repair them, of course) is the true mark of a professional, because many leaks are caused by something small and can be repaired rather easily.

Many old American slate roofs have been ruined by having tar spread all over them. Apparently, the repair person didn't know how to look for and locate the source of a leak. Instead, he got out the tar bucket and the brush and went wild, hoping he'd hit the leak, which was probably the size of a pinhole. There is no greater folly than painting

Emergency Valley Repair This will stop the valley from leaking until the valley flashing is replaced.

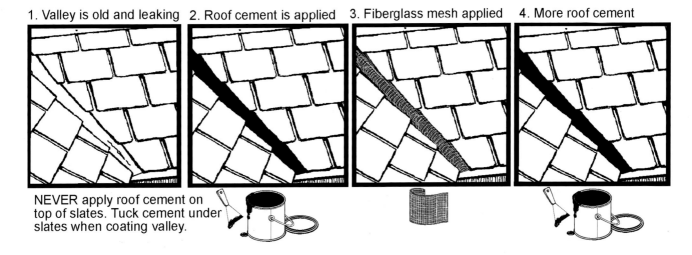

1. Valley is old and leaking   2. Roof cement is applied   3. Fiberglass mesh applied   4. More roof cement

NEVER apply roof cement on top of slates. Tuck cement under slates when coating valley.

a slate roof with tar, yet some contractors *advertise* the "service" of coating slate roofs with brush-on roof tar! The tar doesn't totally ruin the roof functionally, but it does totally ruin the roof aesthetically, and it takes about 50 years to wear off. In the meantime, besides having to look at an ugly roof for the rest of your life, if the roof does need repaired (and it will — that's why it was coated — it leaked!) the job of repairing it properly is much more difficult when the roof is all glued together with tar. If you run into a contractor offering to tar your slate roof, smile, speak gently, humor him, and get rid of him as quickly as possible.

Some buildings, particularly barns, are exposed to excessive wind, and the windward side repeatedly loses pieces of roof slate along the gable end during gales, leaving a ragged and damaged roof edge. When these slates are replaced, they should be bedded in trowel grade roof cement, so that the cement is sandwiched between the slates and is not visible. This simple technique tightens up the windward edge and prevents the slates from rattling loose again.

HOLES IN SLATE

One common source of leaks on slate roofs is a hole in a slate. Holes often result from nails that weren't nailed down far enough when the roof was installed, or which backed out of the roof as the sheathing boards dried. The nail heads then worked against the overlying slate eventually to wear a hole right through. If you examine an old slate roof closely, you'll likely find these holes, and the guilty nail heads will be happily peeking through them. They're easy to fix. Simply tap the nail head down where it's supposed to be, then slide a piece of metal (copper or painted aluminum — brown side facing out) under the hole and over the nail. A good size for the metal is 4" wide and 7" long, and it should have a slight bend in it lengthwise so it doesn't slide back out. A dab of clear lifetime silicon caulk in the hole after the bib has been slid into place will forever hold it there. You can fix a lot of holes on a slate roof quickly and inexpensively this way. On very old large roofs, there could be fifty or more holes like this. Alternatively, the slate with the hole can be completely removed and replaced.

Holes in slate can be caused by other things too. One farmer thought it was a good idea to shoot the pigeons that roosted in the rafters of his barn. His .22 caliber rifle might not have killed many pigeons, but it sure made a lot of neat little holes in his slate roof. No wonder it leaked!

Then there will always be holes in slate roofs that have no obvious explanation. Maybe someone fired a gun in the air a mile away and the bullet hit someone's roof, or maybe it was a miniature meteorite, or maybe a kid threw his dad's screwdriver at

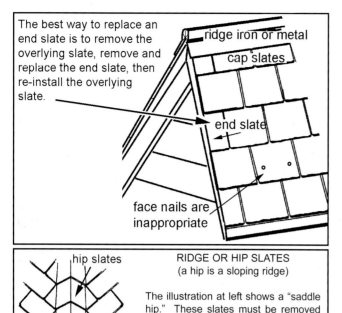

The best way to replace an end slate is to remove the overlying slate, remove and replace the end slate, then re-install the overlying slate.

ridge iron or metal

cap slates

end slate

face nails are inappropriate

hip slates

RIDGE OR HIP SLATES
(a hip is a sloping ridge)

The illustration at left shows a "saddle hip." These slates must be removed when replacing the underlying slates.

[From Slate Roofs, 1926, p. 19]

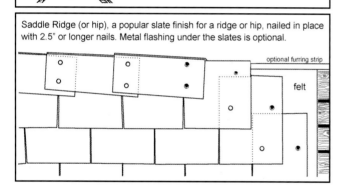

Saddle Ridge (or hip), a popular slate finish for a ridge or hip, nailed in place with 2.5" or longer nails. Metal flashing under the slates is optional.

optional furring strip

felt

a bird on the rain gutter, or maybe . . . well, who knows why, but there are sometimes holes in slate roofs that are totally inexplicable. Again, a piece of flashing slid *under* the slate will solve the problem quickly and easily. Do not tar over the hole unless you really like looking at ugly things.

HIDDEN LEAKS

This is where the fun begins. Hidden leaks are leaks that roofing contractors "fix" and they still leak. Most American roofing contractors make their living tearing off and replacing asphalt shingles and aren't very well versed in the art of finding hidden leaks on slate roofs; after two or three attempts they just give up, leaving the roof owner pulling his or her hair out and fuming. Ironically, most hidden leaks are *caused by* roofing contractors. That's probably why they can't find them.

Numerous examples spring to mind. One homeowner who was building his own house sub-

contracted the roofing to a roofing contractor who also flashed the chimney that protruded through the roof. After the flashing job, the roof leaked around the chimney. The owner called the contractor back and he "fixed" it. It still leaked. He called him back again, he "fixed" it again, and, of course, it still leaked. At this point a competent roofer was called in to have a look at the problem and the leak was immediately obvious — the chimney wasn't flashed properly. No amount of tinkering was going to make the chimney flashing waterproof if it wasn't installed properly to begin with. The only solution was to tear it out and do it again. In this case, the roofing contractor didn't know the secret to flashing a chimney (folding the corners, which is explained in Chapter 19 of this book). In the end, the contractor agreed to let the homeowner get someone else to reflash the chimney and he promised to pay the bill, neither of which had been done a full year and a half after the original flashing job.

Bad flashing jobs on chimneys are widespread, but other common causes of hidden leaks on slate roofs are *faulty old repairs*. If you've been reading this book and paying attention, you know by now that when you nail a replacement slate in place, you hide the nail in the overlying slot and cover it with non-corrosive flashing, or else you use a slate hook. Unfortunately, many roof repair people don't do either. Instead they *create* leaks:

1) Problem: they face-nail slates. These are nails driven through the face of the slate and left exposed to the weather. They'll eventually leak, if they don't leak right away. Sometimes the face-nails are gasketed, sometimes they're cemented over, sometimes they're caulked. Eventually they'll all leak. Face-nailed slates are obvious, however, and therefore easy to find and remove, although a lot of people don't realize that they're the cause of leaks. The heads of very old face-nails may completely rust away and the nails will then be very hard to find unless the roof is examined closely. Ironically, some old *soft-slate* roofs are so aged and rapidly falling apart that face-nails are the only thing holding the roof together. When an old soft-slate roof only has five or ten years left, face-nailing may actually keep it from falling apart so quickly, provided the nail heads are kept caulked or gasketed.

Solution: On hard slate roofs remove the face-nails and the ruined slate that was face-nailed and replace the slate. Fasten the new slate properly. On old, dilapidated *soft slate roofs* (especially PA ribbon slate), caulk the nail heads with lifetime clear silicon caulk and don't try to take the slates out. Chances are that if you start taking a slate or two out, the entire roof will start falling apart and you'll have created a major headache for yourself.

2) Problem: the repair person used corrosive flashing as a bib. When nail heads are covered by flashing that will rust, eventually it does rust away and the exposed nail head, even though tucked down in the slot, will leak.

Solution: Use your ripper to remove the corroded flashing and replace it with copper or painted aluminum or other non-corrodible bib material. When using painted aluminum, use a brown color and leave the brown facing out.

3) Problem: the repair person didn't cover the nail in the slot with any piece of bib flashing. It's not uncommon for a roofer to replace a slate, nail it in the overlying slot like he should, then leave it like that without covering the nail with flashing, as if his brain was working fine and then just shut down before the job could be finished. This is one of the hardest hidden leaks to find, because the nail is tucked in right at the top of the slot where it's hardest to see, especially from above, which is the most common vantage point for a roofing contractor. Then the repair nail rusts away until it matches the color of the roof, and becomes invisible.

Solution: When examining the leaking area, always look in the slots for hidden nails. Look closely. If you find one, renail the slate in the slot and install a bib over the nail.

4) Problem: the repair person covered the repair nail head with roof cement. Some roofers nail the replacement slate in the slot, then put a dab of roof cement over the nail head instead of flashing. Besides being ugly, the roof cement eventually wears away and the exposed nail then leaks. These leaks can be hard to find because there isn't much to see, and like the previous example, close examination of the slots in the leaking area will flush the culprit out. *Solution: renail and flash.*

As long as we're on the topic of leaks hidden in the slot, it is worth mentioning that sometimes slates crack widthwise; the crack is up underneath the overlying slate where it can't be seen, and a leak develops where the crack crosses the slot. This kind of leak can only be found by closely examining the slots, while maybe wiggling the slate a bit. You must get your face right down close to see such a crack.

CAPILLARY ACTION (OR ATTRACTION)

When liquids are in contact with solids, an adhesive force exists between them that may cause the liquid to creep in one direction or another along the surface of the solid, independent of the force of gravity. Water will run sideways or even uphill. This is an extremely important principle to be aware of when trying to understand leaks on roofs and it pertains particularly to leaks around flashing. The rule of thumb is that water has the capacity to run sideways nearly 2", which is why the bib flashing (the flashing that covers the nail on a replacement slate) is 4" wide. If you have a situation on your roof where water can run sideways and can find a place to leak within that 2" limit, you may have a leak by capillary action. If you do have a leak by capillary action, your roof will probably look like there is absolutely nothing wrong with it, and you will be driven crazy trying to find the source of the leak. The solution is to examine the roof closely where it's leaking and look for a place where the lateral overlap on the roof is less than 2". This may be a slate overlapping a piece of flashing, or a slate overlapping another slate. If you find such a situation, *slide more flashing metal under the guilty area to reinforce it.* That should easily solve the problem.

WATER TRAVELS DOWNHILL (USUALLY)

A house with a T-shaped slate roof had a chimney near the center of the roof. A leak developed in a bedroom ceiling, but the slate roof above the bedroom showed no sign of a leak. A close examination in the attic revealed that the water was leaking in around the chimney, running down a hip rafter, detouring down a jack rafter toward the bedroom, dripping on the ceiling's edge, then pooling in the center of the ceiling because the old ceiling joists were sagging slightly. The water then dripped annoyingly into the room. Repair of the chimney flashing, located on the other side of the house, stopped the leak. If a leak shows up in your roof, you must understand that the water can be entering the roof from any point above the leak.

I'VE TRIED EVERYTHING AND
I STILL CAN'T FIND THE LEAK

Don't despair. The author boasts that he can find any leak on any slate roof, and every now and then he runs into a tough one. The solution is not convenient but it's simple: wait until it's raining and get up into the attic of the house, look for the leak, trace it back to its origin, mark the spot with a piece of chalk or a thumbtack and rag, or anything prominent, then call your roofing contractor. He'll come in

NEVER TAR THE SLOTS ON A SLATE ROOF
as shown above on an historic house in Mercer, PA, with excellent sea green slate. This is a popular way to ruin a roof aesthetically, and the tar does not repair the roof. If you think there may be a leak in a slot, *slide a piece of metal under the slates, do not tar over them*! In the photo above, someone in the past had walked on the slates, evidently cracking one, and it subsequently leaked in a slot. The repairperson's inane "solution" — tar <u>all</u> the slots!

Photo by Barry Smith.

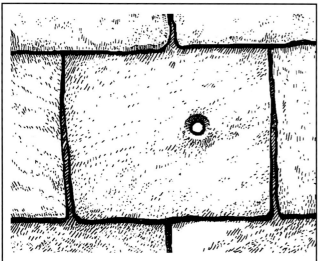

NAIL HOLE IN SLATE

This is a common sight on old slate roofs, especially on soft slate roofs from Pennsylvania's Lehigh-Northampton district. To repair, simply tap the nail down a bit with a bolt, and slide a piece of copper or aluminum flashing under the slate and over the nail.

dry weather, climb into your attic, find the source of the leak, go up on the roof and fix it. This works.

The author was called to a house where the chimney suffered from chronic leaking around the flashing. He asked the homeowner if the chimney was used for anything, and when he replied that it was not, the suggestion was made that the chimney be removed to below the roof line — a routine job. The owner agreed, and the chimney top was taken down, the hole was closed up with matching rough sawn lumber, and the area was slated to match the rest of the roof. When the job was done, you couldn't tell by looking at the outside of the roof that there had ever been a chimney there. The homeowner was assured that, with the chimney gone, the chronic leak had been brought to an end.

A few weeks later the homeowner called and said it was still leaking. He said he could hear the water dripping in the roof as he lay in bed during a rain. So the next day the roof was examined closely. Nothing could be seen that might be leaking, but all was double checked and reinforced with some bib flashings just in case. "I don't see any leaks, and I've never had a chimney removal job leak before," the author informed the homeowner, "but it should be OK now."

A couple weeks later the owner called again, madder than a wet hornet. "It's still leaking! I heard it dripping in the roof last night in the rain!" He wanted the author to drop everything and get over there right away; after all, he had already paid to fix his leak and he was insinuating that he had been ripped off. So the next day the roof was examined again, but nothing was seen that could be contributing to a leak. Finally the owner was informed, "You have a *hidden leak*. There are no signs of anything that would be causing a leak. The only way to fix this thing is to wait for it to rain again, get up *inside* your roof, go to where the chimney is (it was still standing in the attic), and mark the spot on the underside of the roof sheathing where it's wet. Then call me. The spot you've marked will indicate where the leak is, and *then* we'll be able to fix it!"

Well, it wasn't more than a week later that we had torrential rains. The rain just beat down in sheets for hours, and you could really hear it pounding on the roof. The homeowner would certainly be calling, or so it seemed. He didn't call, however, and more rain came, another week passed, and he still didn't call.

Then, a month later the author ran into the guy at the grocery store. The owner looked at him sheepishly as if he hoped he wouldn't be seen. The

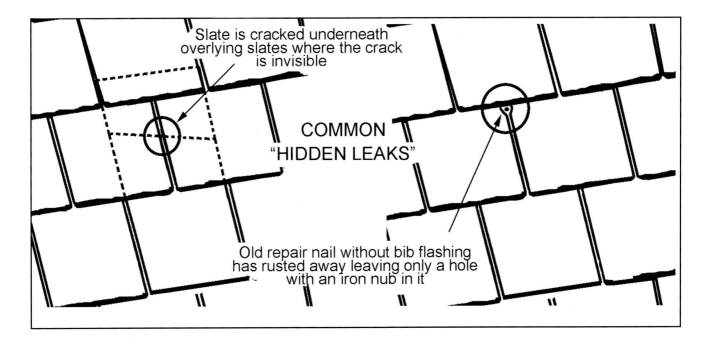

Slate is cracked underneath overlying slates where the crack is invisible

COMMON "HIDDEN LEAKS"

Old repair nail without bib flashing has rusted away leaving only a hole with an iron nub in it

inevitable greetings were exchanged, then the owner took the author aside and said, "You know that leak I had?" "Yes." "You know those real heavy rains we had a few weeks ago, when the rain just beat down like hell?" "Yes." "Well, I did what you said, I went up into the roof with a flashlight." "Yes?" "And it was just raining like *hell*, the sound of the rain beating on the roof was deafening." "Yeah? Did you find the leak?" "Well, you know what, I watched that spot on the roof above the chimney for a full *hour* with my flashlight, and guess what — *it didn't leak a drop*. It was just as dry as a bone."

At this point, the author remembered the two extra trips he had made to the top of this guy's roof, and unseemly thoughts began to form in his mind, but he remained polite and made a mental note to avoid the fellow in the future. The job had been done right in the first place, but the owner heard some dripping during a rainstorm and immediately concluded that it was dripping where the roof had been repaired. It was probably dripping outside, but nothing would convince the man until he got up into the roof during a rain and saw for himself.

The moral of the story is this: if you have a leak and can't for the life of you figure out where it's coming from, get into your attic, if possible, and look at the underside of your roof on a rainy day. Who knows, you may find that the leak is a figment of your imagination! Otherwise, you'll pinpoint the location of the leak and, if you've marked it well enough so it can be found again on a dry day, it can be readily repaired.

On another occasion a fellow called the

author to complain that he had a leak in the same place on his ceiling for fifteen years. He said he'd had a whole slew of roofers try to fix it and no one ever succeeded. Of course, nothing interests the author more than a challenge like this, so he stopped by the man's house and took a look at his roof. Sure enough, the roof area directly above the leak appeared OK. There was nothing that looked like it would be causing water to get through the roof. After fifteen years of roofers trying to find this one, it was decided that this leak be placed directly into the "hidden leak" category, to be found from the inside of the house.

This was a bit of a problem because a dropped ceiling had to be removed and a wooden ceiling had to be cut through in order to get at the underside of the roof. The homeowner was instructed to call the next time he saw any leaking occurring, and soon thereafter, on a rainy day, he did. A drive through the rain to his house and a climb up into the ceiling with a flashlight revealed a trail of water that proceeded up the roof about ten feet and sideways about four feet to the point of origin. After measuring the distances and writing them down, the leak was mapped. Later, during dry weather, the author climbed on the roof and went to the leaking spot like finding treasure with a treasure map. The water was coming in at a roof hip because the roof slates were cut too short against the hip when the roof was installed, and the hip slates covering them didn't have enough overlap (remember capillary action). All of this was impossible to see until the roof was taken apart. The water traveled from the hip down a jack rafter, then ran sideways on a

Leaks are often the result of poor repair work. Many U.S. slate roofs have been abused by unqualified roofing contractors and amateur repair people.

Wrong type of slate, ugly work

Should have used a bib flashing, must have been blind.

Nail in the slot is a common hidden leak.

The site of a Neanderthal convention?

Face nails by unscrupulous contractors are often a problem.

Roof repair by an auto mechanic?

Nail heads that have worked holes through the overlying slates are common.

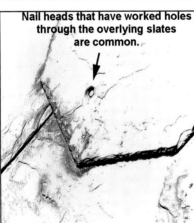

Repair work by a freshly lobotomized roofer?

Missing slates will leak

Slate sliding out, needs refastened

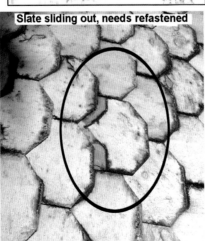

Roof cement art of the Cro Magnon variety?

sheathing board, crossed two rafters and finally dripped through the ceiling far from the point of origin. This orncry little leak confounded many a roofer and drove the homeowner crazy for fifteen years, but was easily repaired once the mystery had been solved.

And by the way, taking a garden hose up on the roof and hosing it down while someone watches on the inside is *not* the way to find a leak on a slate roof. Such a stunt creates an artificial flow of water that does not accurately approximate rainfall. It also creates unnecessary hazards and impositions on both the roofer and the property owner. A good slater experienced in slate roof repair and restoration will *never* have to do such a thing.

CHIMNEYS

Chimneys are a common source of leaks on a slate roof. They leak around the flashing, which is the metal that seals the gap between the roof and the chimney itself. The flashing eventually rusts, develops holes, then leaks. The solution is to replace the flashing, which is a routine job on old slate roofs, a job that should be done once every 75 years or so. A good material to use when replacing chimney flashing is 16 ounce copper. We will not go into the reflashing of chimneys in any depth here because we focus on chimneys in Chapter 19.

However, there are people who don't want to, or can't afford to pay to have their chimney reflashed, or they want to stop it from leaking in an emergency. Although reflashing is the only permanent cure, the chimney can be sealed at the flashing with roof cement and fiberglass membrane as a temporary measure. Such a seal could last about ten years, then will have to be redone, or the chimney will have to be reflashed at that time.

Sealing a chimney at the flashing is done in the same manner as sealing a valley. The three step process of applying trowel-grade, plastic roof cement over the flashing, working in a layer of fiberglass mesh, then applying another layer of roof cement, will create a waterproof seal that will last for quite some time. The problem with this process, however, is that it makes the reflashing job considerably more difficult when that job is finally done because all that roof cement must be removed. On the other hand, such a seal is quick and cheap, so a lot of people do it anyway, and if you're going to do it, you might as well do it right. The fiberglass membrane is the essential ingredient any time a leak is being repaired with roof cement.

A roof cement and fiberglass seal doesn't need to be ugly; it can be neatly done, and the cement can even be painted with tinner's red to make it look like real flashing from a distance. When applying the cement, use a three-inch square-edged trowel preferably with a flexible blade. Apply the cement over the flashing on the sides of the chimney *and* on the slate roof a distance of 3" from the chimney. Use at least a 6"-wide roll of roofing fiberglass, placing the fiberglass so it lies on the roof 3" and runs up the side of the chimney another 3" (minimum). It's critical that the fiberglass be folded at a 90 degree angle to lie on the roof and on the side of the chimney at the same time in a single piece, because the area most likely to leak is the crack where the chimney and roof join. Roof cement will dry and crack over the years and water will leak right through it if you don't reinforce the joints with the fiberglass. You want to run both the cement and the fiberglass up the side of the chimney far enough to cover the existing flashing, so you may have to slightly overlap the fiberglass on the chimney sides by applying a second layer. If the flashing on the side of the chimney is loose, remove it before sealing the chimney.

Use the roof cement liberally, but don't spread it anywhere it isn't necessary. It's hard to remove from bricks, and excess roof cement will look ugly when the chimney is finally reflashed. Be aware that when you put roof cement on top of slate, those slates are aesthetically ruined and should be replaced when the chimney is finally reflashed.

LEAKS AT RIDGES

The ridge is the horizontal peak of the roof, also sometimes called the "comb" by old-timers. There are several types of ridges commonly used on American slate roofs, but these can be lumped into two basic categories: metal and slate.

Metal ridges are far more common than slate or ceramic tile ridges in the U.S., and are typically made of galvanized steel, although both copper and aluminum ridges are also available. Galvanized ridge metal, also known as "ridge iron" by slaters, and "ridge roll" by manufacturers, will rust and deteriorate if not kept painted. The quality of ridge iron has progressively dropped over the years, and it's hard to find anything worth putting on a slate roof. Galvanized ridge iron should be 26 gauge, although 28 gauge is more common; 30 gauge should be avoided. Galvanized ridge iron must be painted approximately every five years with tinner's red or tinner's green oxide, the traditional American roofing paints.

When ridge iron is neglected, it rusts and will eventually leak. When that happens, the solution is to replace the ridge iron with new, 26 gauge, galvanized, or 16 to 20 ounce copper ridge (always watch out for wasps, bees and bats under ridge metal). Traditionally, the ridge metal is nailed in place with 8 penny common, galvanized or copper nails (always use compatible metals), and the nail heads are cemented with roof cement, caulked with lifetime silicon, or gasketed. Rubber or neoprene gaskets can be used on the nail heads when installing ridge, but these tend to leak in time.

The ridge iron is nailed in every slot between the slates underneath it. Don't use screws on ridge metal because old rusty, painted screws are *extremely* difficult to remove without damaging slates when it's time to replace the ridge metal. Copper ridge metal is preferable to galvanized metal because it doesn't need to be painted, it lasts longer, and it looks better. Use copper nails when nailing copper ridge; both copper nails and copper ridge metal are available through some of the suppliers listed at the back of this book (or jenkinsslate.com).

It's often a good idea to paint galvanized ridge iron before it's installed, although new galvanized metal will not take paint unless it's first primed with a *special galvanized metal primer*. When painting new ridge iron, put a coat of galvanized metal primer on first, then a coat of paint. Neither copper nor aluminum need to be painted.

Always remove the old ridge iron before installing new ridge iron. It's a bad practice to install new ridge over old, as the old ridge prevents the new ridge from laying well on the roof. A common flat pry bar ("wonderbar") and slate hammer work well to remove old ridge iron. Straddle the ridge and pry the nails out one at a time by prying against the metal ridge, *not* against the slate.

Often, when nailing new ridge iron on old buildings, it may seem like the nails aren't hitting any wood, and they may not be. When this happens, simply nail closer to the top of the ridge, instead of near the bottom edge. It's important that the nails all hit something solid, as ridge iron, being metal, is subject to expansion and contraction and the nails will work loose if not nailed tightly.

It's also very common to find old nail holes in the slate along the ridges, and it's critical when replacing ridge iron to examine the roof closely for any exposed nail holes or heads and to caulk them when caulking the new nail heads on the new ridge. One small nail hole near a ridge will cause water to drip through a bedroom ceiling. One dab of silicon will put a stop to it.

It's a good practice to crimp any *exposed* ends of the ridge metal (at each gable end of a gable roof, and at the bottom of each hip on a hip roof) before installing it. The end is crimped to keep out bats and bees, and to improve the appearance of the job.

Aluminum ridge metal is not recommended for use on slate roofs. Aluminum is a bit flimsy and tends to come loose (usually because it's nailed with iron nails which react galvanically and rust). True, it doesn't need to be painted, and it can be nailed with aluminum nails thereby avoiding a galvanic reaction, but aluminum ridge just doesn't hold up like a heavy gauge galvanized or copper ridge. The ridge is one of the main "walkways" on a slate roof. Roofers will walk along the ridge to get from one place to another, and light gauge metals like aluminum will buckle under the weight of a person. This is why ventilated ridges aren't recommended on slate roofs either. The ventilation should be out the gable ends of the roof, or out roof vents, but not out the ridges

---

EMERGENCY REPAIR OF CHIMNEY FLASHING

1. Old flashing is leaking.
2. Roof cement is troweled over the leaky flashing.
3. Fiberglass mesh is worked into the cement.
4. Another layer of roof cement is troweled on. Slates with cement on them will have to be replaced when the chimney is reflashed.

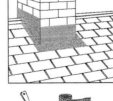

if it can be avoided. When a person works on slate roofs day in and day out, he'll cringe when he sees a slate roof with aluminum ventilated ridge, because he knows how difficult it will be to get around on the roof without damaging the ridge if the roof needs work. Although ventilated ridge will perform well on a slate roof, it can interfere with the routine upkeep and maintenance of the roof by obstructing hook ladders as well as roofers. Slate roofs are made to last for centuries, and by design they should allow for routine maintenance and upkeep. People who put unusual ridges on slate roofs usually aren't thinking about having to work on the roof over the next century.

Leaks at the ridge are often caused by exposed nail heads, both on the ridge iron and on the roof itself adjacent to the ridge. A little roof cement, silicon caulk, or a well placed bib flashing will quickly put an end to these leaks. Additionally, the slates directly under the ridge iron sometimes need to be replaced, and in order to do so the ridge must be lifted by prying the nails out. With the ridge lifted up a little, the nails can be removed from the cap slates using a wonderbar, claw hammer, pointed end of a slate hammer, or ripper, then the cap slates or the slates underneath them can be replaced.

Ridges that are made of slate are simply repaired by removing and replacing the slate. Some older roofing publications recommend cementing ridge slates into place as well as nailing them. Don't do this. In fact, it's almost never a good idea to routinely cement or caulk slates into place when installing them. Why? Because the guy who has to repair the roof years later will have a big mess on his hands when he discovers that the slates have been glued together. Use roof cement or caulk under slates only when absolutely necessary — never use it routinely. Nothing holds a slate in place better than a couple of good nails, and nails are relatively easy

to remove or punch through a piece of slate if the slate ever has to be removed.

Old mitered slate hips that spread open over the years and leak can be repaired by covering with ridge iron or copper.

RESTORATION

What is the difference between repair and restoration? Often, restoration amounts to repairs done in such a manner as to not be visible to a layperson, restoring the original appearance of the roof. In addition, restoration can involve the removal of an entire original slate roof due to it having reached the end of its life, and the subsequent replacement with new slates to restore the roof to its original condition.

Slate roofs are restorable when the slates are still good, no matter how old or how bad the roof looks due to abuse or neglect. On the other hand, if the slates have worn out, no prayer will save the roof.

Following are four examples of restoration projects. The first (page 227) is a common residential job where the valleys had been so abused by "roofing contractors" that the entire roof was almost given up on. The average home-owner would have torn this roof off, but the young lady who owned the house knew a good thing when she saw it and decided to have the roof restored instead. For a fraction of the cost of replacing the entire roof, she kept her good slate roof with the knowledge that it will still last her lifetime.

The Cathedral of St. Andrew in Little Rock, Arkansas (pages 228-229), provides another good example of the proper way to approach an aging slate roof. In this case, the roof was not repairable due to the worn out Pennsylvania black slates, 120 years old, even though the Vermont green slates were still good. The proper solution was to replace the

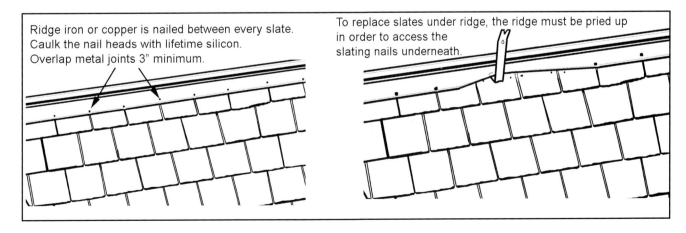

Ridge iron or copper is nailed between every slate. Caulk the nail heads with lifetime silicon. Overlap metal joints 3" minimum.

To replace slates under ridge, the ridge must be pried up in order to access the slating nails underneath.

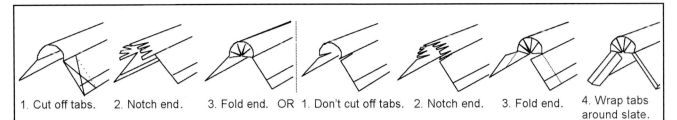

1. Cut off tabs.   2. Notch end.   3. Fold end. OR 1. Don't cut off tabs.   2. Notch end.   3. Fold end.   4. Wrap tabs around slate.

Metal ridge roll should be crimped at the outer ends before installation

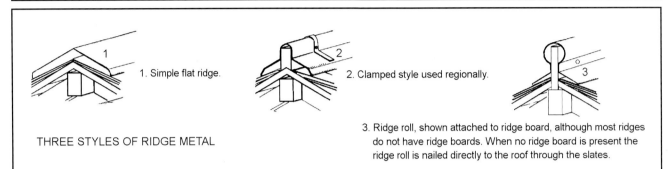

1. Simple flat ridge.

2. Clamped style used regionally.

THREE STYLES OF RIDGE METAL

3. Ridge roll, shown attached to ridge board, although most ridges do not have ridge boards. When no ridge board is present the ridge roll is nailed directly to the roof through the slates.

original slate roof with a new slate roof. But that costs too much money to be practical in most situations, doesn't it? In fact, the roof, installed in 1880, cost $1,951.50 at that time, including labor and materials. That works out to a little less than $16 per year for a beautiful slate roof that is a work of art, on a cathedral no less. Face it, you can't get any cheaper than that. When the entire life of the roof is taken into consideration, nothing beats a slate roof for cost effectiveness. Of course, the Diocese did the right thing and replaced the old slate roof with a new one, saving themselves a heck of a lot of money in the long run.

The dilapidated barn roof illustrated here on page 231 was also one that, at first glance, looked hopeless. However, *because the slate was still good*, the roof was whipped back into shape in just a few days. The west gable end had been almost stripped clean of slates due to excessive exposure to high winds over many years coupled with the inability of the owners to maintain the roof. The west end rafters had to be scabbed over with new rafters and all of the sheathing along the west gable end had to be replaced back to the second rafter. Once this was done, matching slates were installed to finish the job.

The Glenridge Hall in Atlanta, Georgia, illustrates another type of restoration job. Originally installed in 1929, it was a beautiful example of slate roofing craftsmanship done in the "old world" style with a random slating pattern, eyebrow dormers, rounded valleys, graduated slate lengths, and a mix of Vermont slate colors. Fifty years later, however,

for unknown reasons, it developed some leakage and instead of conducting proper repairs, a contractor convinced the owner to pay a huge sum of money to have the entire slate roof removed and then reinstalled in the infamous and discredited "economy" method — a method that eliminates headlap, ignores sidelaps, and instead utilizes felt paper between each course of slates to prevent water penetration. Of course, a scant twenty years after this expensive roofing travesty, the felt paper began to wear thin, rendering the roof a time bomb about to spring leaks in every conceivable place. The unfortunate but wise owner decided to restore the roof back to its original condition by again removing all of the slates and reinstalling them in the proper manner, with sufficient headlaps and sidelaps — an example of restoration done right, illustrated on page 230.

If you have an historic building and want to find funding to help pay for its restoration, try contacting the National Trust for Historic Preservation, 1785 Massachusetts Ave., NW, Washington, DC 20036 or browse their publications at www.preservationbooks.org.

ROUTINE RESIDENTIAL RESTORATION

The valley on a residence in western Pennsylvania at right has been completely tarred and tarred again, as have the slates on either side of the valley (a roofing practice some call the "re-tarred" method). This is a common sight on older American slate roofs and it looks so bad that one may think the roof is, at this point, hopeless. The valley still leaks like a sieve despite the fact that the "repairs" cost the home owner a fair amount of money. This is what the author would call a severe case of Neanderthal syndrome.

The way to deal with the situation is simple enough. All of the tarred slates, as well as the entire tarred valley, are removed. The valley metal is replaced with new copper, in this case 20 ounce partially hardened "red" (uncoated) copper. The key to the success of this approach is simply to have a stock of replacement slates that will match the existing roof. These slates must be the same type, in this case Vermont "sea green" slates, the same size, the same shape (in this case mostly with scalloped corners), *and about the same age*. This is the critical factor in most restoration projects — you usually cannot use new slates to restore existing old roofs. There are some exceptions to this, but not many. This is perhaps the main reason more contractors do not do restoration work on slate roofs (either in Europe or in the U.S.). They would have to maintain a stockpile of many varieties, sizes, and shapes of salvaged slates, which is a job unto itself. A reputable slate roof restoration professional will have these replacement slates or know where to get them.

With the proper tools, supplies, know-how, and replacement slates on hand, a severely abused roof such as this can be put back into its proper shape. At right is the valley replacement completed. The ridge metal still needs to be replaced on this project as well as some of the other tarred slates and other abuse scattered around the roof surface. Once completed, the roof looks like it never had anything wrong with it and it will not leak a drop anywhere. This valley replacement was an afternoon's job for one man. No underlayment of any kind was used on this job — it is totally unnecessary

Photos by author.

AN EXAMPLE OF SLATE ROOF RESTORATION DONE RIGHT

The ornate black and green slate roof on the Cathedral of St. Andrew in Little Rock, Arkansas, reached the age of 120 years before it became a concern to the Diocese (left, shown before re-roofing). A professional roof inspection revealed that the Pennsylvania black slate on the roof had reached the end of its life, although the VT unfading green slate was still in good condition. The only way to properly restore this roof was to remove the original slates and reslate it with new slates following the original pattern. Welsh black slates from Cwt-y-bugail were combined with Vermont unfading green slates to restore the original look of the roof. Copper flashings were replaced with terne coated stainless steel in the critical areas, including the built-in gutters, the valleys and ridges. Four pound sheet lead was used as flashing against irregular stone surfaces. Stainless steel nails were used to fasten the new slates to the original southern yellow pine sheathing.

The ingenious fundraising efforts of Monsignor Scott Marczuk (right), which helped to finance the roof project, included salvaging the re-usable original slates as they came off the roof during the re-roofing project, silk-screening an image of the cathedral onto each slate, and selling them as mementos. He also cut salvaged slates into squares to make coasters and sold them in packs of four, each wrapped in a ribbon. He even had slate clocks and slate crosses made from the original slates, raising a considerable amount of money.

HOW MANY BAKE SALES DOES IT TAKE TO PAY FOR AN ORNATE CATHEDRAL ROOF?

One of the most common complaints about slate roofs is their cost. They are just too expensive to consider installing on most buildings, or so many people believe. The cathedral roof shown at left was installed in 1880 for $1,951.50, including labor and materials. That works out to a little less than $16 per year for a beautiful slate roof that is a work of art on a cathedral. Still think slate roofs are expensive? When you consider the life of the roof, there is no less expensive roof worth buying.

Right: the Cathedral spire is scaffolded for reslating by Midland Engineering. Below right: Lyle Bandurski, foreman on the project. Bottom, middle: Steve Kurtz, project manager for Midland Engineering, photographs the new roof; Bottom, left: Ken Sage, also of Midland Engineering, inspects the new slate on the rear dome of the Cathedral. Opposite page: Bandurski proudly displays his work where both the old slates and the new ones can be see side-by-side.

Photos by author (except aerial photo of old roof at immediate left — photographer unknown).

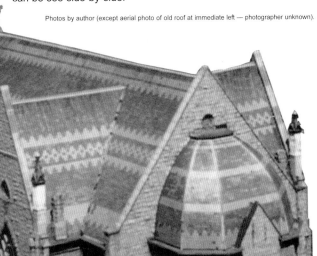

A RESTORATION JOB OF MONUMENTAL PROPORTIONS would best describe the re-slating of Glenridge Hall in Sandy Springs, Georgia, near Atlanta. The original, mixed-color, Vermont graduated slate roof had been removed and re-installed in the 1970s by a roofing contractor who used the discredited "economy" method of slating, which eliminated all headlap, instead relying on felt paper between each course of slate to prevent leakage. Twenty-five years later, the roof had to be entirely removed and reslated with proper headlap, a job successfully undertaken by The R.W. Stokes Company of Atlanta. At left is Victor Manuel Avila, job foreman, and above is roofing technician Sergio Avila.

Bottom photo by Ron Stokes, other photos by author.

SLATE BARN ROOFS such as this one in Pennsylvania (top) are often considered beyond hope after a century of battering by Mother Nature, coupled with owner neglect. The Vermont sea green roof slates were still good, however, and a few days of expert restoration work brought the roof back almost to its original condition (above). Some woodwork was required on the windward gable end (below).

Photos by author.

Chapter Eighteen

FLASHINGS

Flashings are pieces of sheet metal that join a slate roof to objects that protrude through the roof, such as dormers, chimneys, pipes or skylights, in such a manner as to prevent any water from leaking through the joint. Flashings also act as a joint between different roof planes — valleys join two planes that slope into each other, while top flashings on shed-roof dormers join two parallel planes of different slopes. Some flashings act as a joint between the roof and a wall abutting either the side, top, or bottom of the roof.

TERNE

The most commonly used flashing on old American slate roofs is terne metal flashing, often referred to as "tin." Terne is a copper-bearing steel coated on both sides with a lead-tin alloy (80% lead, 20% tin). The typical coating weights of 20 lbs. and 40 lbs. refer to the total weight of lead-tin alloy on both sides of 112 sheets that are 20" X 28". In fact, 20 lb. terne has .047 lbs. per square foot of coating,

while 40 lb. terne has .092 lbs. per square foot. The manufacturer of terne metals, Follansbee Steel, also makes a lead-free terne metal called "Terne II" coated with zinc and tin ("ZT Alloy").

Terne metal makes a good flashing that will last for many years — 90-year-old terne metal valleys on slate roofs are not uncommon, but terne metal must be kept painted or its longevity is substantially reduced. In fact, unpainted terne will begin to show rust in a year or so. Terne metal has an affinity for paint, however, and Tinner's Red or Green Oxide are traditional paints for painting terne on roofs and are available at better hardware and paint stores or from professional roofing supply outlets. Follansbee sells primers and paints specially formulated for terne. Terne metal is also available pre-painted.

Now for some good news and some bad news. The bad news is that most older American slate roofs have been neglected, and the original flashings have deteriorated to the point that they need replacing. The good news is that almost any slate roof with

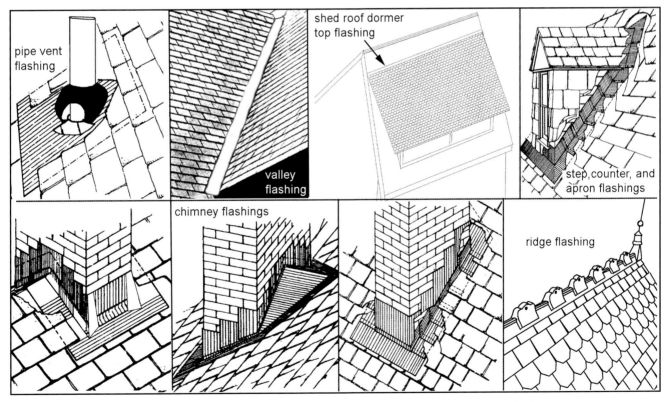

Opposite page: German house is clad with German slate near Fredeburg, Germany, where Magog slate is mined.   Photo by author.

good slates can be reflashed, and in most situations no special tools or equipment are required, except for an additional hand tool or two that is easily obtainable (see jenkinsslate.com).

When removing terne metal flashing from slate roofs, it's advisable to replace it with a metal that doesn't need painted, such as copper, stainless steel, lead coated copper, terne coated stainless, or aluminum. Terne can be used to replace terne, of course, but its use may best be limited to situations where maintenance painting will be done regularly. Experience has shown that many more roofs go neglected than are well maintained, which is the primary incentive for the use of non-corrodible metal.

STAINLESS STEEL

Stainless steel is an extremely durable roof flashing and can hardly be surpassed for longevity. Uncoated, it remains shiny indefinitely, which may make it aesthetically unsuitable for some applications. Stainless falls into several categories: the 300 series typical of flashings are alloys of steel incorporating chromium, nickel and manganese; Type 302 consists of 18% chromium and 8% nickel and is most often specified for flashing. It is used interchangeably with Type 304, a lower carbon variation of 302; Type 316 also contains molybdenum; Type 318 contains 2-3% molybdenum and is more corrosion resistant than the other types of stainless steel; the 400 series is used in interior applications as it contains no nickel and is therefore less resistant to corrosion. Stainless steel also has the characteristic of being non-magnetic in most cases.

Stainless steel used for detailed flashings is best fully annealed and dead soft. Stainless also comes in a variety of finishes ranging from non-reflective to a mirror finish. Terne-coated stainless is type 304 stainless coated with terne and is a popular flashing material. The terne coating eliminates the shiny surface gloss of pure stainless while making the normally difficult stainless easier to solder, requiring no pre-tinning or special preparation for soldering (although pre-tinning is recommended). Prolonged handling of this material with bare hands may pose a health risk because of the lead content. However, a terne-coated stainless alternative known as TCSII, produced by Follansbee, contains no lead at all. The terne coating is on dead soft Type 304 stainless steel and consists of zinc and tin ("ZT Alloy"). TCSII comes in three gauges ranging from 28 ga. (.015"), to 26 ga. (.018") and 24 ga. (.024"). When soldered with pure tin solder, TCSII creates an environmentally friendly, durable roofing system, with no lead content, that never needs painting.

LEAD

Sheet lead is a very popular flashing material in Europe and is used there for valleys, ridges, and just about everything else on slate and tile roofs. Again, the prolonged handling of lead can pose a health risk to the people using it. Lead poisoning, caused by the absorption of lead into the body, can result in anemia and paralysis, so gloves should be worn when handling lead or lead-bearing metals.

Sheet lead, like copper, is specified according to the weight per square foot and is available in grades ranging from one to eight pounds per square foot. Four pounds per square foot of sheet lead yields a sheet that is 1/16 inch thick. Three or four pound sheet lead is suitable for such things as lead pipe flashings, for example. Heavier lead sheet may be used for lead roofing.

One positive characteristic of lead flashing is its workability — it bends easily and can be made to conform to irregular surfaces. It's also very durable and highly resistant to atmospheric corrosion, and will not stain a roof like rusting terne metal or galvanized steel will. A drawback to lead is its high expansion rate; allowances must be made for expansion and contraction of lead when used in larger sheets.

GALVANIZED STEEL

Galvanized metal flashing has been commonly used in roofing applications. It is composed of steel or iron coated with zinc on both sides, either hot-dipped or electro-plated. It's fairly inexpensive and long-lasting if kept painted, although it will rust quickly if neglected and is therefore not recommended for slate roofs. Galvanized metal reacts galvanically with copper, and these two metals should not be allowed to contact each other in the presence of moisture.

ALUMINUM

Aluminum is currently perhaps the most popular flashing material in America for use on asphalt roofs. It comes in a natural silver-gray color (mill finish), or coated with paint. The material is readily available at most building supply centers because it's used for soffit, fascia and siding, as well as flashing, and can be bought in 50-foot rolls ("coil-stock"). It is light in weight, the thinner gauges are easily bendable, and it is highly resistant to corrosion due to the protective oxide that forms on the surface of the metal. Much of the mill finish aluminum available in rolls is too thin (.015") for any use on a slate roof, and even the painted aluminum (.019"/.020") should be avoided for any long term flashing jobs when copper or stainless steel are available. Painted aluminum does, however, make excellent bib flashings for covering the nails of replacement slates (see Chapter 17). Heavier gauge aluminum (.032" - .040") will make a suitable valley on a slate roof, although care must be taken to use aluminum nails whenever nailing aluminum valleys into place.

COPPER

Copper is a non-magnetic, malleable, corrosion resistant metal which shines like brass when new, but soon turns reddish brown upon exposure to the elements, and finally green. The most common flashing material recommended today for use on slate roofs is 16 ounce copper, which gets its name from the fact that it weighs 16 ounces per square foot. An even more durable copper flashing is 20 ounce, which, of course, weighs 20 ounces per square foot, although the 16 ounce may be more workable with hand tools when detailed work is required.

Copper has long been the slate roof flashing of choice in America because it's durable without needing painted; it develops a green patina with age that many people find aesthetically pleasing, and which is said to form a protection against corrosion; and most importantly perhaps, it's easy to work with. Sixteen ounce copper is easy to bend, and can be bent by hand in most cases, or with a pair of sheet metal hand tongs. Additionally, pure copper doesn't present the toxic hazards that lead-bearing metals such as sheet lead, lead-coated copper, terne, and terne-coated stainless do, exposure to which can have an adverse effect upon the health of the people who handle it regularly for prolonged periods. The mining of copper certainly does its share of environmental degradation, yet roof flashing must rank as one of the more honorable uses of the metal, and surely its costs are justified.

Copper, like all metals, does benefit from paint, which will prolong its life indefinitely. It must also be nailed with copper, brass, or stainless steel nails, as galvanized nails will corrode in contact with copper flashing. Copper reacts galvanically with aluminum, steel, zinc, and galvanized steel, and should

MECHANICAL PROPERTIES OF COPPER

| TEMPER | TENSILE STRENGTH (KSI) | | YIELD STR. |
	MIN.	MAX.	(KSI)
060 soft	30	38	?
H00 cold rolled 1/8 hard	32	40	20
H01 cold rolled 1/4 hard	34	42	28
H02 half hard	37	46	30
H03 3/4 hard	41	50	32
H04 hard	43	52	35

Source: Copper Development Association, Palatine, IL (1 ksi=1000 pounds/square inch)

not be placed in contact with these metals in the presence of moisture.

Copper flashing may be sold as either "sheet" or "strip" — sheet copper according to the Copper Development Association in England (1951) being of exact length, over .0006 inches thick, but not over 3/8" thick, and over 18" wide; while strip copper falls into the same thickness range but may be of any width and usually comes in coils.

According to the American Copper Development Association (www.copper.org), architectural copper usually falls under ASTM B370, consisting of 99.9% copper available in six tempers as follows: 1) 060 (soft); 2) H00 (cold rolled, 1/8 hard); 3) H01 (cold rolled, high yield); 4) H02 (half hard); 5) H03 (three quarter hard); and 6) H04 (hard). The harder the copper, the stronger it is. Dead soft copper (060) is not recommended for most building projects; instead, cold rolled (H01) is preferred. Half hard copper is superior in valleys and other applications where extra strength and durability is required. Copper can be softened by annealing, or heating to a dull red with a propane torch and either quenching in water or allowing to cool naturally in the atmosphere.

Lead-coated copper is copper coated on both sides with lead and is popular among architects because it remains dull gray in color, is durable, and is easy to solder.

GALVANIC CORROSION

"Galvanic action," simply stated, is the corrosive reaction between incompatible metals. The corrosion is expedited by an "electrolyte," which is a non-metal substance which will conduct an electric current — water, for example, makes a good electrolyte, and salt water is even better. Metals that are more electro-positive (anodic) corrode more easily, and metals that are more electro-negative (cathodic) are more corrosion resistant.

The farther apart metals are on the galvanic

scale, the more they will react with each other. For example, zinc and aluminum will react with copper much more than they will with each other. Therefore, it's important to avoid using incompatible metals in contact with each other when installing flashing. If such contact seems unavoidable, place a layer of roofing paper, roof cement, paint, or other non-absorbent, non-conductive material between the metals.

USING COPPER

According to the Copper Development Association (CDA) in England,

"The aim of the sheet copper worker must be to achieve the necessary shape or form with as little actual working of the metal as possible. Thus it will be seen that the working techniques of other metals are not always applicable."

Copper expands and contracts under varying temperature conditions; sheet copper expands 1/8" per ten feet of length for a hundred degree temperature change. The amount of movement that takes place for a given temperature variation is about 40%

less, however, than with either lead or zinc. Nevertheless, the expansion and contraction of copper creates limitations in its use, and the Copper Development Association warns, *"It cannot be too strongly emphasized that the superficial area of each individual piece of copper sheet or strip must not be greater than 14 square feet for thicknesses up to and including .022 inches . . ."* When large sheets are used in flashing situations, the copper sheets must be fastened to the roof with copper cleats to allow for movement of the metal.

The *Architectural Sheet Metal Manual* (1993) recommends the use of 50/50 solder (50% lead and 50% tin) when soldering copper, and recommends that the flux be neutralized (flushed with soda water) after soldering. Many flashing jobs on typical slate roofs may not require soldering; nevertheless a section on soldering is included later in this chapter for those situations where soldering is required or preferred. See Chapter 21 for a section on soldered-seam (also called "lock-seam," "flat-seam" or "flat-lock") copper roofing.

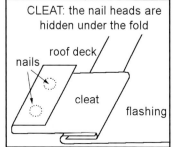

CLEAT: the nail heads are hidden under the fold

nails
roof deck
cleat
flashing

BASIC FLASHING PROBLEMS

Older slate roofs develop two primary flashing problems: deteriorated valley flashing (see Chapter 20) and deteriorated chimney flashing (see Chapter 19). A third, less serious, but still common flashing problem involves deteriorated sewer vent pipe flashings (illustrated in this chapter). Other flashings that wear out on slate roofs include dormer step flashings (sides of dormers), dormer top flashings on shed roof dormers, skylight flashing, and flashings that connect the slate roof to other roofs, or to the outside walls of the building.

SKYLIGHTS

Skylights fall into two general categories as far as flashing is concerned: those that protrude above the surface of the roof, and those that do not. The ones that protrude above the roof should be flashed in a manner similar to a chimney — the sides should be step flashed either with one-piece step flashing or step and counter flashings; the bottom should have an apron flashing; the top should have the largest piece of flashing which extends well up under the slate (at least 12"); and all four corners should be folded (or soldered) to prevent leaking.

SKYLIGHTS
Flush skylights, as above, are single panes of tempered safety glass worked into the roof like giant slates, then finished on the interior with another double pane of safety glass to create a triple pane roof window. Standard commercial skylights, as below, are flashed into a slate roof in the same manner as a chimney. Photos by author.

Some of the standard commercial skylights come with pre-formed flashings which may need to be modified to fit your particular situation. Flashings are easy to make, so it may be advisable to make your own rather than buying expensive, generically pre-formed skylight flashing. Remember this rule when making step flashings: *the bottom of the step flashing should line up with the bottom of the slate above it, and the top of the step flashing should line up with the top of the slate below it.* If the step flashing must be fastened (nailed) to the roof, then allow an inch or two of flashing to extend above the top of the slate below it for nailing. If it fastens to the skylight, that won't be necessary.

Skylights that lay flush with the roof usually fall into the "home-made" category, which means that they're relatively inexpensive, and if done right, work well. The outer piece of glass on the skylight is laid into the roof as if it were a big shingle. It's flashed only on the bottom edge, while the sides and top are overlapped by slate. This piece of glass should be tempered safety glass, and should overlap the roof sheathing by three inches minimum on all four sides. It is laid directly on the felted roof sheathing on a *generous* bed of roof cement, and held in place by the heads of roofing nails tacked in around it on the corners. Wider pieces of glass may require a wooden support or two (a horizontal 1x1 for example) to prevent collapse under unusual snow weight, or when a cat's up there walking on it. It's important to know that a single piece of glass, when used as a skylight, will condense moisture on its interior surface, and drip water when used on a heated building in cold weather. The solution to this is simply to install another piece of glass interior to the skylight, and a double paned glass panel is even better (as far as heat retention is concerned). The drawbacks to "home-made" flush skylights are that they don't open and can't be used for ventilation, and may be difficult to clean. However, they will greatly increase the amount of natural light entering a room and therefore reduce the amount of electricity needed for lighting.

SEWER VENT PIPES

Nearly every home and building has at least one sewer vent pipe protruding through the roof. The flashing on these pipes wears out eventually and must be replaced. Most of the old flashings were made of either copper or lead, lead being the more common. Today they can be replaced either with pre-fabricated lead, aluminum/neoprene, rubber, or made-to-order lead or copper flashings. The most durable are the lead and copper flashings, although lead carries with it the hazard of handling a toxic metal. The most widely available flashings are either aluminum/neoprene or rubber, because these are typically used on asphalt shingle roofs and can be bought at most building supply centers. Lead and/or copper vent pipe flashings may be available through your local plumbing contractor, building supply outlet, or roofing contractor. See the illustrated lead vent pipe flashing fabrication sequence in this chapter.

Vent pipes are found in a variety of diameters, the more typical ranging between 1" and 4"; it is therefore necessary to measure the pipe diameter, as well as the roof angle, and often the length of the pipe as well, before ordering a pre-fabricated flashing. If no suitable flashing is available, try having one made at a local sheet metal shop.

When replacing old vent pipe flashings, the same flashing rules apply: remove enough slates from the roof to completely expose the old flashing, remove the old flashing and replace it with new, then re-slate around it making sure that the slates and flashing overlap in the direction of water flow. On steep roofs less expensive store-bought flashing may have to be forcibly bent to align properly with the slope of the roof.

STEP FLASHINGS

Step flashings wear out on dormer sides on old slate roofs after a century or so and must be replaced. Often the sides of the dormers as well as the roof are covered with slate, so replacing the step flashing is simply a matter of removing the slates from the sides of the dormers (after numbering them), then removing enough slates from the roof to completely expose the pieces of step flashing, one at a time. The step flashings are then pried off one at a time until all are removed; then they're replaced one at a time with new material as the slates are nailed back in place. After all the step flashings are replaced, the slates on the side of the dormer are nailed back on — an easy job if they were numbered ahead of time (e.g., row A1, A2, A3, B1, B2 etc). The flashing can be nailed either to the side of the dormer or to the roof when being installed.

When installing step flashings, remember the simple rule: line up the bottom of the step flashing with the bottom of the slate on top of it, and the top of the step flashing with the top of the slate underneath it. Extend the top of the step flashing up

another inch or two for nailing purposes if desired. Lay the step flashing about 3"-4" horizontally on the roof and about 2"-4" vertically up whatever it is you're flashing.

Some dormers don't have slate on their sides and there is no way to get the step flashing underneath the siding. In that case you may have to be content with tightly nailing the step flashing to the side of the dormer against a backing of silicon caulk or roof cement, leaving the vertical section of the step flashing exposed, or covered by a piece of wood trim. This is a last resort, however, as both vertical and horizontal sections of the step flashing should be covered by slate and siding. Counter flashing can also be used to cover the vertical sections of step flashings, and the counter flashing can be bent and set into a reglet (groove) cut into the side of the vertical wall.

TOP FLASHING ON A SHED ROOF DORMER

Flat roofed (shed roofed) dormers on slate roofs can sometimes cause problems because they don't have the slope that the rest of the roof does, and therefore rain and snow don't drain off of them fast enough. This becomes a real problem when the flashing that ties the shed roof to the main roof becomes deteriorated and begins to leak. However, this flashing can routinely be replaced, and this is the beauty of a slate roof — it can be taken apart, the flashing replaced, and the roof put back together again whenever necessary.

The procedure for replacing the top flashing on a shed roof dormer is basically the same as for replacing any flashing: remove the slates to completely expose the flashing, pry the old flashing off the roof, replace it with new metal, then replace the slates. When replacing the slates, substitute good ones for any that are cracked, broken, or tarred.

Number the slates with a nail when random width slates are used (numbering isn't necessary when uniform slates are used), keep the slates nearby after removing them so they'll be handy when you replace them, use the same slates whenever possible when re-slating, and put them back in the same spot (when they're random widths). A top flashing replacement sequence on a random width slate roof is illustrated on pages 242-243.

SOLDERING

Soldered flashings are not typically required when doing general slate roof restoration. However,

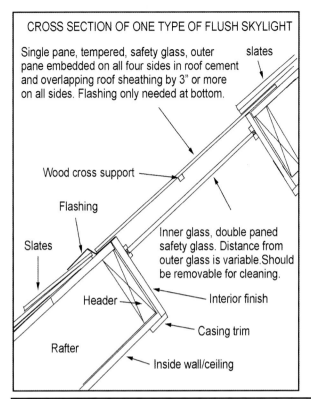

CROSS SECTION OF ONE TYPE OF FLUSH SKYLIGHT

Single pane, tempered, safety glass, outer pane embedded on all four sides in roof cement and overlapping roof sheathing by 3" or more on all sides. Flashing only needed at bottom.

slates

Wood cross support

Flashing

Slates

Inner glass, double paned safety glass. Distance from outer glass is variable. Should be removable for cleaning.

Interior finish

Header

Casing trim

Rafter

Inside wall/ceiling

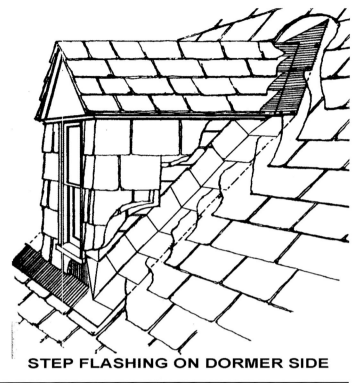

STEP FLASHING ON DORMER SIDE

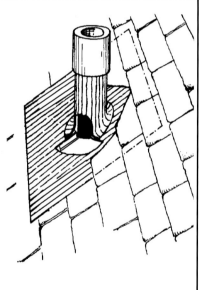

SEWER VENT FLASHINGS

on some occasions one may wish to solder something (e.g. a finial, a built in gutter, a cupola, or a shop-fabricated vent pipe flashing) so a brief review of soldering principles is appropriate.

Simply stated, soldering is the act of joining two pieces of metal with a molten metal or metal alloy. For our purposes, soldering involves joining copper to copper, or lead-coated copper to lead coated copper, terne to terne, terne coated stainless to TCS, or stainless to stainless, using lead/tin or pure tin solder. There are enough problems involved in soldering aluminum that the Architectural Sheet Metal Manual states that it cannot be soldered.

The official definition of soldering is *"a group of welding processes which produces coalescence of materials by heating them to a suitable temperature and by using a filler metal having a liquidus not exceeding 450 degrees C. (840 degrees F.) and below the solidus of the base materials. The filler metal is distributed between the closely fitted surfaces by capillary attraction* [American Welding Society Soldering Manual, 1978, p. ix]." The solidus temperature is the highest temperature at which a metal or solder is completely solid, while the liquidus temperature is the lowest temperature at which a metal or solder is completely liquid.

Fabricating a Pipe Vent Flashing from Sheet Lead

Plumbing vent pipe flashings tend to wear out and leak on older buildings. Institutions often have pipes that are unusually high or large in diameter, and replacement flashings must be custom made. Sheet lead is a common material for this purpose, although 16 ounce copper is often used as well and can be used following the same procedures with some minor variations. Lead is a very malleable metal and can be easily bent or formed. It is advisable to wear gloves when working with lead.

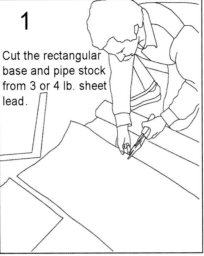
1 Cut the rectangular base and pipe stock from 3 or 4 lb. sheet lead.

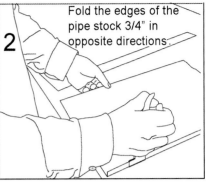
2 Fold the edges of the pipe stock 3/4" in opposite directions.

3 Clean folds with steel wool and flux with paste flux before folding completely.

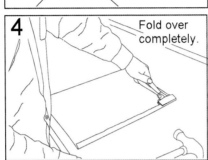
4 Fold over completely.

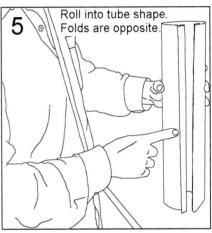
5 Roll into tube shape. Folds are opposite.

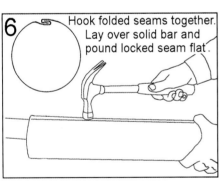
6 Hook folded seams together. Lay over solid bar and pound locked seam flat.

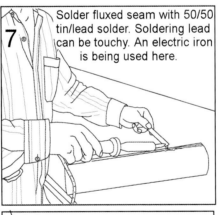
7 Solder fluxed seam with 50/50 tin/lead solder. Soldering lead can be touchy. An electric iron is being used here.

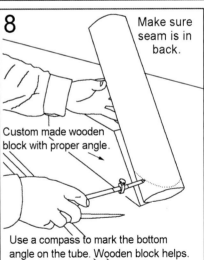
8 Make sure seam is in back.

Custom made wooden block with proper angle.

Use a compass to mark the bottom angle on the tube. Wooden block helps.

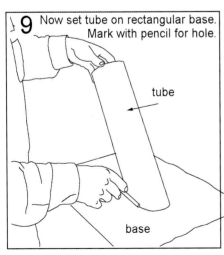
9 Now set tube on rectangular base. Mark with pencil for hole.

tube

base

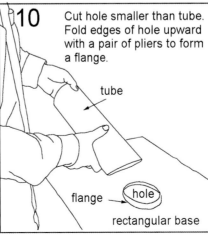
10 Cut hole smaller than tube. Fold edges of hole upward with a pair of pliers to form a flange.

tube

flange   hole

rectangular base

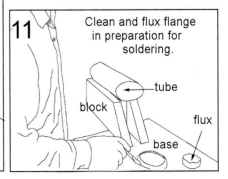
11 Clean and flux flange in preparation for soldering.

tube

block

base   flux

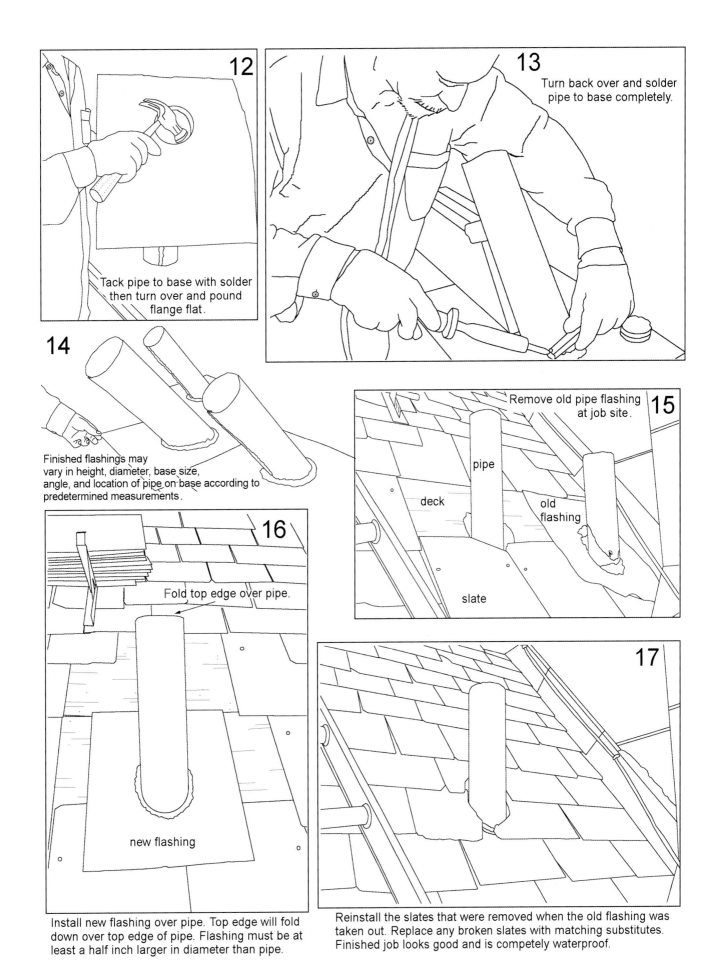

12

Tack pipe to base with solder then turn over and pound flange flat.

13

Turn back over and solder pipe to base completely.

14

Finished flashings may vary in height, diameter, base size, angle, and location of pipe on base according to predetermined measurements.

15

Remove old pipe flashing at job site.

pipe

deck

old flashing

slate

16

Fold top edge over pipe.

new flashing

17

Install new flashing over pipe. Top edge will fold down over top edge of pipe. Flashing must be at least a half inch larger in diameter than pipe.

Reinstall the slates that were removed when the old flashing was taken out. Replace any broken slates with matching substitutes. Finished job looks good and is competely waterproof.

▼ BEFORE - The entire length of the old copper top flashing is pitted, leaking, face-nailed and spot-tarred.

▼ AFTER - New 20 ounce copper top flashing on this dormer will last for generations.

▼ 1 - Number 3 rows of slates 1A, 1B, 1C..., 2A, 2B, 2C... etc - (number random width roofs only), remove top row with slate ripper.

▼ 2 - Carefully remove lower slates by pulling out the nails with a flat pry bar (continued on next page).

If your mind is still functioning after reading that, let's continue. "Wetting" is when a molten solder leaves a continuous, permanent film on the base metal surface, often referred to as "tinning" or "pre-tinning." Wetting makes the soldering action possible, and although pure lead doesn't readily wet (or adhere to) either copper or steel, a tin/lead solder readily wets both, which is why solder contains tin.

The solder is drawn into the joints by capillary attraction, and a space between the joints of about .15mm or .005 inches is suitable for most work. A clean surface free of oxidation is critical to ensuring a sound soldered joint, and some fine steel wool will clean most surfaces adequately prior to soldering. Solder that is 50% tin and 50% lead is suitable for use on copper, lead sheet, galvanized steel, stainless steel, terne metal, and terne coated stainless (although when soldering Terne II or TCSII, a 100% tin solder is recommended if a lead free project is desired).

Prior to soldering, a flux is applied to the surfaces for the purposes of additionally cleaning away oxides, preventing re-oxidation, and promoting wetting of the surface. Fluxes fall into three general categories: corrosive, intermediate, and non-corrosive. The mildest flux that works should be the one chosen for the job. Corrosive fluxes remain chemically active after soldering and can cause corrosion to the joint, which is why they're not used for electrical applications (for example).

Highly active and corrosive fluxes must be used on stainless steel, high alloy steels, and galvanized steel because they have hard oxide films, although rosin (non-corrosive) fluxes are satisfactory in most cases for soldering terne metal. Fluxes such as Ruby Fluid contain zinc chloride and are considered corrosive fluxes. The soldered areas where a corrosive flux was used should be washed after soldering using water with baking soda added.

Copper tends to form less tenacious oxide films, and therefore mild (non-corrosive) fluxes are suitable when soldering it. Non-corrosive fluxes all have rosin as a common ingredient.

After the flux is applied, heat must be applied to the metal to be soldered. This is typically done by heating a soldering iron and placing the sol-

▲ 3 - Work from one end toward the other when removing the slates. Starter slates are underneath 1st row and here have no shim.

▲ 4 - Expose and remove the old flashing. Work off planks if the roof is not too steep, otherwise use roof jacks and planks.

▲ 5 - Remove any bad slates that were under the flashing.

▲ 6 - Replace the bad slates and sweep the roof.

▲ 7 - Nail in the new flashing (in this case 20 ounce copper bent over the edge of a ladder). Underbend flashing so it will fit tightly.

▲ 8 - Begin re-slating in the reverse order, putting same slates in original spots. Align with weather marks. Avoid nailing through metal.

▲ 9 - This flashing is bent 6" on dormer and 10" on main roof, and is nailed along top edge only.

▲ 10 - Top row of slates are fastened with nail-in-slot and bib flashing technique. Photos by author.

dering iron on the metal with the solder in close proximity. The solder will then flow into the cleaned, fluxed joint as the iron and solder are slowly moved along the joint. It may help to wet (pre-tin) the metal with solder before soldering, although wetting and soldering can take place in one operation. This is a part of the process that has a few variables, such as the amount of heat, type and size of soldering iron, etc, that will vary according to the type of metal being soldered, and these details must be ascertained by the artisan doing the work. For example, copper may require a hotter iron than steel because it has a greater rate of conductivity, while stainless steel may require a cooler soldering iron and longer contact. Practice soldering on a piece of scrap before doing the actual work if you need to figure out these details.

After soldering, the soldered joint should be encouraged to cool as rapidly as possible without cooling so quickly as to warp. Non-corrosive, rosin-based flux residues do not need to be removed after soldering. Zinc chloride fluxes can be removed by flushing with a 2% hydrochloric acid solution, then with hot water containing some sodium carbonate, then with clean water.

It's important to remember that all soldering fluxes give off fumes and smoke when heat is applied, and these may be toxic. Metals such as cadmium, lead, zinc and their oxides may also be toxic when present in the atmosphere as fumes or dust. Therefore, adequate ventilation is imperative when soldering any metal. Soldering fluxes containing zinc chloride may produce severe burns and dermatitis if allowed to remain on the skin for any period of time.

Two typical solder joints related to flashing are the lap seam and the lock seam (below). The lap seam can be joined with rivets of a compatible metal before soldering, although the joint as well as the rivets must be soldered. The flat lock seam can be cleated to an underlying structure before soldering, if needed (see Chapter 21).

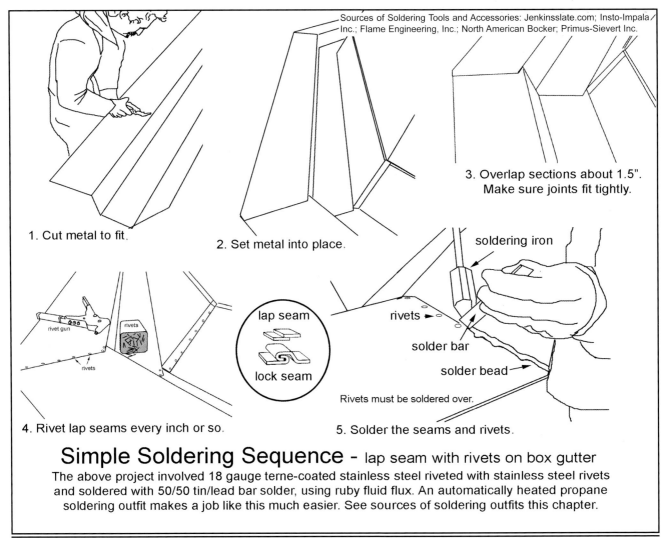

Sources of Soldering Tools and Accessories: Jenkinsslate.com; Insto-Impala Inc.; Flame Engineering, Inc.; North American Bocker; Primus-Sievert Inc.

1. Cut metal to fit.

2. Set metal into place.

3. Overlap sections about 1.5". Make sure joints fit tightly.

rivet gun
rivets
rivets

lap seam
lock seam

soldering iron
rivets
solder bar
solder bead
Rivets must be soldered over.

4. Rivet lap seams every inch or so.

5. Solder the seams and rivets.

Simple Soldering Sequence - lap seam with rivets on box gutter
The above project involved 18 gauge terne-coated stainless steel riveted with stainless steel rivets and soldered with 50/50 tin/lead bar solder, using ruby fluid flux. An automatically heated propane soldering outfit makes a job like this much easier. See sources of soldering outfits this chapter.

COMMON STEP FLASHING REPLACEMENT AGAINST THE SIDE OF A DORMER

1. Old Flashing is worn out, tarred, and leaking.

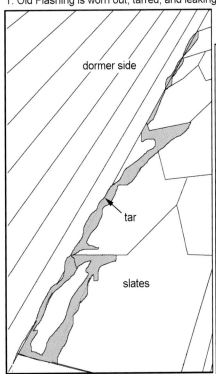

2. Top slate is removed to expose flashing.

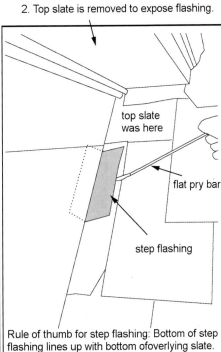

Rule of thumb for step flashing: Bottom of step flashing lines up with bottom of overlying slate. Top of step flashing lines up with top of underlying slate (can extend above it).

3. Pull old step flashing out, then remove half slate underneath it.

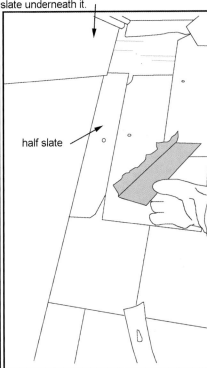

4. Remove next piece of step flashing. Flashing will be behind siding, so siding may have to be removed or pried loose in order to get the step flashing out.

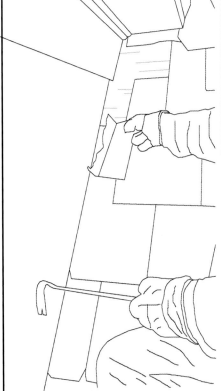

5. Continue to remove slates using a slate ripper, exposing and removing each step flashing. Remove all step flashings before replacing any with new ones.

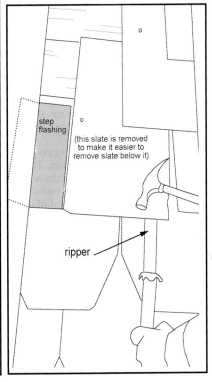

(this slate is removed to make it easier to remove slate below it)

Getting the step flashing to go under the siding is often the biggest challenge. Cut a reglet and install counterflashing if you can't go under the siding. Then caulk the reglet groove.

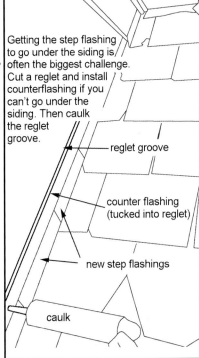

6. Once all old step flashings are removed, the roof can be put back together in the reverse fashion using new copper or stainless steel step flashings. The step flashings should extend up under the siding by about two inches and on the roof 3" to 4". The siding may have to be pried loose to get underneath it. If brick, cut the mortar joints. If siding is slate, number and remove before replacing step flashings, then replace.

TYPICAL FLASHING SCENARIOS ON AN OLDER, RESTORABLE SLATE ROOF.

Above are photos of a brick bell tower on a church in Mercer, Pennsylvania. The photo above left shows a common sight — old flashing buried in tar, and broken slates where the roofers walked on and damaged them over the years. The repairs were faulty and it should be no surprise that a persistent leak occurred there. The only permanent solution was to remove all of the offending materials including the old tar, flashings and broken slates, and replace them. The photo above right shows the job completed in copper, a day's work for an experienced slate roofer and a helper. The slate was Pennsylvania black, 104 years old and still in good condition.

The same church, shown at left, also had deteriorated valley flashings. The old metal has been removed and new 20 ounce half-hard copper valley flashing has been installed. The slates are now being nailed back into place. Note the board roof deck and the absence of underlayment (felt paper). Although felt paper had been installed on this roof over a century ago, it was now deteriorated to the extent of being practically non-existent. The new valley was installed with no underlayment whatsoever and was, of course, leak free. Although felt paper is recommended when installing a new slate roof, it is the slate and the flashings that permanently keep the water out of the roof, not the underlayment. For more on this issue, see Chapter 13. For more about valley flashings, see Chapter 20.

Photos both pages by author.

NEGLECTED CHIMNEY FLASHINGS ARE NOTORIOUS PROBLEM AREAS ON OLD SLATE ROOFS (ABOVE)

Chimney flashings will deteriorate and leak after many years, and they must be replaced. This is routine maintenance on old slate roofs as the metal flashings simply do not last as long as the slates, in many cases. The photo above left shows an old chimney with deteriorated flashings and some tar patches, a candidate for reflashing. The same chimney, above right, has been reflashed with new copper using the "folded corners" flashing method described in Chapter 19.

BUILT-IN GUTTER REPLACEMENT BENEFITS FROM STAINLESS STEEL (BELOW)

Built in or "box" gutters are notorious for wearing out on older slate roofs. The gutters on the Cathedral of St. Andrew in Little Rock, Arkansas, were originally copper and had already been replaced once in the 125-year history of the roof. When the cathedral was reslated with new slate in 2002-2003, terne coated stainless steel gutter flashing, instead of copper, was installed in order to add an extra element of longevity to the job. The photo below shows the job in progress. The scaffolding platform, collecting slate dust, allows easy access to the lofty job.

## Chapter Nineteen

CHIMNEYS

Chimneys on old slate roofs present a variety of maintenance and repair issues. First, the flashing wears out and must be replaced. Secondly, the mortar becomes soft, especially above the roof-line, and the top of the chimney must be rebuilt periodically. Third, many old chimneys no longer serve a functional purpose in the house and are no longer used. These chimneys should be removed to below the roof-line and the roof closed up. In some cases, chimneys must be rebuilt from the attic floor, rather than from the roof-line. These are all routine maintenance jobs on old slate roofs.

An old style of chimney construction sometimes found in old farm houses involves building the chimney on a wooden platform elevated off the floor of the house. These chimneys may not even have a wooden pedestal underneath them, but instead may be supported only by a wooden shelf propping the massive brick structure five feet off the floor! Most chimneys, however, are built from the basement floor up through the house and on through the roof.

Older chimneys were typically built of 4" x 8" standard thickness brick that were solid, with no holes in them. Therefore, when rebuilding an old chimney, the same size and type of bricks are recommended. They go by the name of 4x8 "pavers," and are a little more expensive than bricks with holes, but they offer greater fireproofing protection because they're solid brick.

The chimneys found in most older American houses with slate roofs are almost universally devoid of ceramic chimney liners, which are additions to chimneys that became popular in more modern times, when bricks began to be made with holes in them to save on material costs and cut down on their weight. The liners added an additional firewall for protection. In the old days, chimneys were used to vent smoke and gasses from coal and wood burning stoves, which tend to burn hot and inefficiently. Today, most chimneys vent either the more efficient "air tight" stoves, or gas/oil furnaces, both of which burn cooler than the old-time heaters. Cool burning woodstoves are notorious producers of creosote, which can leak through brick chimneys and even ceramic liners.

Opposite page: German Chimney Photo by author.

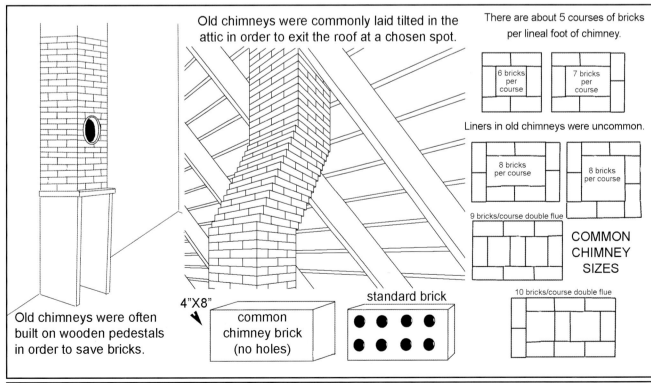

▲ New York red slate blocks and shingles create an elegant Vermont chimney that is a work of art.

Photo by author.

▼ Strings anchored to the floor and ceiling make great plumb guides when laying a chimney.

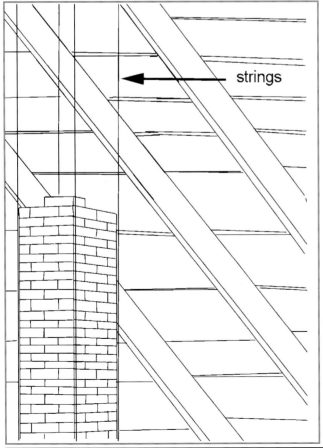

strings

When rebuilding an old chimney or part of it, the original style should be imitated, unless the entire chimney will be rebuilt; then any style can be used. However, the old style of using solid bricks laid up without a ceramic liner is a tried and proven style, and many thousands of these old chimneys are still in service after a century of use. Nevertheless, old chimneys can be dangerous, especially if the mortar and/or bricks become soft enough to crumble and allow flames to escape to a flammable surface during a chimney fire. Chimney fires are unlikely when the chimney vents only a gas or oil furnace, because these don't create an accumulation of creosote as a wood stove does. When using an old chimney to vent a wood stove, the chimney should be inspected closely beforehand to make sure it's safe. If it doesn't appear to be safe, it should be completely rebuilt or lined (with a stainless steel liner, for example).

Another peculiar style of chimney construction that was once common involved the practice of building a section of the chimney, usually in the attic, at an angle. Once through the attic floor, the chimney was angled so the top would exit the roof directly through the center of the ridge — a style adhered to by certain ethnic builders. It's these leaning chimneys that must often be rebuilt from the attic floor because of deteriorating bricks or mortar, which can create quite a hazard if the chimney should begin to collapse. When rebuilding a leaning chimney, it should be built straight and plumb, after which it will exit through the roof slightly to one side of the original hole. New roof sheathing that matches the existing sheathing is simply added to fill in the hole after suitable framing material has been nailed into place. The area is then reslated as the chimney is flashed. See Chapter 13 for more information on roof sheathing.

When building a chimney for a new building, be aware that there are best and worst locations to build them. The best location for a chimney is inside the structure and through the peak of the roof. Inside chimneys stay warmer, radiate more heat throughout the building, and develop less creosote problems than cooler, outdoor chimneys. The emergence of the chimney at the peak of the roof prevents any water from pooling up against the chimney. From a roofing point of view, a chimney built up against the outside of the building through the peak of the roof works just as well as one built inside the building. However, from a heating point of view the inside chimney is preferable.

A less desirable place to situate a chimney is

WHERE TO LOCATE A CHIMNEY: The best place for a chimney is inside the house, through the peak of the roof. A wide chimney at the eaves is not recommended because it creates a dam and blocks the flow of water and snow off the roof, although a cricket will usually solve that problem. Don't build a chimney at the bottom of a valley unless you enjoy grief.

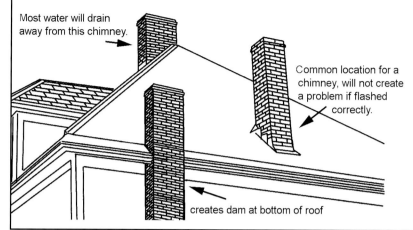

Most water will drain away from this chimney.

Common location for a chimney, will not create a problem if flashed correctly.

creates dam at bottom of roof

This is a bad idea.

REMOVING AN OLD CHIMNEY

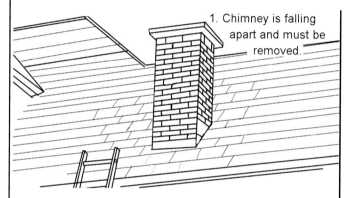

1. Chimney is falling apart and must be removed.

2. Cap stone and bricks are removed

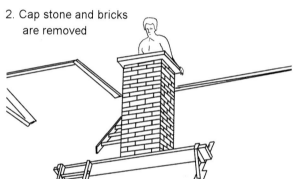

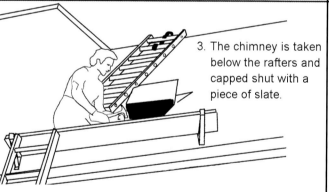

3. The chimney is taken below the rafters and capped shut with a piece of slate.

4. 2x4s are nailed to the rafter sides and the hole is boarded shut with lumber that matches the existing roof sheathing.

5. Replacement slates that match in size, shape, and color (type) are used to slate over the patch.

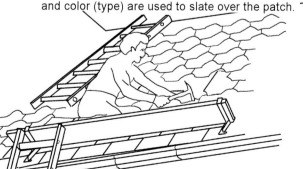

6. Properly done, the finished job will look like there never was a chimney there.

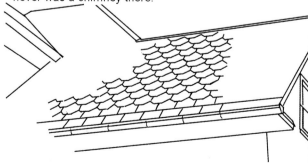

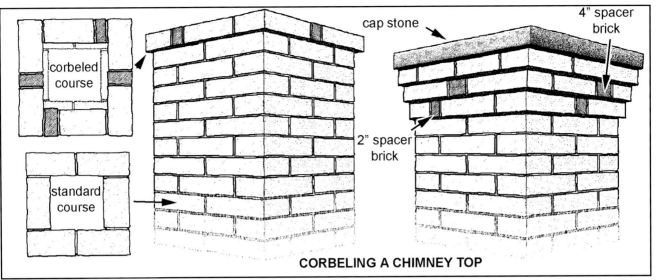

CORBELING A CHIMNEY TOP

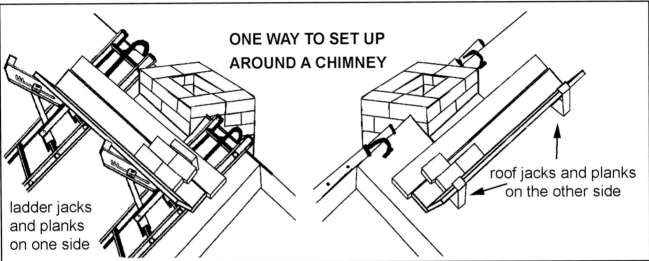

ONE WAY TO SET UP AROUND A CHIMNEY

ladder jacks and planks on one side

roof jacks and planks on the other side

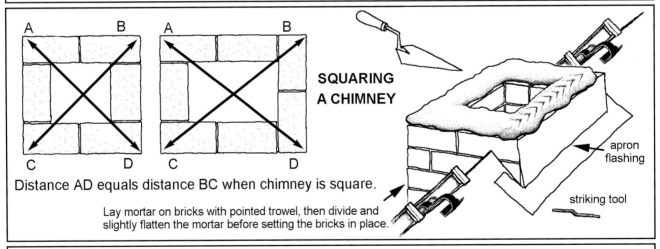

SQUARING A CHIMNEY

Distance AD equals distance BC when chimney is square.

Lay mortar on bricks with pointed trowel, then divide and slightly flatten the mortar before setting the bricks in place.

apron flashing

striking tool

REBUILDING THE TOP OF A CHIMNEY

TOOLS YOU WILL NEED: Mortar box and hoe for mixing mortar, small mortar box for use on the roof to hold the mortar (a plastic dish pan will do); pointed masonry trowel; 5-gallon plastic buckets for carrying mortar and possibly bricks up onto the roof; brick tongs for carrying brick (optional but recommended); 2-foot and 4-foot levels; striking tool for striking mortar joints after bricks are laid; slate ripper; slate cutter; slate hammer; flat pry bar; small sledge hammer for knocking old top apart (may not be needed); bucket of clean water and rags for wiping chimney clean after bricks are laid (important); tin snips; flashing (sheet metal) tongs (optional when using 16 ounce soft copper).

EQUIPMENT: Ground ladders, hook ladders, roof jacks, scaffold planks.

MATERIALS: Regular masonry cement (mortar); mason's sand; new bricks; flashing material (16-ounce or 20-ounce soft copper or terne-coated stainless steel recommended); water; slates that match the roof.

MORTAR MIX RATIO: One part regular masonry cement to three parts masonry sand. Mix dry first, then add water slowly to consistency of buttery mud.

---

at the eaves of the building, especially if the chimney is wide, as it creates a dam on the roof which will eventually turn into a leak. The worst place to put a chimney is at the eaves *and* at the bottom of a valley. This sort of chimney falls into the "roofer's nightmare" category, although they can be made leakproof with a carefully constructed cricket.

REBUILDING THE TOP OF A CHIMNEY

The first step in rebuilding a chimney top is setting up roof scaffolding. This can be easily and quickly done by using hook ladders and ladder jacks. When the chimney is located halfway down the roof instead of at the peak, hook ladders alone with two pairs of ladder jacks (one pair above and one below the chimney) will work. If hook ladders can't be used because there's no ridge above the chimney, roof jacks and planks will have to do. The danger of using any roof scaffold is in overloading it, with bricks for example, and/or carelessly installing it. Double check everything before any weight goes on the scaffold, then make sure it isn't overloaded.

Once the roof scaffold has been set up, the old chimney top should be removed and then rebuilt with new bricks. Old chimney tops can usually be taken apart by hand because the mortar becomes soft after a century of weathering, although some of the more stubborn chimneys may require a small sledge hammer and a flat pry bar to get the bricks apart. The top should be taken down to just below the roof line, far enough to remove all of the old flashing. All old mortar should be removed from the top of the last remaining course of bricks, then the bricks brushed clean of dust in preparation for new mortar.

When a chimney top is rebuilt, it makes sense to reflash it too, because it's much easier to flash a chimney when rebuilding it than it is to try to flash it after it's been built. The old flashing, then, is completely removed as are the slates surrounding the chimney covering the flashing, as well as any bad slates (broken or tarred). It's always easier to completely clean all the bad stuff out of the way before beginning to put the roof and chimney back together.

When building a chimney indoors, such as through an attic, run strings where the four corners of the chimney will be and use them as a guide when laying the brick. This saves quite a bit of time because the sides don't need to be plumbed with a level as the chimney is built, although the courses still need to be checked for levelness as each course is laid. Furthermore, when building a chimney from the bottom up, make sure there is a clean-out door somewhere near the base of the chimney in the basement and try to leave a drain in the bottom of the chimney so any rain water that accumulates down there will have a place to go. Heavy downpours can allow a lot of water to enter a chimney, even when capped.

When you have the bricks and tools in place on your roof scaffold, the first batch of mortar can be mixed. It's usually a good idea to have one person mixing mortar and hauling materials and another laying the bricks. The mortar is mixed according to a ratio of one part masonry cement to three parts masonry sand. These materials are always mixed together dry first until the mix is a uniform grey color, then water is slowly added until the correct consistency is obtained. If the mix is too wet the mortar will run all over the chimney and make a mess, but if it's too dry it will make the brick-laying job more difficult. If it is too dry, however, a little water can be added while on the roof scaffold in order to get the mortar to the right consistency, which should be like thick pudding (or "butter" as the contractors say).

The mortar is first laid on the clean brick surface with a pointed trowel, then cut into two halves and slightly flattened with the point of the trowel. The bricks are laid in place, one at a time. After the first brick is laid, the next must be buttered on the end that will abut the first brick, before setting it into place. Every course of bricks must be staggered in relation to the one below it so the joints don't align. The courses must be checked with a level as they're laid, for levelness and for plumbness. A two-foot level can be used to plumb the first two feet of the chimney, then a four foot level is necessary. The chimney should also be checked for squareness, either by using a framing square, or by measuring the diagonals (diagonal corner to diagonal corner), which will measure exactly the same if the chimney is square. Smooth the excess mortar against the bricks on the inside of the chimney with the trowel.

Fold the chimney flashing into the mortar joints a distance of about two inches. Start with the apron flashing, then the lower corners, which are folded around the corners of the chimney, then the sides. Sixteen-ounce soft copper is recommended as flashing material because it's long lasting and it can be bent by hand. You'll have to set the apron flashing in place, then lay another course of bricks (maybe two) on top of it, then install the corner flashings,

DRIPSTONES
Stone shelves built into the Welsh chimney shown above divert water away from the bottom of the chimney, acting much the same as flashing. Dripstones are a common sight in Wales where both the chimney and the building are constructed of similar materials.

Photo by author.

then lay a course or two of bricks, etc. This is a tedious job that benefits greatly from experience.

After the bricks have been laid and the mortar begins to set, the joints must be struck with a striking tool. The joints must be struck before the mortar gets too hard, so it's usually a good idea to strike the joints as you go along. As soon as the job is completed (most single or double flue chimney tops can be reflashed and relaid in a day by two workers), the bricks should be carefully washed with clean water and a clean rag. There is no easy way to do this because if the mortar gets wet, it will run over the face of the bricks and make a mess, so the wet rag must be wrung out and carefully wiped across the face of *each and every* brick to remove any traces of mortar. Otherwise, when the mortar dries, it will appear as white splotches or streaks on the bricks and no matter how well the bricks were laid, it will forever look bad. Once the mortar does dry out it's almost impossible to clean it off without using an acid solution, which is not advisable on a slate roof due to the effect of the acid on the flashing. Take a little extra time to carefully clean *each* brick when the job is done, before the mortar dries, and the chimney will be beautiful.

Finally, after the chimney top rebuilding job is done, you must go down into the basement of the house and find the clean-out door at the bottom of the chimney (and you better hope there is one — otherwise you may have to make one). There you will need to remove any mortar or other debris that has fallen down the chimney when the top was rebuilt. Otherwise you risk plugging the chimney, and a plugged chimney is dangerous because it can allow carbon monoxide, an odorless and lethal gas, to accumulate in the house. If you have any doubts about whether your chimney is venting properly, install a carbon monoxide detector in your home, available at most hardware stores.

One last thing — when rebuilding a chimney top, take an extra hour or two and do a fancy job. The extra time is well worth it when one realizes that the new chimney top will be visible for generations.

BUILDING A FANCY TOP

Chimneys were often traditionally flared at the top in order to accommodate a chimney cap stone, which was usually a single piece of sandstone laid over the top course of bricks with a square hole cut in the middle to allow the chimney to exhaust. Bricks can be easily stepped out, flared, or *corbeled*, by simply adding spacer bricks in the proper courses. A one-inch spacer brick on every side of the chimney course will space that course out 1/2 inch on all four sides. A two-inch spacer will space the course out an inch, etc. By successively increasing the size of the spacer brick by two inches, each subsequent course of bricks will be stepped out an inch further than the one below it. A course can be stepped out, then stepped back in, then stepped out again. There are an endless number of variations related to chimney top designs, and the design one settles upon for his/her own work is only limited by one's creativity. A good chimney builder may build a "signature" top on his/her chimneys, such as the one shown on the next page.

REMOVING AN OLD CHIMNEY

Old chimneys no longer used are best removed from the roof. Once gone, they cannot leak, and they will require no further expense or upkeep. They can be rebuilt if needed if they're only removed to just below the roof line. Such a removal job is routine and can be done so that the roof appears never to have had a chimney at all.

This old style fancy chimney top is not difficult to build, and the procedure is illustrated on the next page. Below left is a double flue chimney on a historic house in Harrisville, PA, with an asbestos shingle roof. The original chimney was a leaning chimney and was rebuilt plumb from the attic floor. Below middle is a rebuilt double flue chimney top, and right is a chimney built new through a sea green slate roof. All three chimneys are flashed with 16-ounce soft copper. (Top) A similar style on a more massive chimney in Crediton, Devon, England.

Bottom photos by author, top photo by Dave Starkie.

1. Lay about 10 courses up from the ridge, then come out an inch, then lay two regular courses, then come out an inch again.

2. Set full bricks in a bed of mortar diagonally on the corners, flush on the ends. Level and straighten them.

3. Fit in the remaining diagonal bricks, lining up their outside edges and making sure they're level. Some will need to be cut short. Use a brick hammer to cut them, or use a masonry or diamond wheel on a grinding tool or electric saw.

4. Lay all the diagonal bricks in place, corners flush with the top square course, outside edges lined up, and everything level.

5. Lay a square course on the top, outside edge flush with the tips of the diagonal bricks.

6. Lay a second square course on top of the last one, then lay a bed of mortar on top of it as a final protective layer. Level it and dress it up.

7. Strike the joints, then carefully clean the chimney, brick by brick, with clean water and a rag.

Building a Fancier Chimney Top

The trick to making a decent chimney top is to take your time, level and plumb everything, use plenty of mortar and make sure it isn't too dry. Slightly wet mortar makes laying small pieces easier too. A top like this will only add an hour or two to the job, and since the top might last a hundred years it's worth the extra time!

The job is simple and can be briefly described. As always, most of the job is in setting up a safe working platform. Again, roof jacks and planks, along with hook ladders, will be sufficient in most cases to get the job done. Once you've made your way to the chimney top, you may find that the bricks come apart in your hand, which is common on old chimneys. If the chimney is stubborn, a small sledge hammer will convince it that its days have come to an end. After the bricks are knocked apart and the chimney is lowered enough to clear the rafters, framing lumber is nailed to the rafters to create a place to nail the sheathing boards that will be patched in to cover the chimney hole. The sheathing patch should be made of the same sort of lumber that already exists on the roof, such as one-inch-thick boards.

Sometimes a rafter is set too far from the chimney hole to be able to scab it out enough to nail sheathing on to it. In this case, one will have to take a saw up on the roof and cut the old sheathing back to the rafter center, then add the new sheathing nailed directly to the old rafter. A cordless circular saw is a great tool to have when removing old chimneys.

When the chimney is being taken down through the roof, the old flashing will need to be removed, as will the slates around the chimney, especially any that are tarred or broken. Bats often live in old chimney flashings, so keep your eyes open for them. When the chimney has been taken out of the way and the mess on the roof cleaned up, there will be only a hole and good slates around the spot where the original chimney used to be. Once the hole is closed up with the new sheathing, matching replacement slates are nailed in place, and the job is done.

FLASHING A CHIMNEY

First off, no two chimneys are alike. Therefore, it is a good idea to understand the fundamentals of flashing in order to understand how to flash a chimney on a roof.

Imagine, if you will, a flat plane such as a floor, and through the plane protrudes a rectangular column such as a chimney. The sides of the column can be easily flashed by simply bending long flat pieces of sheet metal to ninety degree angles and fitting them against the four sides of the column and against the floor. If the column were an actual chimney, the top edges of the flashing would be bent and fit into a mortar joint to prevent water getting

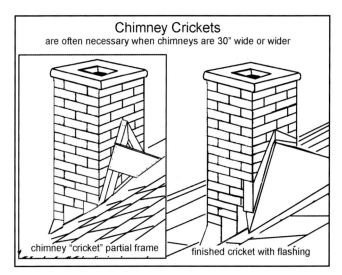

Chimney Crickets
are often necessary when chimneys are 30" wide or wider

chimney "cricket" partial frame   finished cricket with flashing

behind the flashing.

What's wrong with this picture? Two things: a) on a roof, the plane is not flat, it's sloped, and b) the four corners of the column will leak because they're not flashed, only the sides are flashed, leaving a tiny gap at each corner where water can penetrate unless the corners are either soldered or folded appropriately. Corner leakage is the single most likely problem encountered when trying to flash rectangular objects protruding through a roof.

Let's look at the same scenario with a sloped plane. Now we have an advantage, because the sloping of the plane forces water to flow in one direction, and because we know the direction of flow, we can control the water and therefore eliminate any leakage.

The sloping plane also adds a new factor to the situation: the top edge of the flashing is no longer parallel to the mortar joints on the sides of the column and cannot be folded into a joint as easily. Instead, the top of the flashing must be cut to follow the stepped joints, and folded accordingly.

Next, we must take into consideration that several rows of slates are going to butt against the side of the chimney column. Since the slates overlap each other, the side flashing on the chimney must be cut into sections that overlap each other as well, otherwise one long piece of side flashing would allow water to run under the edge of the slates (see illustration, page 259).

The flashing is then installed in overlapping sections and "stepped" up the side of the chimney, each step lining up with a row of slates. This is known as "step flashing." To make step flashing easier to install, it is often not folded into the mortar joints at all, but instead it's simply folded up the side of the chimney two or three inches, then covered with "counter flashing," which, in this case, is also installed in overlapping sections (the pieces of counterflashing overlap themselves *and* the step flashing). Step flashing is also known as "base flashing," and counter flashing is also known as "cap flashing."

The side flashing pieces need not be installed as separate step and counter flashings, but can be installed as single pieces. This single side-piece system works well when re-flashing smaller chimneys, especially on older buildings that have already settled. Very large brick or stone chimneys may be easier to flash using separate step and counter flashings. Chimneys on new construction also benefit from a combination of step and counter flashing as the roofs will settle slightly over time while the chimney will not. This causes the step flashing, attached to the roof, to drop in relation to the counter flashing, attached to the chimney, which is not a problem if the two flashings are separate.

We have now demonstrated how to flash the *sides* of a column protruding through a sloped plane, but what about the corners? What about the top side and bottom side of the column?

All four corners are rendered leakproof by building little "roofs" over them with folded corner flashing. The bottom corners of the chimney are covered with the bottom piece of side flashing, and the top corners of the chimney are covered with the uppermost section of chimney flashing, also called the "top" or "back" flashing. In both cases this is simply achieved by extending the flashing beyond the chimney corner about four inches, then folding the flashing around the corner and tucking it into the mortar joint. Extending the flashing beyond the corner forces the water to flow away from the corner, while folding the flashing around the corner and into a mortar joint prevents water running vertically down the outside of the chimney from finding its way into the corner.

This simple flashing technique can be used for any rectangle protruding through the roof, such as a skylight, and it requires no soldering, adhesives, roof cement, caulk, or fancy sheet metal work. Most importantly, if done correctly, it's foolproof. The size of the "roofs" built over each corner is variable and may be made larger

or smaller according to the situation or to the personal preference of the person doing the work.

An alternative to the folded-corner method of flashing involves fitting the apron flashing (the lowest piece on the chimney) with soldered gusset pieces, thereby eliminating any holes in the lower corners. This style of flashing was once popular among roofers who were involved in a lot of sheet metal work, and for whom the soldering or brazing of flashing was part of the normal routine. A similar technique is the use of folded soldered seams at the corners of the chimneys to provide a waterproof flashing job. This technique is often limited to the sheet metal specialist who has proper sheet metal tools and soldering equipment (see illustration below).

The *apron* flashing on the chimney is the piece that fits at the lowermost front of the chimney. Often this is installed as two pieces, a base piece and a cap piece; however, one piece is often sufficient and is easier to install, especially when reflashing an old chimney. The apron should extend down the roof far enough to cover any exposed nail heads, as well as to provide a headlap on the second course of slate below it. Four to six inches of apron flashing extending down the roof is common. The apron need not extend beyond the sides of the chimney, although it can be allowed to do so and then can be laid on the roof alongside the chimney to allow a place to nail it.

After the apron is installed, the side pieces are installed. They should lie on the roof about four inches and run up the side of the chimney 2-4" for step and counter flashings, and until they reach and tuck into a mortar joint for single piece flashings. Single piece side flashings should not be cut on the ground based on drawings and measurements, but instead should be laid against the chimney with the 4" fold against the roof and marked in place using a level and a marker, then cut and installed. This ensures a good, tight fit, both against the chimney and against the roof. If step flashings and counter flashings are used, the *bottom* edges of the counter flashings can be folded under about half an inch for stiffness.

The uppermost piece of flashing on a chimney (top flashing or back flashing) is very impor-

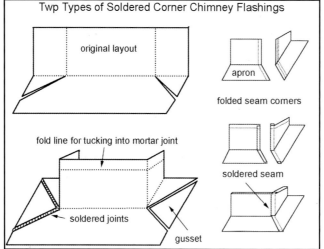

Two Types of Soldered Corner Chimney Flashings

original layout

fold line for tucking into mortar joint

soldered joints

apron

folded seam corners

soldered seam

gusset

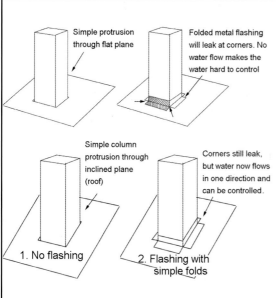

Simple protrusion through flat plane

Folded metal flashing will leak at corners. No water flow makes the water hard to control

Simple column protrusion through inclined plane (roof)

Corners still leak, but water now flows in one direction and can be controlled.

1. No flashing

2. Flashing with simple folds

Preliminary fundamental steps to understand when flashing a chimney.

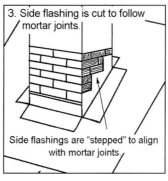

3. Side flashing is cut to follow mortar joints.

Side flashings are "stepped" to align with mortar joints.

4. Side flashing is stepped and overlapped.

Stepped flashing is cut so that it is installed in overlapping sections, one section for each course of slate.

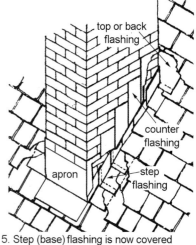

top or back flashing

counter flashing

apron

step flashing

5. Step (base) flashing is now covered with counter (cap) flashing. Step flashings can also be installed as a single piece tucked into mortar joint, more appropriate for re-flashing on buildings that have already settled.

Top edge of flashing always tucks into mortar joints at least a half inch.

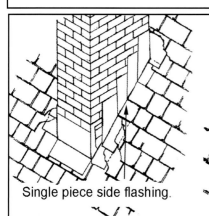

Single piece side flashing.

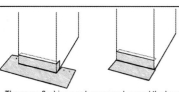

The apron flashing can be wrapped around the base of the chimney as at left, above, or simply cut off even with the edges of the chimney, as at right. Neither will leak once the flashing corners have been wrapped. Left procedure allows more nailing options.

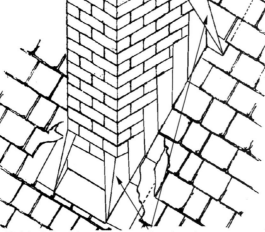

Folded flashing corners divert water from the leak-prone chimney corners.

Folded Flashings

Folding flashing properly at corners and other critical areas is a trick of the trade that will make a flashing job permanently leakproof without the need for solder, roof cement, or caulk. Note that on chimneys both the top and bottom corners must be folded.

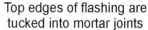

Top edges of flashing are tucked into mortar joints

top side

extends 3" laterally

Folded flashing corners must divert the water that's both running down the roof and down the chimney.

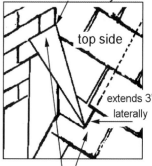

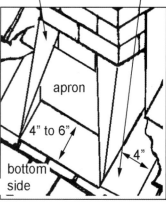

apron

4" to 6"

4"

bottom side

Height of apron and bottom corners is variable.

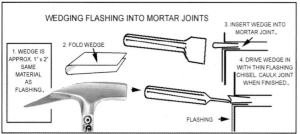

WEDGING FLASHING INTO MORTAR JOINTS

1. WEDGE IS APPROX. 1" x 2" SAME MATERIAL AS FLASHING.

2. FOLD WEDGE

3. INSERT WEDGE INTO MORTAR JOINT.

4. DRIVE WEDGE IN WITH THIN FLASHING CHISEL. CAULK JOINT WHEN FINISHED.

FLASHING

COMMON CHIMNEY REFLASHING JOB — BEFORE AND AFTER
Single flue chimneys exiting through the ridge of a roof are common, and unfortunately so are the ugly messes that are evident in the photo at left. In about five hours a single worker cleaned up the roof and reflashed the chimney, as shown at right. The step-by-step procedure is illustrated on the following page.

Photos by author.

tant as it catches the brunt of the rain and snow (this is not found on chimneys that protrude through the peak of a building). A wide stock of copper should be used; 24" stock is usually sufficient, and need not be run up the back of the chimney more than eight inches (three courses of bricks). That leaves about 16" or more to run up the roof, and care should be taken not to nail through this section of flashing when laying the slate over it, except within one inch of the top edge. The roof slate should not be brought down tightly against the back of the chimney, rather, the flashing on the roof at the base of the chimney should be left bare for at least a few inches to allow the water to quickly run off. It's very important that this flashing be folded tightly against the base of the chimney, otherwise capillary action may allow water to creep under the side wings and make its way into the chimney corners. A tight fit can be achieved by over-bending the flashing to make a tight crease, then opening it back up to its proper position before laying it in place.

The top or back flashing should extend beyond the sides of the chimney about 4" and be folded around the chimney corner and tucked into a mortar joint. In order to do this, *one must allow some extra flashing length where the metal is going to tuck into the mortar joint on the back of the chimney;* so if you're going to tuck 3/4" into the mortar joint, leave 2" of metal instead. The extra length is trimmed off with a pair of tin snips *only* where the flashing tucks into the mortar joint, but <u>not</u> where it extends beyond the sides of the chimney. This should leave enough flashing material on the sides to reach around the corner and tuck into a side mortar joint.

There are two places where hidden leaks may occur after installing top flashing on a chimney. The first can occur at the top edge of the flashing, which is under the slate, but may be exposed in the slots because the flashing did not extend up the roof far enough. If so, slide extra pieces of flashing under the slates at the slots, but over the flashing metal, in order to cover the exposed edges. Secondly, the side edges of the flashing (where the water runs off) may fall on or near a slot between the underlying slates. If so, reinforce the slot with flashing slid underneath the slates.

Flashing a Single Flue Chimney at the Ridge Using Folded Corners

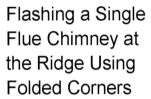

1. Remove bad slates and old flashing.

2. Grind out mortar joints with a diamond saw.

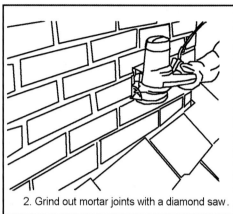

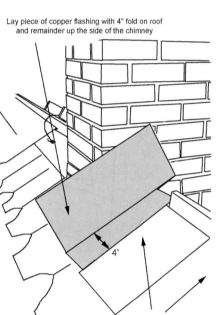

3. Replace slate up to chimney and then install apron flashings.

Lay piece of copper flashing with 4" fold on roof and remainder up the side of the chimney

4"

4. Prepare corner flashing.

Mark the mortar line cut with a pencil and level.

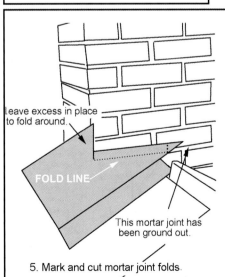

Leave excess in place to fold around.

FOLD LINE

This mortar joint has been ground out.

5. Mark and cut mortar joint folds.

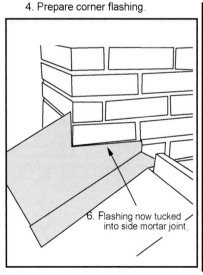

6. Flashing now tucked into side mortar joint.

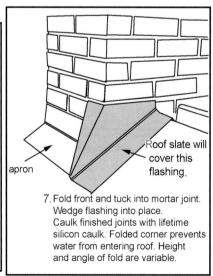

apron

Roof slate will cover this flashing.

7. Fold front and tuck into mortar joint. Wedge flashing into place. Caulk finished joints with lifetime silicon caulk. Folded corner prevents water from entering roof. Height and angle of fold are variable.

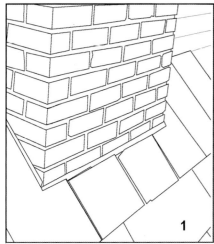

1. Remove all old slates and flashing. Install new slates up to bottom of chimney.

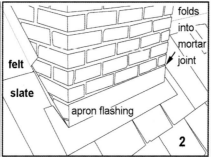

2. Install copper apron flashing after cutting out mortar joint with 1/8" diamond blade.*

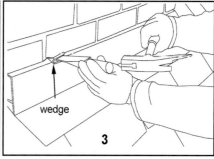

3. Wedge apron flashing in place using lead or copper wedge material.

REFLASHING AN EXISTING DOUBLE FLUE CHIMNEY
(16 ILLUSTRATED STEPS)

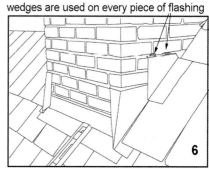

6. Once corner flashings are installed, lay the next course of slate in place and then install the next piece of flashing (step flashing) as shown above. The top edge of the step flashing also tucks into the mortar joint by about 3/4 inches. Joints are caulked with lifetime silicon after flashing is wedged into place.*

7. Install the next course of slate on top of the step flashing.

8. Continue this procedure until both sides of chimney are flashed to top edge as in #8 and #9 below, and #10 on next page.

4. Install first corner flashing (four inches wide where it lays on roof). Top edges tuck into mortar joints after it is folded around corner.

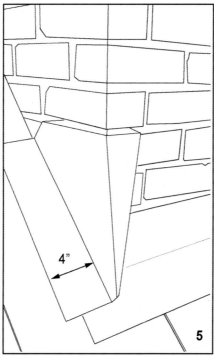

5. Procedure for making a folded corner flashing is shown at beginning of flashing section in this book.

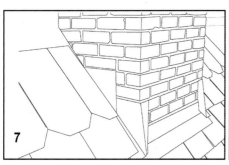

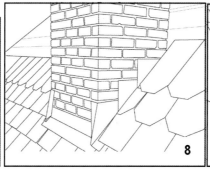

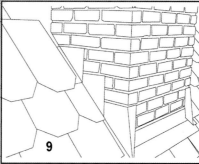

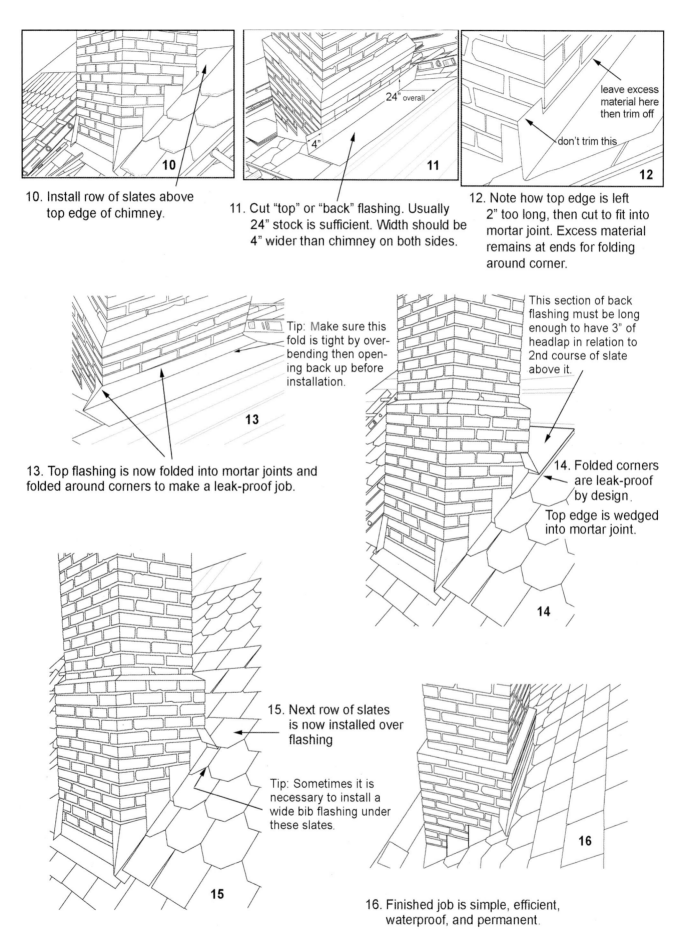

10. Install row of slates above top edge of chimney.

11. Cut "top" or "back" flashing. Usually 24" stock is sufficient. Width should be 4" wider than chimney on both sides.

24" overall

4"

leave excess material here then trim off

don't trim this

12. Note how top edge is left 2" too long, then cut to fit into mortar joint. Excess material remains at ends for folding around corner.

Tip: Make sure this fold is tight by over-bending then opening back up before installation.

13. Top flashing is now folded into mortar joints and folded around corners to make a leak-proof job.

This section of back flashing must be long enough to have 3" of headlap in relation to 2nd course of slate above it.

14. Folded corners are leak-proof by design.

Top edge is wedged into mortar joint.

15. Next row of slates is now installed over flashing

Tip: Sometimes it is necessary to install a wide bib flashing under these slates.

16. Finished job is simple, efficient, waterproof, and permanent.

*Note: Mortar joints can be completely cut out with a 1/4" diamond blade then filled with mortar cement after the flashing is installed.

Reflashing a Single Flue Chinmey Near the Ridge

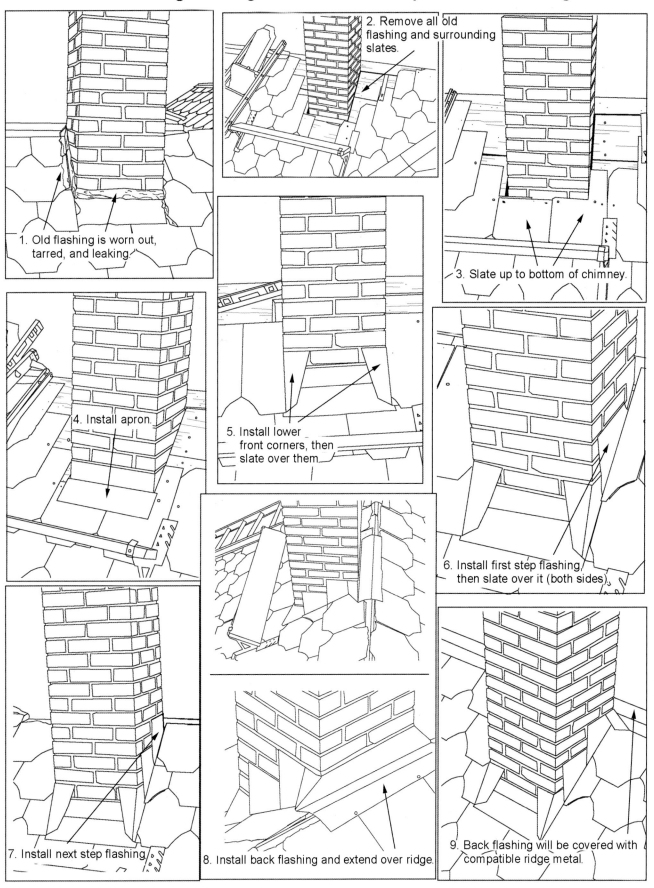

1. Old flashing is worn out, tarred, and leaking.

2. Remove all old flashing and surrounding slates.

3. Slate up to bottom of chimney.

4. Install apron.

5. Install lower front corners, then slate over them.

6. Install first step flashing, then slate over it (both sides).

7. Install next step flashing.

8. Install back flashing and extend over ridge.

9. Back flashing will be covered with compatible ridge metal.

Reflashing a Double-Flue Chimney at the Peak of the Roof

7 steps, using folded corners

1. Remove old slates and flashing around chimney.
2. Once the area is cleaned, start replacing slates.
3. Install slates up to bottom of chimney as shown.
4. Install apron flashing after cutting out mortar joint.
5. Install corner flashings, again tucking into mortar joints.
6. Corner flashings overlap at peak, then are covered by ridge.
7. Thoroughly caulk all cut out mortar joints with high quality sealant such as lifetime silicon. Caulk ridge ends as well.

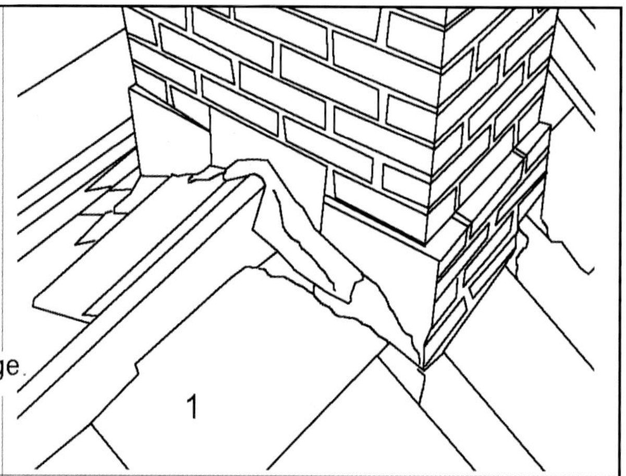

1

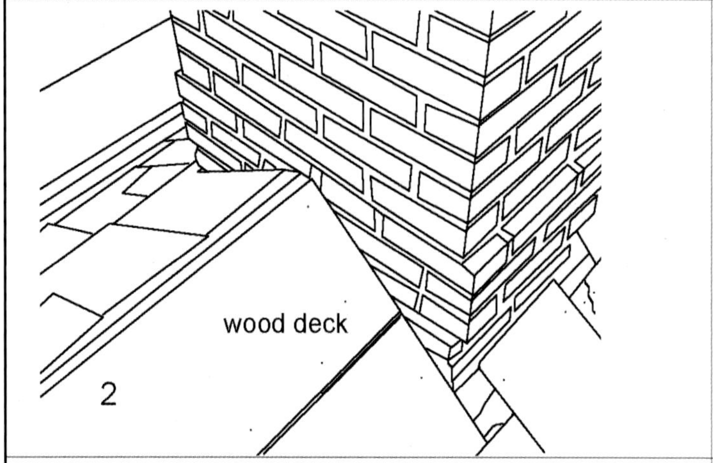

wood deck

2

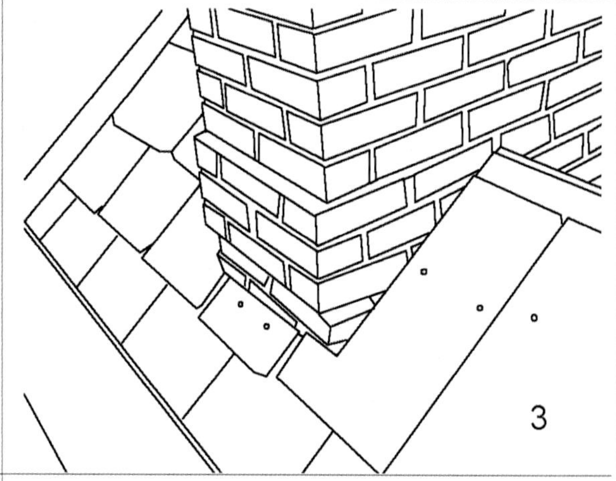

3

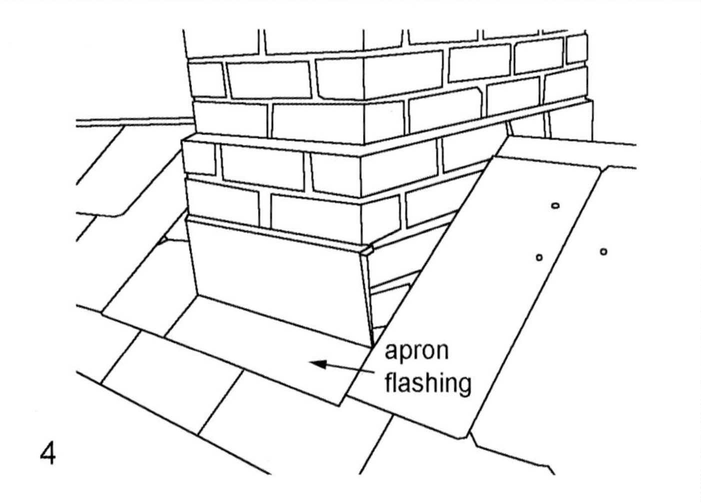

apron flashing

4

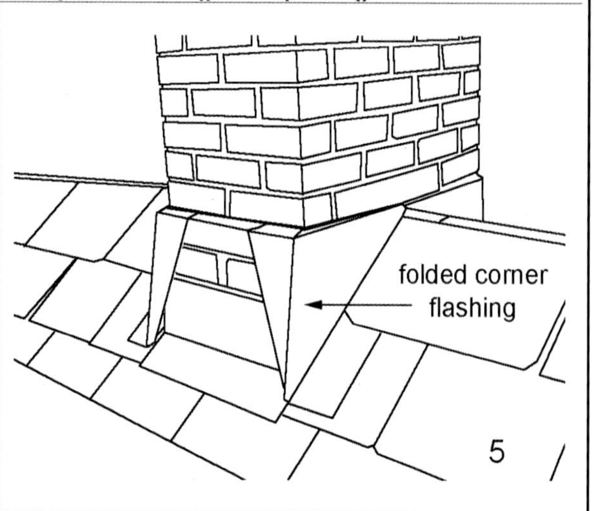

folded corner flashing

5

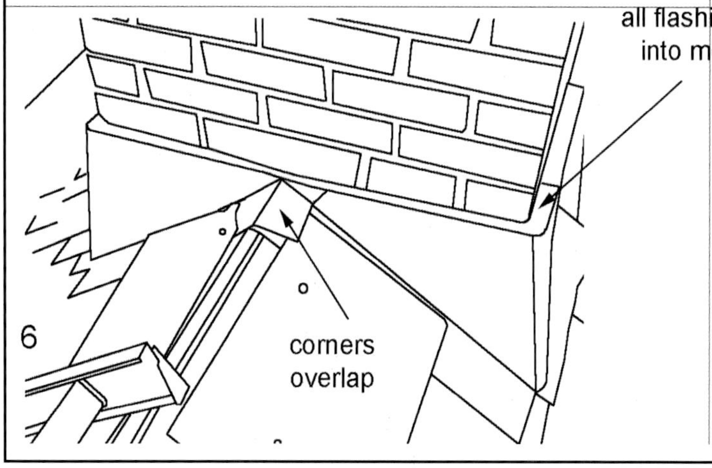

all flashing is tucked into mortar joints

corners overlap

6

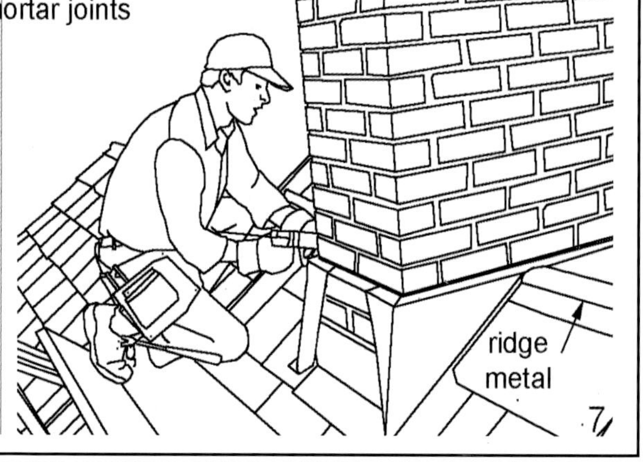

ridge metal

7

Finally, when a chimney is very wide (usually over 30 inches) and creates a large dam on a roof and therefore also creates a potentially chronic problem, a "cricket" should be installed between the top side of the chimney and the roof.

Practice makes perfect. Flashing work is a skill that benefits from practice, and the more one does of it, the easier it gets. Nevertheless, a person with no prior experience should be able to flash or reflash a chimney, although they may have to take their time and pay close attention to detail to get it just right. Tools required for most flashing work include a pair of tin-snips (straight cutting aviation snips), sheet metal hand tongs for bending the edges that fit into the mortar joints, a thin-bladed masonry chisel for cleaning out mortar joints and for forc-

ing in wedges to hold the flashing (when reflashing only — new flashing is not wedged into place, it is simply placed into the mortar joints as the masonry is being laid), and a two-foot level. An electric grinder with a masonry wheel or diamond blade is very useful for cleaning out old mortar joints in preparation for reflashing.

Large chimneys, such as this massive one on a flat shingle-tile (not slate) roof on a dormitory at Grove City College, Grove City, Pennsylvania USA, are reflashed according to the same principles as a smaller chimney. More scaffolding equipment is needed, however, to gain access to the job site. The flashing shown here is terne-coated stainless steel, installed with counter flashings.

Photos by author.

Chapter Twenty

VALLEYS

Valleys are sections of the roof where two sloping planes intersect. Typically, metal valley flashing is installed over the felted sheathing before any slating begins. The felt paper is optional and can be left out completely during valley replacement jobs where the old felt has already disintegrated. The felt paper on new construction need not overlap the valley flashing — the valley metal can be laid right on top of the felt if it is going to be slated immediately. Once the flashing and slate are installed, the felt will be entirely obsolete as a waterproofing material. It is much more important to avoid nailing anything into the valley (such as felt paper) than it is to have the felt paper overlap the valley metal. Before the metal is installed, a chalk line is struck up the edge of one side of the roof deck on the felt paper to indicate where the edge of the metal valley flashing should be. The metal is then nailed in place, one side at a time, with a nail of compatible metal. The nails are kept to within one inch of the edge of the valley metal.

Valleys don't need to be creased on a brake, leaving them with a bend line down the center, except in cases where the two roof surfaces are very steep or the valley is very narrow. Otherwise, a creased valley is simply a matter of personal style and creates no advantage in the function of the valley. Instead, the valley metal can be nailed along one edge, then carefully forced into the roof with the pressure of a knee as the other side is nailed.

Valleys should be laid in sections not to exceed twelve feet in length, although a ten foot maximum length is recommended, due to the adverse effect of expansion and contraction that can cause long pieces of metal to buckle and develop a leak over time. The valley sections should simply be overlapped by six inches — no soldering is necessary or recommended, as it's the old solder joints on the old valleys that tend to leak first due to expansion and contraction. Do not use roof cement or other adhesives along the edges of a valley, as these make later repairs of the roof unpleasant and difficult while adding no advantage to the functioning of the roof.

Furthermore, valleys do not need to be cleat-ed along the edges — that is, fastened with bent pieces of metal (called cleats) instead of nails. It was once recommended that valleys be cleated in place for fear of expansion and contraction wearing any nails loose along the edges, but most valleys were nailed in place anyway, and time has shown that this procedure is much simpler and works just as well.

American valleys are typically laid "open," with approximately six inches of metal exposed. The overall width of the valley metal can vary from 12" to 20" or wider, depending on the width of the exposed valley. The rule of thumb is that the slates must overlap the valley by a minimum of five inches, so for a six-inch exposed valley, 16" stock, or wider, must be used.

Open valleys typically have parallel sides running from bottom to top, although some roofers prefer open valleys that gradually widen toward the bottom. When laying slate into a valley, chalk a line the length of the valley metal on both sides to indicate the edges of the slate, then draw over the chalk lines with a permanent felt-tipped pen, as the chalk lines will wear off the metal almost immediately. When nailing slate over the valley metal, be careful to nail only along the edge of the metal, and not anywhere near the center. If a small, triangular piece of slate cannot be nailed over the valley at the end of a row without nailing too close to the center, you can probably eliminate that piece of slate — chances are you won't need it.

Valleys can also be laid "closed," with no metal exposed, and "rounded" — a style more common in Europe, in which no metal flashing need be used at all, as is discussed later in this chapter.

REPLACING VALLEYS

Worn out valleys on old slate roofs create one of the biggest problems any roof owner could have. Valleys collect water from two or more roof planes and therefore channel a greater concentration of water than any other part of the roof. A hole the diameter of a match stick in a valley can cause buckets of water to leak into a building during heavy rains. Worse, the bottom ends of the valleys often wear out first, and they also bear the greatest amount of water, so serious structural water damage can happen to the area at the bottom of the valley if left leaking and neglected for a long time.

Fortunately, valley replacement is a routine job when restoring old slate roofs. When valleys leak they should be replaced in their entirety. That means the old valley flashing must be completely removed and new valley flashing installed in its place. Some roofing contractors will try to slip sections of flashing over the old metal without removing the old valley. This is a temporary solution that doesn't work very well in the long run. Others will tar the old valley and often the roof on the valley sides. This is a huge mistake. Valleys can be sealed as a temporary measure to prevent leaking until they're replaced, and the procedure for doing so is illustrated on page 217; but if the job is to be done correctly and permanently, the old valley must be removed and a new one installed.

In order to remove an old valley one must simply remove enough slates on both sides of the valley to completely expose the old valley metal, which is usually about 14"- 20" wide overall. The metal is then pried loose and removed, the exposed sheathing underneath swept clean with a broom, and a line is chalked for the new valley. The new metal is nailed into place, and the original slates are replaced in the same spots from which they were taken. This is facilitated by first numbering the slates before removal, as illustrated in this chapter. Any bad slates (broken, cracked, perforated on the face, or tarred) are simply replaced *with matching slates* when the valley is reslated. If the slate along the valley is in good shape, a master slater can expect to replace 34' of valley in a 10-hour day, working alone. If the slate is not good (i.e. tarred and/or broken), a master slater and a competent helper can expect to replace a 34' valley in a long (10 or 12 hour) day. The slates along a valley are often broken because people who have attempted to repair the valley in the past have walked on them; they have also spread tar all over the valley and adjacent slates, thereby making it very difficult to take the roof apart. It's preferable to replace a valley in a single day, if possible, as a roof should not be torn apart, left open and subject to a potentially disastrous cloud burst.

TAKING THE OLD VALLEY OUT

The single most important step in removing old valleys is the roof set up. Typically, a hook ladder is positioned on either side of the valley, set back far enough to allow access to the slates that must be removed (about 3-4 feet back from the valley center). If the roof is steep, the bottom of the hook ladders will try to slide into the valley and must be tied to a roof jack nailed to the roof for that purpose. If the valley is long, such as the 34' length referred to above, a 20' hook ladder (which is the longest practical hook ladder available) will be too short to extend the entire length of the valley. Then, two ladder sections can be bolted together end-to-end with aluminum plates and stainless steel bolts to create a custom-made, longer hook ladder.

The bottom of a hook ladder *will* slide into a valley on a steep roof, because the slates, when removed, are stacked on the hook ladders, which weighs the ladder down. It also, however, makes it very easy to install the slates back on the roof after the new valley metal has been installed, because the slates are waiting adjacent to where they were removed.

The trick-of-the-trade that makes it even easier to put the roof back together is *numbering* the slates before they're removed. Starting at the top row, a nail or other sharp instrument is used to scratch the number "1" on the first row, the number being scratched on the first 4 or 5 slates out from the valley, on both sides of the valley. Then the second row is scratched "2," the third row "3" and so on, until the entire length of the valley is marked, row by row, about 4 or 5 slates out from the valley center. Mark the starter slates at the bottom with an "ST." This only takes a few minutes, but saves an incredible amount of time when the valley must be reslated. Measure the exposed width of the valley (usually between 4" and 8") and make a note of it before beginning to remove the slates.

Pry the ridge iron, if any, loose enough to remove the nails from the cap slates underneath. The numbered slates are then removed, starting at the top, using a slate ripper and hammer, but mostly a flat pry bar, and the slates are propped on the hook ladders as they're removed. *All the slates from the right side of the valley must be propped on the right hook ladder, and all the slates on the left side of the valley must be propped on the left hook ladder.* Do not overlook this important detail if you want to put the valley back together quickly and efficiently. Also, *remove the slates from one side of the valley at a time* for maximum speed and efficiency (expose one entire side of the valley before exposing the other). If two people are working on the same valley, then both sides may be removed at the same time, one worker on one side and the other worker on the other side,

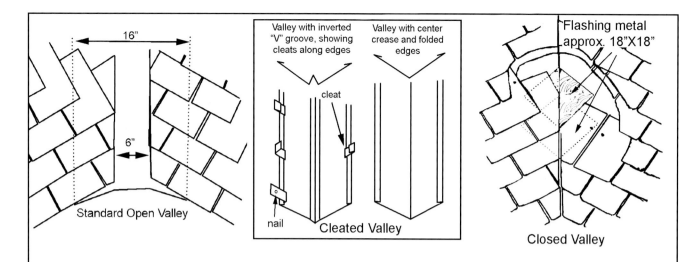

16"

6"

Standard Open Valley

Valley with inverted "V" groove, showing cleats along edges

Valley with center crease and folded edges

cleat

nail

Cleated Valley

Flashing metal approx. 18"X18"

Closed Valley

METAL VALLEYS

• STANDARD OPEN VALLEY (above, left). Most metal valleys on slate roofs do not require a center crease, nor do they need to be cleated into place. A standard open valley has about six inches of exposed metal, has no center crimp, and is nailed along the edges directly through the valley metal; total width ranges from about 14" to about 20" or more, with 16" being practical. Lengths of sections should not exceed twelve feet, with ten feet being the maximum recommended length due to expansion and contraction issues. The ten-foot-long valley sections can simply be overlapped six inches — no soldering is needed. No roof cement or adhesives are used under or over the valleys or on the slate along the valleys. No cant strip is used alongside a valley. Some roofers prefer the exposed valley to widen toward the bottom, although this is mostly a stylistic issue.

• CLEATED VALLEY showing an inverted "V" groove (center, left), and a center crease (center, right).Cleated valley installations are preferred by some roofers, although the cleating of valleys has been proven to be unnecessary. Flashing is cleated in order to minimize the stress on solder joints from expansion and contraction. Valley installations typically do not require any soldering at all, and therefore cleating is not needed. The inverted "V" groove is used when a steep roof pitch drains onto a shallow one — the "V" prevents water from forcing its way under the slates on the shallow side. A center crease is preferable in closed valleys, or very narrow valleys, or on very steep slopes, or it can be added simply for style.

• CLOSED VALLEYS (above, right) show no exposed metal in the valley, since the slates butt against each other in the valley center. Typically, they are installed with step flashing. The bottom of the step flashing lines up with the bottom of the overlying slate; the top of the step flashing lines up with the top of the underlying slate. The top of the step flashing may extend above the underlying slate for nailing purposes.

QUICK REFERENCE - VALLEY INSTALLATION

Make sure the board roof sheathing is covered with 30 lb. felt paper. The felt does not need to be overlapping the valley if the valley is to be slated immediately (the valley metal can be installed directly on top of the felt). Run a chalk line up one side of the valley felt to indicate where the flashing metal should be positioned. For example, for 16"-wide valley flashing, measure approximately 8" on the roof from the valley center and chalk a line parallel to the length of the valley.Use 16- or 20-ounce copper, or stainless steel (18 gauge or heavier) sheet metal, 16 inches wide for a six-inch exposed valley, 18-20 inches wide for an eight-inch exposed valley (more common on churches and institutions). The rule of thumb is that the slates must overlap the valley metal by at least five inches, and that the valley flashing should only be nailed within an inch of its outer edges.

Valley flashing typically does not need to be cleated, crimped or creased, but should have an inverted V groove one inch high if a steep roof slope is running into a shallow one. The valley metal should have a center crease if the valley is especially narrow or on a very steep roof, or closed. Lay the valley metal onto the roof, carefully forcing it into place using your knee (if the metal has not been creased on a brake — if creased, it will lay into the valley easily), and then nail within 1" of the edges of the metal, using nails of compatible metal. Nail one side into place first, then the other. Keep metal valley sections to a maximum length of 10', overlapping 6" at ends; don't use roof cement or adhesives. Chalk lines the length of the valley flashing to indicate where the edges of the slates will be (3" out from the center of the metal for a 6" exposed valley), then immediately draw over the chalk lines with a permanent felt-tipped marker as the chalk will quickly rub off. The drawn lines will act as guides for the exposed edges of the valley slates. Leave six inches of metal exposed for a standard open valley; chalk lines at a slight angle to create an open valley that widens near the bottom, if desired. After valley metal is in place, chalk remainder of the roof horizontally for the slates, if needed, chalking over and onto the valley metal. When nailing slates over the valley metal, hold nails back to within an inch of the edge of the metal if at all possible.

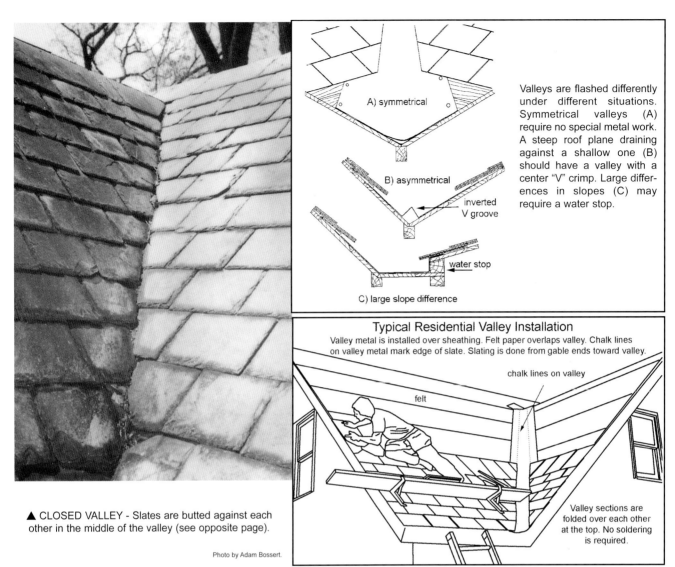

A) symmetrical

B) asymmetrical

inverted
V groove

water stop

C) large slope difference

Valleys are flashed differently under different situations. Symmetrical valleys (A) require no special metal work. A steep roof plane draining against a shallow one (B) should have a valley with a center "V" crimp. Large differences in slopes (C) may require a water stop.

▲ CLOSED VALLEY - Slates are butted against each other in the middle of the valley (see opposite page).

Photo by Adam Bossert.

Typical Residential Valley Installation

Valley metal is installed over sheathing. Felt paper overlaps valley. Chalk lines on valley metal mark edge of slate. Slating is done from gable ends toward valley.

chalk lines on valley

felt

Valley sections are folded over each other at the top. No soldering is required.

Before Removing Slates When Replacing a Valley

Always number them first by scratching with a nail as indicated below. The roof will go back together much more quickly.

Valley

Note: On random width slate roofs, number the rows 1A, 1B, 1C etc, 2A, 2B, 2C, etc.

Numbers will not be visible from the ground once the slates are replaced.

▼ Roof jacks (brackets) are used to stack slates on when a hook ladder can't be used.

Photo by author.

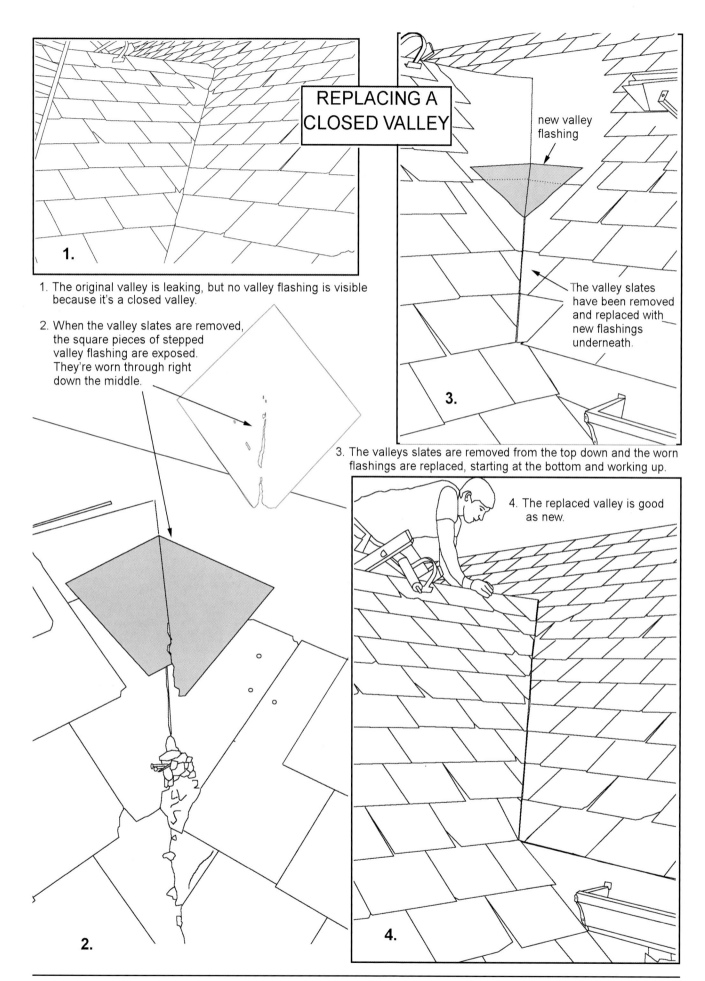

REPLACING A CLOSED VALLEY

new valley flashing

1.

1. The original valley is leaking, but no valley flashing is visible because it's a closed valley.

2. When the valley slates are removed, the square pieces of stepped valley flashing are exposed. They're worn through right down the middle.

3.

The valley slates have been removed and replaced with new flashings underneath.

3. The valleys slates are removed from the top down and the worn flashings are replaced, starting at the bottom and working up.

4. The replaced valley is good as new.

2.

4.

Slates are removed from either side of this 20'-long valley on a house in Sandy Lake, PA, and propped on the hook ladders, which are prevented from sliding into the valley by a roof jack tied to the ladder.

After the new metal is installed (in this case, 20-ounce half-hard copper), the numbered slates are nailed back in place, and any bad slates are replaced in the process.

Photos by author.

one above the other.

If the slates have been tarred into place, try to carefully pry them loose without breaking them. In some cases you may have to cut along the tarred valley edge with a utility knife to get the slates to come loose in one piece. Heavily tarred slates that can be pried loose can usually be cleaned of most of the tar and re-used. The tar is knocked off by tapping directly against the edge of the slate with a hammer, or by using the tip of a slate ripper or flat pry bar to chip it off.

If you cannot place a hook ladder on either side of the valley because there isn't a ridge to hook onto, then you must nail roof jacks to the roof and prop the slates on them (see Chapter 12 for instructions on nailing roofs jacks). Again, the slates from one side of the valley must be propped on *that* side of the roof.

Once all the slates are removed and the old valley flashing is completely exposed, the old metal can be pried out with the flat pry bar. This is a seriously dirty job, and a dust mask is a good idea when taking out old valleys. Also, be on the lookout for wasp nests underneath the bottom of the old valley when you roll it out. Pry the old metal loose, roll it

down the roof starting at the top, and drop it over the edge, if possible. Then sweep the roof clean with a broom, and examine it closely for nails and any sharp objects that might rub against the new flashing and eventually penetrate it. *Use a gloved hand* and brush it over the roof boards — it will catch on any sharp nub. Pound the nub in or pull it out with a claw hammer. When you're sure you have the old sheathing clean, chalk a line on one side of the valley to indicate where the edge of the new metal will be. If the new metal is 16" wide (for example), snap your chalk line at 8" from the center of the valley.

Often there will be a small spot on the sheathing that has rotted due to a prolonged leak. If the new flashing is going to cover it, don't worry about it — put the new metal in right over top of it. It's very rare that sheathing in a valley will be so bad that it needs to be replaced. It's more important that the valley be replaced and the roof be closed up as quickly as possible, so minor imperfections in the valley boards can be tolerated. They'll do no harm.

Umberto Perlino, a slate roofing contractor in Pennsylvania, works at replacing a 34'-long valley on a white oak lath-roofed barn. Although the old terne metal valley was completely deteriorated and the valley slates were tarred and broken, this replacement job was a day's work for a master slater and a helper. Notice that there is no felt underlayment on this 100-year-old roof. There never was. The repaired roof is 100% leakproof. Felt paper is totally unnecessary for the proper functioning of most slate roofs.

Photo by author.

Routine Valley Reflashing on a Residential Job

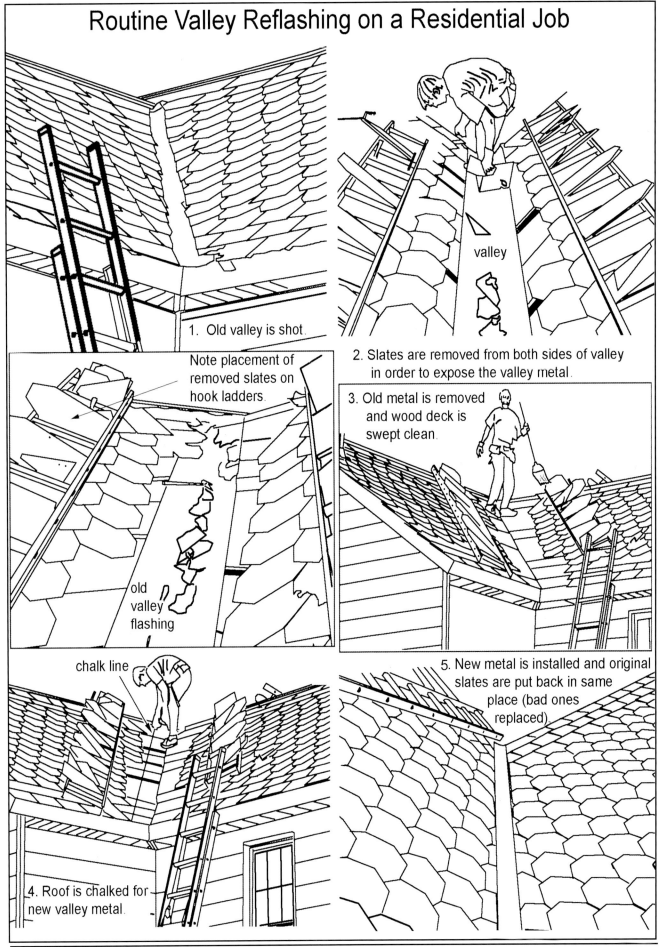

1. Old valley is shot.

2. Slates are removed from both sides of valley in order to expose the valley metal.

valley

Note placement of removed slates on hook ladders.

old valley flashing

3. Old metal is removed and wood deck is swept clean.

chalk line

4. Roof is chalked for new valley metal.

5. New metal is installed and original slates are put back in same place (bad ones replaced).

▲ Hook ladders are positioned alongside this 34'-long church valley. The valley was tarred, but the sea green slate was quite good.

▲ The slates are pried off, stacked on the hook ladders, and the old valley metal removed. The roof is swept and chalked.

▲ A 12' section of .025" stainless steel, 16" wide, is forced into place and nailed along the edges, aligned with the chalk line.

▲ A second section of stainless over laps the first by 6". 10' lengths are preferable, 12' is maximum.

▲ The third and final section of stainless is installed.

▲ The slates are nailed back where they came from. The new valley should last the life of the roof and require no maintenance.

General Valley Replacement Sequence

Opposite page: 1) The slate rows are numbered by scratching with a nail. 2) Hook ladders make the job easy. Slates are pried from each side of the valley and placed on the adjacent hook ladder. The original felt may be dried up and no longer functioning as a waterproof barrier. 3) The metal is completely removed, and the exposed wood deck is swept clean (wear a face mask). All protruding nails are removed or pounded down. 4) The wood is chalked for the new flashing. There is no need to install felt or any underlayment under the new metal — you can simply nail it directly to the wood with compatible metal nails (no side cleating of the valley metal is necessary either). 5) The numbered slates can then be replaced in their original positions. Broken slates are replaced with reclaimed slates matching in size, shape and type (color). The finished job is permanently waterproof and looks good. Virtually any valley can be replaced in a single day by experienced slaters. The church valley above was a 10-hour job for one experienced slate roof restoration professional. The stainless steel had been pre-drilled and almost all of the slates were good enough to be reused.

Photos by author.

▲ A carefully placed knee forces valley metal into position without buckling it (above left). After nailing the valley in place, one side is re-slated first while working off the bare roof (above, middle). Then the hook ladder (now empty) on that side can be moved in close to the valley to sit on while slating the other side, thereby avoiding damaging the newly replaced slate. Top photo at right shows finished valley. Center photo shows one side slated (the left side).

Photos by author.

INSTALLING THE NEW VALLEY FLASHING

When laying the new metal in place, remember that it doesn't have to be creased down the center on a sheet metal brake, nor does it have to be "cleated" in place. Instead, the metal can simply be carefully forced into the contour of the valley into a rounded valley shape. This won't work, however, if the *exposed* valley is too *narrow* (under 4") or on a very *steep* roof (such as valleys on some dormers or on mansard roofs)— then the valley needs to be creased. However, narrow valleys on steep roofs are uncommon, and ninety-nine times out of a hundred a force-fit valley works fine.

One disadvantage of using creased valleys lies in the fact that old valleys on old roofs are often not straight. Over the years the roof may have settled or even bowed a bit. Creased valley metal is stubbornly straight and may not form to the contour of the roof. This presents a problem that can only be solved by cutting the creased valley metal into shorter lengths, or by using non-creased valley stock. One advantage to using creased valley stock, on the other hand, is that the valley metal lies down tight against the roof boards, preventing birds and other nuisances from creeping under the metal and establishing nests.

It's important to remember that the valley metal should not be laid in lengths longer than about ten feet due to possible expansion and contraction of the metal and consequent long-term

harm. Twelve-foot-long valleys may be necessary in some very unusual instances but twelve feet should be the limit. The sections should be overlapped six inches in the direction of water flow. One side of the valley should be nailed first along its entire length (the chalked side), then one end forced into place (usually with a knee) and nailed, then the other. The end you want to have the tightest fit — usually the bottom end because it is visible from the ground — should be forced in and nailed *last*. Be very careful when forcing a copper valley into place, as too much force can cause the soft metal to kink. However, get the valley metal to conform to the shape of the roof as tightly as possible without improperly bending it.

When replacing two valleys that converge at the top (a common situation), carefully fold the tops over each other at the ridge. This can be done without the need for solder, although a little lifetime silicon under the metal in a couple of critical spots doesn't hurt. Caulk or no caulk, a properly folded valley top will not leak. Remember that there will probably be a segment of ridge slate or metal that will cover some of that valley top and keep the rain out. What doesn't fold over the ridge will lay flat on the roof and should also be overlapped. If there is not adequate overlap, an extra piece of metal can be installed at the top convergence of the valleys to make it leak proof before it is re-slated. Practice makes perfect.

The bottom of the valley usually has to be cut to conform to the corner of the roof, leaving

---

enough overhang to channel the water into the gutter. If you're not sure how much overhang to leave, better to leave too much — it can be trimmed off or bent over later. Make sure the valley metal is *under* the shim or cant strip at the drip edge and not on top of it (assuming a shim strip exists). Keep the shim well back from the valley center — it need only overlap the valley metal an inch or two. The bottom of the valley metal at the drip edge can also be folded down about an inch to create its own drip edge, thereby preventing capillary attraction up the valley bottom as well as obstructing birds and other unwanted visitors who may try to crawl under the valley metal after installation.

Some roofers insist upon installing felt paper under replacement valley metal. Others insist (religiously) upon installing self-adhesive underlayment under all valley metal. The author is not a big fan of self-adhesive underlayment as the metal can stick to it, thereby inhibiting expansion and contraction; plus, it's an unnecessary material which eats up time and money, both paid for by the property owner. Felt paper under a replacement valley doesn't hurt anything, and may even act as a cushion for the valley metal. Nevertheless, when replacing valleys, the valley metal can usually be laid directly on the cleaned bare wood with no ill effect. The original felt paper is typically dried up and non-functional by that time anyway, and the contact the valley metal has against the bare roof deck does not harm it in any way. There is no reason to believe that felt paper under valley metal will prolong the life of the valley, although the felt can act as a nominal insulating thermal barrier where one is needed. More often than not, roofing contractors rely on underlayment to conceal shoddy workmanship. If they think their valley job will leak, they pour on the underlayment before installing it. A properly installed valley will not leak, underlayment or no underlayment.

Once the valley metal has been installed on the roof, you must chalk a line up either side of the metal to indicate where the edges of the slate will be. Some foresight is necessary here, and although most valleys have parallel exposed edges, not all exposed valleys are the same width. Six inches is average for an open valley, but they'll vary from 4" to 8" or 10" if open, or they may be closed instead. If the original valley *exposure* was 6" wide, measure 3" from the center on both sides and chalk two lines the length of the valley on the metal, 6" apart. Then immediately go over those lines with a permanent felt tip marker, pushing the chalk out of the way with the felt tip as you go (the ink will eventually wear off the valley, although it will not be visible from the ground at any time). When re-slating, align *the inside edges of the slates with those ink lines,* and don't worry about the *outer sides* of the slates being exactly parallel to each other if you want the *valley* to look neat.

RESLATING THE VALLEY

When reslating the valley, start at the bottom. Use the same slates that were removed from the roof as much as possible, for two reasons. First, they'll match that roof better than any other slates because they came from that roof. Secondly, they have weather marks on them showing where they were overlapped by the other slates, and can therefore be placed back in the same spot they came from by simply lining up the weather marks with the slates already on the roof. With numbered slates positioned on the proper sides of the valley, and weather marks to guide you, you can zipper a valley back together in less time than you may think.

When you start reslating at the bottom of the valley, make sure you have a piece of wood shim (or cant) in place under the starter slates, about 1/2"-5/8" thick (for standard thickness slates) and an inch or so wide, positioned right at the bottom of the sheathing (see Chapter 13). If the original cant strip is still there and is usable, fine. The cant strip lies *on top* of the valley metal, but should not extend into the center of the valley as far as the slate, and usually *need not extend over the valley metal very far at all.*

When re-slating, simply nail the slates back into place in the same spots from which they were removed — *one side of the valley at a time,* if working alone. If two people are reslating a valley at the same time, both sides can be reslated at once with one person working above the other. If working alone, after one side has been reslated, slide the hook ladder from the finished side closer to the valley center and work off the repositioned ladder while slating the other side. Be careful that the foot of the hook ladder is not digging into the valley metal.

You'll find that many of the slates you're nailing back into place will be covered on one side by a nailed overlying slate, and therefore the replacement slate can be nailed only on one side. Punch an extra hole with a slate hammer on the exposed side a few inches above the existing hole so that you nail every slate with two nails. If the old nail holes are still good, use them, if not, punch new ones *near the outer edge of the slate.* Make sure your nails are 1 1/2 inches long (for standard thickness slates). Occasionally, you'll run into a slate that

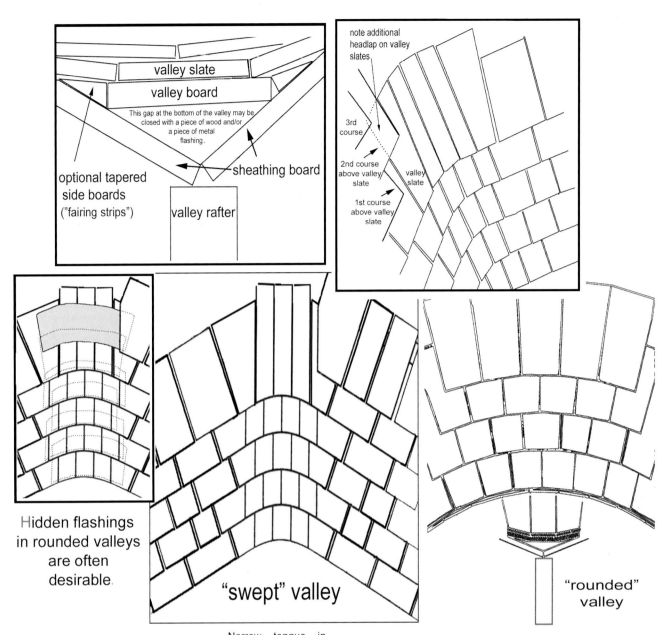

valley slate

valley board

This gap at the bottom of the valley may be closed with a piece of wood and/or a piece of metal flashing.

optional tapered side boards ("fairing strips")

sheathing board

valley rafter

note additional headlap on valley slates

3rd course

2nd course above valley slate

valley slate

1st course above valley slate

Hidden flashings in rounded valleys are often desirable.

"swept" valley

"rounded" valley

Narrow tongue in groove boards were originally used to round the valleys at Glenridge Hall in Sandy Springs, Georgia (left). When reslated, 20 ounce copper strips were laid in with each course of valley slates to ensure a water tight installation (right). Alternatively, longer valley slates with double headlap can be used (top right), or the valley slates can be installed with twice as many courses as the field slates (pages 281-282).

Rounded and Swept Valleys

Photos this page by Ron Stokes.

SLATED VALLEYS

What you can't see: Valley boards underneath the valley slates extend above the ridge and meet in the center, creating a sweeping curve where the gable ridge meets the main roof at the top of the valleys. The valley board is completely hidden in these photos, of course, as is the fact that the valley slates are longer than the slates on the main roof.

The valleys shown at left and at top left have hidden copper step flashings installed under each course of valley slates. The slates in these valleys are six inches wide.

Photos by author.

slides into place and has no exposed side on which to nail. Nail it with the nail and bib technique that is described in Chapter 17, or use a slate hook, if possible. Cut replacement slates as you need them with a slate cutter, and always use matching slates. Make your cuts right there in the valley; don't go down onto the ground to cut the slates.

Be very careful when using soft copper valley flashing to not drop any sharp object on the valley metal. Soft copper punctures easily, and a dropped slate hammer or ripper (or slate) can punch a hole in it. Furthermore, be careful when working over a soft copper valley. If you sit or stand on it and put too much weight on it in the wrong way, you can permanently buckle it. Stainless steel valleys and heavy aluminum (.040") are much stronger than soft copper and can take a lot more punishment. Partially hardened copper is preferable to soft copper in valleys, and one of the best valley flashings available is 20-ounce half-hard copper.

ROUNDED VALLEYS

Perhaps the most aesthetically appealing of all valleys are those made of slate with no exposed metal. These are generally referred to as *rounded* or *slated* valleys and they exist in many different styles. Some rounded valleys utilize valley slates of uniform sizes, others are completely random. The various styles of slated valleys are illustrated in some European texts, such as *Architecture and Techniques of Slate Roofing,* by Jose Luis Menendez Seigas, and others.

Slated valleys can be found on very old buildings in Europe. They were probably first developed in order to avoid the use of metal on the roof, as suitable metal flashings may have been hard to obtain a couple hundred years ago. It is reported, therefore, that valleys can be constructed utilizing slate only and no metal, although slated valleys can also employ hidden metal step flashings.

VALLEY BOARD

There are a few primary differences between open metal valleys and rounded valleys. The first is the valley board, which is a board that runs parallel to the valley up its center, used on rounded valleys but not on other standard open valleys. This board may have an additional tapered board on either side of the valley board (a "fairing strip") in order to help round the valley smoothly. The width of the valley board can vary somewhat depending on the size of

the slates and the slope of the roof, although a width of 9" to 11" seems to be most effective.

EXTRA HEADLAP

The slates in the valley must be longer than the standard slates on the roof if step flashing is not being used in the valley. The headlap should extend not just three inches under the second overlying course, which is standard headlap; it should extend completely up to and under the *third* overlying course by an inch or so. The extra headlap is necessary to prevent water penetration since the valleys carry more water and have a lower slope than the rest of the roof.

ROOF SLOPE, VALLEY LENGTH,
AND OTHER CONSIDERATIONS

Slated valleys cannot be installed on low slope roofs. It is recommended that slated valleys be installed on roofs with slopes no lower than 30^0 (7:12), although much steeper slopes, such as 20:12, are recommended.

The longer the valley, the more water it must carry, especially at the bottom. Therefore it is not recommended that rounded valleys be designed on long runs, with 20' being a recommended maximum length. Also, regardless of valley length, some valleys may be subjected to a great amount of water flow due to the size of the roof surfaces that drain into the valley. This must also be taken into consideration when designing a rounded valley. If the water flow or other characteristics of the roof cast some doubt on the efficacy of a rounded valley installation, hidden flashings may be installed between the valley slate courses.

HIDDEN FLASHINGS

Rounded slate valleys may have a piece of step flashing installed between each course of valley slates. This can overcome problems associated with valley slates that are too short, slopes that are too shallow, or roofs that are too large and drain too much water. Some contractors install hidden flashings in rounded valleys as a matter of course, no matter what. This hidden flashing can be copper, stainless steel (preferably terne-coated in order to dull it down), lead or zinc. This is illustrated on page 278.

Above: German slated valleys, the one on the right is under construction, installed over light colored paper.
Below: Another German valley. Right, top photo: cross section of valley showing valley board and fairing strips.

Below: Rounded ceramic tile valley. Note valley board.

Four examples of German slated valleys

Chapter Twenty-One

CLAY TILE ROOFS, ASBESTOS ROOFS, FLAT SEAM COPPER ROOFS AND MISC. ROOFING

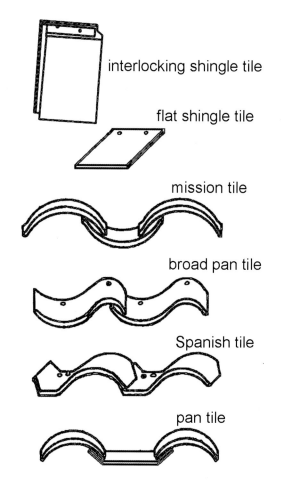

interlocking shingle tile

flat shingle tile

mission tile

broad pan tile

Spanish tile

pan tile

Ceramic tile roofs are mentioned here because slate roofers are often called upon to repair old tile roofs, and sometimes tile roofs are confused with slate roofs by their owners. Many tile roofs are similar to slate roofs in that the tiles are semi-permanent, they're on steep roofs, some are shaped like slates and even look like slates, and their repairs require basically the same tools, equipment and procedures as those used on slate roofs, with some notable exceptions. These include the cutting of the tiles, as a slate cutter will not cut a ceramic tile and a diamond blade saw must be used instead, or a special hand-operated tile cutter must be used, several varieties of which are available in Europe. Tiles cannot be punched for nail holes like slate can — they must be drilled with a masonry drill bit. Also, it is common to hang some tiles in place using copper wires during specific applications, such as along valleys during replacement of the valley metal. Finally, individual tiles, when replaced in the field of the roof, are rehung using special tile hangers, which wire to the back of the tile. Neither slate hooks nor nails and bibs are used with ceramic tiles.

Clay tiles fall into two general categories: flat or *shingle* tiles, and rounded or *pan* tiles. There are many, many different types and styles of clay

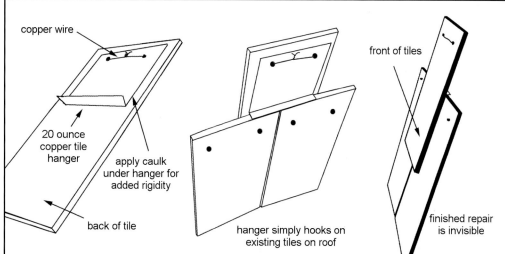

copper wire

20 ounce copper tile hanger

apply caulk under hanger for added rigidity

back of tile

hanger simply hooks on existing tiles on roof

front of tiles

finished repair is invisible

This technique for replacing tiles on a roof will work for just about any shape or type of tile, including Spanish tiles. The tile hanger is simply custom-made to suit the tile being replaced, some longer, some shorter, some wider, some narrower.

The old tile is pulled out using a slate ripper. Although it is common in some roofing circles to use exposed strap hangers to replace clay tiles, these are unsightly and should be avoided. Any finished repair should be invisible.

VALLEY REPLACEMENT ON A FLAT SHINGLE TILE ROOF

Shingle tile roofs are similar to slate roofs, and replacement of a valley follows similar steps. First the tiles are removed from either side of the valley using a slate ripper. Before removal they're numbered by scratching with a nail. The tiles from each side of the valley are propped on hook ladders situated on the same side of the valley. Once the tiles are off, the old valley metal removed, and the valley swept out, new metal is installed as shown at left (#1) where terne-coated stainless steel is being used. Once the metal is installed, the valley is re-tiled, starting on one side with the starter tile (#2). Note that the small triangular tile in the valley center is hung on a piece of copper wire and glued in position with clear silicon caulk for added anchoring. The next course of tiles is then installed (#3) in the same manner. A tile hanger is wired to the back of the tile (bottom left) when there is no way to nail the tile in place.

Photos both pages by author.

1.

lifetime silicon caulk

starter course

2.

copper wire

copper tile hanger

back of tile

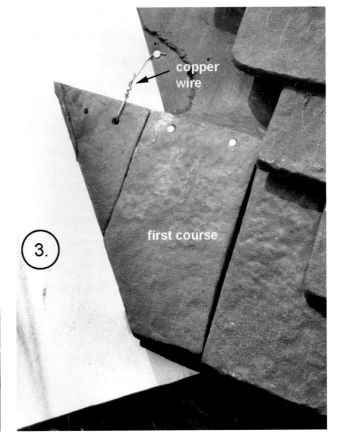

copper wire

first course

3.

The courses of tiles are continued in the same manner using tile hangers, copper nails, and copper wires, up one side of the valley (#4) until that side is completely tiled back in. Then the other side is tiled. Both sides can be tiled at the same time if two men are working on it together; however there is an advantage to having one man simply wiring tile hangers to the back of tiles as the other is installing them (#5). Nails used to install these tiles were 2.5" copper roofing nails.

The photo at the bottom left shows the same procedure being used on an interlocking shingle tile roof where a 20-ounce copper valley is being installed. The tile hangers used on this roof are a different size (longer) in order for the tiles to be positioned correctly once re-installed. A wire attached to a copper nail holds the small triangular piece of tile in the center of the valley (circled).

For more information about tile roofs see Hobson, Vincent H., 2001, <u>Historic and Obsolete Roofing Tile</u>, Remai Publishing Company, Inc., Evergreen, Colorado, USA; www.rooftilebook.com

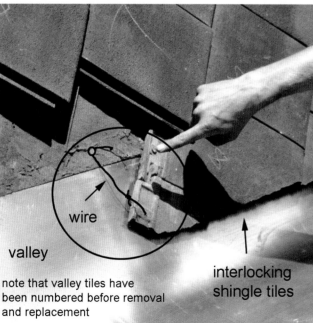

wire

valley

interlocking shingle tiles

note that valley tiles have been numbered before removal and replacement

tiles in existence. Any proper repair work on a tile roof requires the availability of matching replacement tiles, which may be hard to find when working on an old tile roof. Fortunately, there are several roof tile salvage services available in the U.S. (listed in the back of this book), and these will help you find the tiles you need.

As always, when working on a roof with unusual clay tiles, slates or asbestos tiles, and suitable replacement tiles cannot be found, it is advisable to cannibalize a section of the roof and use the shingles taken from that section to repair the rest of the roof. This means removing a section of the roof in a relatively invisible area, such as one half of a rear dormer roof, using the tiles removed to repair the remainder of the roof, then re-tiling the cannibalized area with another tile that does not match perfectly, but is close enough. This strategy will save many a roof while maintaining the aesthetic integrity of the roof overall.

The proper way to replace a ceramic tile in the field of the roof is to carefully remove the broken tile with a ripper, then wire a copper or stainless steel tile hanger to the <u>back</u> of the replacement tile, then slide the replacement tile into place, hooking the tile hanger to the tiles underneath the replacement tile. Visible strap hangers should never be used.

ASBESTOS ROOFS

Asbestos roofs are mentioned here because people who have this type of roof on their buildings often mistake them for slate. Some even call them "asbestos slate" roofs. Asbestos roofing is a human-made material consisting of asbestos fibers and a binding agent. It does not have the characteristics of natural slate roofing, although it can resemble slate in appearance, especially to the layperson. Like slate roofs, asbestos roofs tend to be old, and it can be difficult to find someone to repair them. One complication associated with asbestos roofs is the fact that asbestos is now considered a toxic material because it can cause lung cancer if inhaled, and therefore the removal of asbestos roofing is expected to be done by toxic waste professionals.

The bright side of this dilemma is that asbestos roofs can be maintained for quite some time *as long as replacement shingles are available* to replace any that become broken. If replacement shingles are not available, then use the "cannibal" method described above to get the shingles you need.

Asbestos roofs can be maintained much the same as slate roofs are maintained — with hook ladders to work on the roof, slate rippers to remove the old tiles, and slate cutters to cut them. Asbestos roofing cannot be punched with a slate hammer, but it

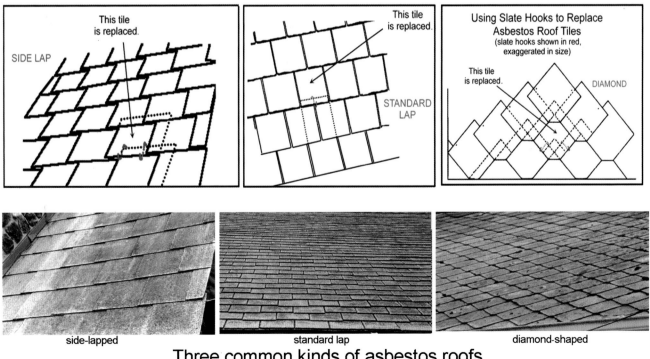

Three common kinds of asbestos roofs
Individual tiles are easily replaced with the use of slate hooks.

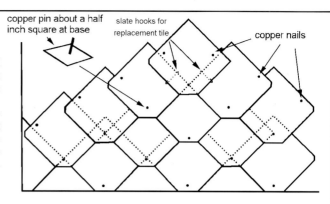
can be drilled, or punched with a slate cutter that has a hole punch on it. Asbestos shingles are often laid in diamond patterns and have no overlying slot, so replacing them usually requires a couple of slate hooks.

The original asbestos shingles are nailed with two nails, and the diamond patterned ones also have a copper clip through the bottom. Sources of asbestos tiles have been listed at jenkinsslate.com.

BOX GUTTERS

Box gutters are very common on buildings that have old slate roofs. These gutters, like soldered-seam or flat-seam metal roofs, were typically made from soldered terne metal. They rust and corrode when not kept properly cleaned and painted, and many of these old gutters leak. They can be replaced, however, by carefully measuring the wooden frame holding the gutter, then having a new metal liner fabricated at a local sheet metal shop in 4-, 8-, 10-, or 12-foot lengths. Use terne-coated stainless, 20-ounce copper, or another metal that won't corrode, overlap the metal sections by an inch and a half or two inches, rivet the pieces together with stainless steel or copper rivets, then solder the joints *and* the rivets (see illustrations, next page).

SOLDERED-SEAM METAL ROOFS

Many old slate roofs have low-slope soldered seam metal roofs abutting or adjoining them. They're usually found on porches, bay windows, and two-story additions. These roofs are usually made of terne metal panels, and sometimes copper panels, soldered at their flat-lock seams, hence the name "soldered-seam," "flat-seam," or "flat-lock" roofs. They're excellent, long-lasting roofs *if they're properly maintained* — most aren't. Proper maintenance means painting regularly if terne metal, or painting

regularly after about 50 years of age if copper. These roofs differ from "standing seam" roofs, which have an obvious metal seam that projects vertically an inch or so, running the length of each metal roofing panel. It's the *soldered-seam* roofs that are most commonly associated with old slate roofs.

The reason they're mentioned here is because they're a maintenance problem for many owners of old slate roofs, since the metal is often run up under the slates where the flat roof abuts the slate roof. If you have a slate roof with a soldered-seam roof associated with it, there are two things you should know. First, terne metal soldered-seam roofs should be kept painted with tinner's red oxide paint (or any good, exterior metal paint). If not, they'll rust and start leaking, usually at the seams. Second, if the soldered-seam roof is already leaking because the roof has long been neglected and now you have to do something about it, the roof can be replaced with new metal. There is also an easy and effective temporary remedy: *liquid asphalt emulsion and fiberglass membrane.*

Liquid asphalt emulsion is an old-fashioned roofing material that works well to preserve old metal roofs when paint won't do. It's a water-based (emulsified) asphalt that must be applied when no rain is imminent as it will wash off if it is subjected to a downpour before it sets up. But once it dries it's insoluble in just about everything, and you'll soon be a believer when you try to wash dry asphalt emulsion off your hands or clothes with water, gasoline, or anything else after applying it to a roof. Liquid asphalt emulsion comes in five gallon buckets and is applied with a long-handled roofing brush. It's liberally brushed onto a clean, dry surface (like painted or rusted metal), then a layer of fiberglass membrane (which is made in rolls to be used with the emulsion) is carefully rolled out over the wet asphalt and brushed in, then a second liberal coat of asphalt is applied over top of the fiberglass. The fiberglass

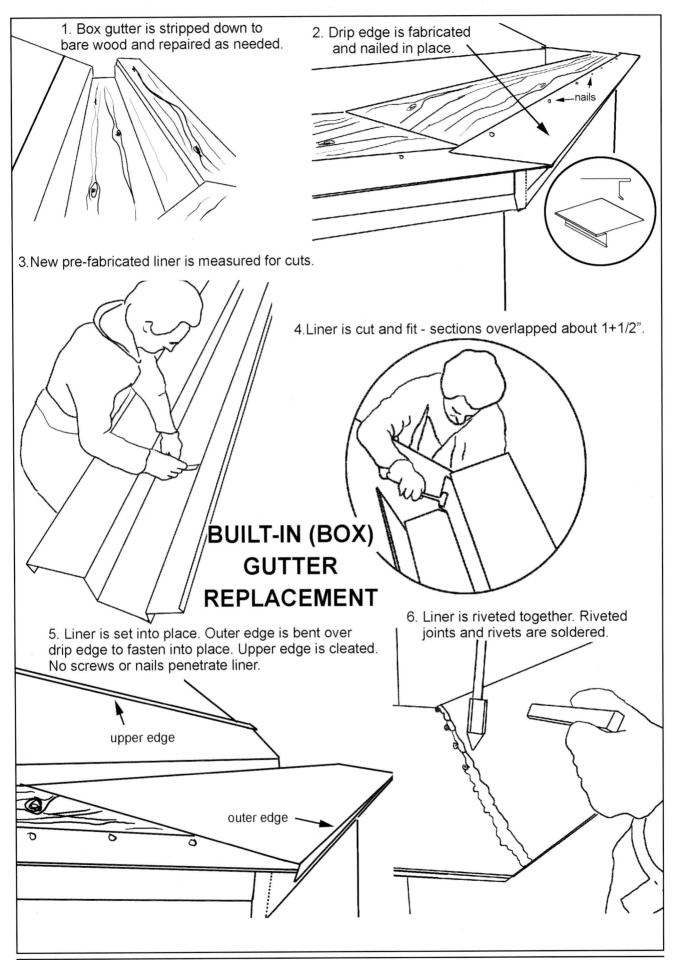

1. Box gutter is stripped down to bare wood and repaired as needed.

2. Drip edge is fabricated and nailed in place.

nails

3. New pre-fabricated liner is measured for cuts.

4. Liner is cut and fit - sections overlapped about 1+1/2".

BUILT-IN (BOX) GUTTER REPLACEMENT

5. Liner is set into place. Outer edge is bent over drip edge to fasten into place. Upper edge is cleated. No screws or nails penetrate liner.

upper edge

outer edge

6. Liner is riveted together. Riveted joints and rivets are soldered.

comes in rolls of various widths, and the three-foot-wide roll is most practical for larger surfaces. The whole thing needs to set up overnight, and maybe longer if the weather is cool or humid Then, a third, final coat of asphalt is applied right over top of the whole shebang.

When all is dry and done, the metal has a thin coating of black asphalt (which can be painted to change the color, if desired) that remains hard even in hot weather, and can be walked on. When the asphalt wears thin a number of years later, brush on another coat and throw in some fiberglass over any weak areas. Be very careful not to apply asphalt over any slates — brush it up *under* the slates when you're coating a metal roof that abuts a slate roof, and make sure the fiberglass is worked up under the slate with a pointed trowel, if necessary.

Liquid asphalt emulsion will seal up your soldered-seam roof so it won't leak, and at a reasonable cost. An alternative covering for old soldered-seam roofs that is gaining popularity is rubber roofing. The problem with rubber roofing is that the rubber must also be worked up under the slates, and most rubber roof guys just slap it over top of the slate and glue it down. This creates big problems when the glue lets loose — and it will. Then try to find your rubber roof guy to come back and repair it (good luck). Properly done, a rubber roof abutting a slate roof requires that the slates are lifted, the rubber installed underneath, then the slates are relaid. Liquid asphalt emulsion is cheaper, easier to work with, available at most roofing supply outlets, and do-able for most do-it-your-selfers. But watch out — it's messy!

The best way to repair a leaking soldered-seam metal roof is the same way flashings are repaired on slate roofs: the old metal is completely removed and then replaced. When the old terne metal is removed, it should be replaced with new terne-coated stainless steel or 20-ounce copper. 16-ounce copper will work too, but for the smaller extra cost and the promise of a metal roof that can last a century with little maintenance, the 20-ounce copper would be preferable (and the stainless the most durable). It has become difficult these days to find roofing contractors who can and will do flat seam metal roof work. Some instructions are illustrated on the following page, and the reader is advised to check jenkinsslate.com for the possible availability of more detailed video and/or print instructions.

The flat-seam roof starts with a drip edge almost exactly the same as the drip edge shown here in the box gutter replacement illustration (page 288). The field of the roof is then covered with metal pans which can be various standard sizes, the most common perhaps today is a panel 24" X 18". Old roofing books describe these standard panels as being 10" X 14" or 10" X 20" (Roofing, 1930); 20" X 28" (Sheet Metal Workers' Manual, 1942, and Architectural Sheet Metal Manual, 1993); 14" X 20" up to 20" X 28" (Roofing Construction and Estimating, 1995); 14" X 20" (Standard Practice in Sheet Metal Work, 1929); and 18" X 24" maximum (Copper Development Association, 1991). So obviously the particulars of flat-seam roofing are not carved in stone.

The flat pans are cut, the corners are snipped off at approximately a forty-five-degree angle, then two adjacent edges are folded up and the opposite two adjacent edges are folded down, the folds being 1/2" to 3/4". The pans are fastened to the roof with cleats, these being made of the same metal, approximately 1.5" X 2", with a fold equal to the folded seam on the pan (1/2" to 3/4"). This fold is fit into the pan seam, and the other end is nailed to the roof deck with two nails, then folded over the nail heads. Two to three cleats are recommended on the long side of the pan and one to two cleats on the short side. The seam is then hammered down with a dead blow hammer and soldered with 50/50 tin/lead solder or other suitable solder. Roof protrusions such as vent pipes are fit closely around the base with a pan, then a sleeve is fabricated with a flared bottom that is riveted to the pan and soldered in place.

ROOF ACCESSORIES

SNOW GUARDS

Snow guards are small, angled metal brackets permanently attached to the roof to prevent ice and snow from cascading off the roof in an avalanche. They fall into two primary categories: those that are nailed to the roof, and those that hook either on the slates themselves, or on the slate nails. The ones that are nailed on must be installed when the roof is installed. Otherwise, slates must be removed from the roof, the snow guards attached, then the slates replaced. On the other hand, the kind that hook on an existing slate or nail may be attached to the roof at any time with much less effort. Some simply have a long, slotted end which is slid up under the slate as if it were a slate ripper, and hooked on a slate nail. Others are slid between the slightly lifted slates until they hook over the top. The snow guard,

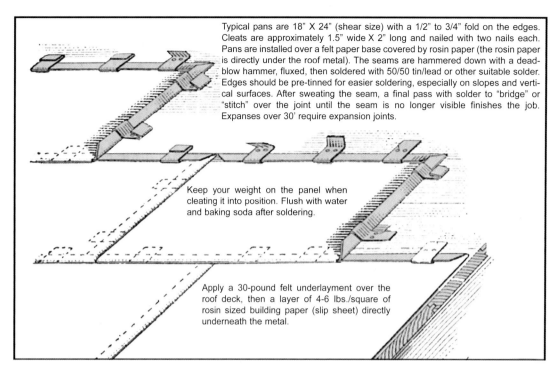

Typical pans are 18" X 24" (shear size) with a 1/2" to 3/4" fold on the edges. Cleats are approximately 1.5" wide X 2" long and nailed with two nails each. Pans are installed over a felt paper base covered by rosin paper (the rosin paper is directly under the roof metal). The seams are hammered down with a dead-blow hammer, fluxed, then soldered with 50/50 tin/lead or other suitable solder. Edges should be pre-tinned for easier soldering, especially on slopes and vertical surfaces. After sweating the seam, a final pass with solder to "bridge" or "stitch" over the joint until the seam is no longer visible finishes the job. Expanses over 30' require expansion joints.

Keep your weight on the panel when cleating it into position. Flush with water and baking soda after soldering.

Apply a 30-pound felt underlayment over the roof deck, then a layer of 4-6 lbs./square of rosin sized building paper (slip sheet) directly underneath the metal.

Flat-Lock Soldered Seam Metal Roofing

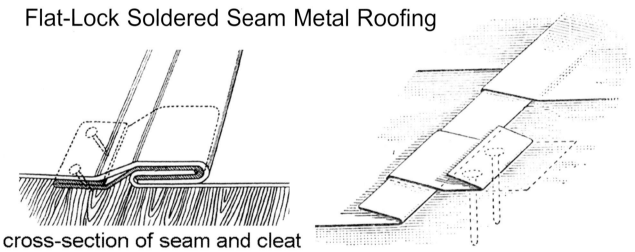

cross-section of seam and cleat

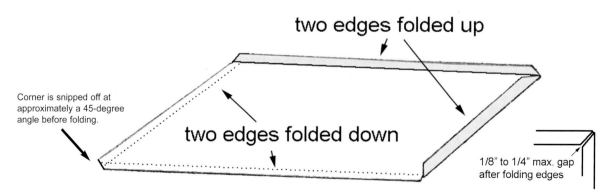

two edges folded up

Corner is snipped off at approximately a 45-degree angle before folding.

two edges folded down

1/8" to 1/4" max. gap after folding edges

Pans are cut (e.g. 18"X24"), then the corners are snipped off at approximately a 45-degree angle so when the edges are folded there is 1/8" to 1/4" clearance between the folded mitered corners. The edges are then folded, two adjacent edges folded one way and the opposite two adjacent edges folded the other.

Check jenkinsslate.com for the availability of an instructional video on flat-lock copper roofing.

20 ounce copper

rosin paper

30 lb felt over
t-i-g board deck

20-ounce flat-lock copper with 18" X 24" pans being installed on a Vermont unfading-green slate roof in New Jersey by Barry Smith Slate Roof Restoration (above). Top left photo shows pans before soldering. Barry Smith installs a terne-coated stainless steel flat-lock roof adjacent to a New York red slate roof located in Troy, PA (below).   Top photos by Barry Smith, bottom photo by Aaron Cable.

once hooked, is left in place.

Snow guards are positioned above doorways, walkways, porches, sidewalks and anywhere where falling ice or snow may present a hazard. They're made of either galvanized steel, painted steel, stainless steel, copper, aluminum or bronze, and may be lead coated. Some are available as "snow railings," an elongated version of the snow guard.

LIGHTNING RODS

Lightning rods are common on old slate roofs. They're often nailed onto the roof right through the slate, usually with rather large (i.e. 16 penny) nails, and the nail heads are caulked or cemented to prevent leakage. The feet of old lightning rods will sometimes cause leaks because the cement over the nails has worn away, and a fresh application of roof cement or silicon caulk will cure

the problem. Lightning rods are designed to divert electric current away from a building by channeling the current through a heavy copper or aluminum cable into the ground. Therefore, a lightning rod must be properly grounded or it won't work at all. Many older homes with old lightning rod systems are now having the rods and cables removed.

Sources of snow guards and lightning rods are listed on the following page.

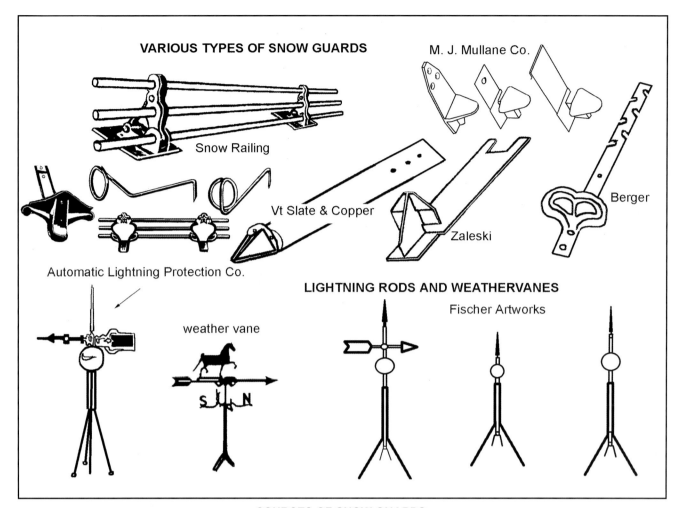

VARIOUS TYPES OF SNOW GUARDS

M. J. Mullane Co.

Snow Railing

Vt Slate & Copper

Berger

Zaleski

Automatic Lightning Protection Co.

LIGHTNING RODS AND WEATHERVANES

Fischer Artworks

weather vane

S — N

SOURCES OF SNOW GUARDS

• **Berger Building Products** — 805 Pennsylvania Boulevard, Feasterville, PA 19053; Ph: 1-800-523-8852 or (215) 355-1200; Fax: (215) 355-7738. They have the handy hook-on kind that can easily be installed on an existing slate roof (in copper, stainless steel, or hot-dipped galvanized). They also have a variety of other styles and types.
• **Gough Snowguards** — 4133 Du Bois Blvd., Brookfield, IL 60513; www.snoguard.com; all types of snowguards
• **M. J. Mullane Company** — 17 Mason Street (P. O. Box 108), Hudson, MA 01749; Ph: (508) 568-0597, Fax: (508) 568-9227. The have a wide variety of snow guards, including bronze and aluminum cast.
• **Vermont Slate and Copper Services** — 55 C Gonyeau Road, Milton, VT 05468; Ph: (802) 893-7703, Fax: (802) 893-1143. They have custom-made copper snow guards, leader boxes and finials.
• **Zaleski** — 11 Alsen Street, New Britain, CT 06053; Ph: (203) 225-1614, Fax: (203) 225-1060. Their slate snow guards protrude between the slates and hook on the top of the slate without the need for nailing.

SOURCES OF LIGHTNING RODS, WEATHER VANES, FINIALS

• **Automatic Lightning Protection** — 11072 W. Mohawk Lane, Sun City, AZ 85373; Ph: 1-800-532-0990; Over 200 styles of lightning rods and weather vanes.
• **Colonial Works** (weather vanes) — PO Box 46457, Hollywood, CA 90046; Ph: 213-460-6838.
• **Crosswinds Weathervanes** - 29 Buttonwood St., Bristol, RI 02809; Ph 401-253-0344
• **Denninger Weather Vanes and Finials** — 77 W. Whipple Road, Middletown, New York 10940-9801; Ph/fax: 914-343-2229.
• **Fischer Artworks** — 6530 S. Windermere St., Littleton, CO 80120; Ph: 303-798-484 or 1/800-441-6067, Fax:795-8805; Copper and cast bronze lightning rods with copper or glass globes.
• **Independent Protection Company., Inc.,** - PO Box 537, Goshen, IN 46527; Ph: 219-533-4116, Fax: 219-534-3719. A variety of lightning rods, including old style (antique).
• **Lehman's Hardware** — One Lehman Circle, Kidron, Ohio 44636; Ph: 330-857-5757, Fax: 330-857-5785; Email: GetLehmans@aol.com; web site: http://www.lehmans.com; A variety of both lightning rods and weathervanes.
• **Vanes and Things** (dog motif weather vanes) - 1112 East C. Street, Commerce, OK 74339; Ph: 918-675-4262.
• **Vermont Slate and Copper Services** — 55 C Gonyeau Road, Milton, VT 05468; Ph: 802-893-7703, Fax: 802-893-1143. Finials.

REFERENCES

CHAPTER 2 — WHAT IS SLATE?

• Behre, Charles E., (1933), <u>Slate in Pennsylvania</u>, PA Geological Survey, Fourth Series, Bulletin M 16. p. 29.
• Bowles, (1934), <u>The Stone Industries</u>, First Edition, McGraw-Hill Book Co., Inc., New York and London, p. 230.
• Levine, Jeffrey S., (1993), *"The Repair, Replacement and Maintenance of Historic Slate Roofs,"* U.S. Department of the Interior, National Park Service, Preservation Briefs # 29.
• Moebs, Noel N., and Marshall, Thomas E., 1986, *Geotechnology in Slate Quarry Operations*, Bureau of Mines.
• Pennsylvania State University, School of Mineral Industries, <u>Properties and New Uses of Pennsylvania Slate</u>, The Pennsylvania State College Bulletin, Volume XLI, No. 30, July 18, 1947, pp 7-25, 134-137.

CHAPTER 3 — IDENTIFYING ROOF SLATE

• Bowles, Oliver, and Coons, A. T., 1930, Slate in 1929, U. S. Department of Commerce, Bureau of Mines, Mineral Resources of the United States, 1929 - Part II (Pages 161-74).

CHAPTER 4 — WALES

• Brigden, John, (date unknown), *"Turning Stone into Bread,"* (publisher unknown). This article was on file at the Buckingham-Virginia Slate Quarry in Arvonia, Virginia.
• Carrington, Douglas C., and Rushworth, T. F., (date unknown), <u>"Slates to Velinheli - The Railways and Tramways of Dinorwic Slate Quarries Llanberis,"</u> found in Green Mountain College, Poultney, Vermont.
• *"Chwareli a Chwarelwyr,"* (Quarries and Quarrymen), 1974, Gwynedd Archives Series, (A booklet to accompany an exhibition prepared to celebrate the centenary of the founding of the North Wales Quarrymen's Union), pp. 47-51.
• Hartmann, George E., (1967), <u>Americans from Wales</u>, The Christopher Publishing House, Boston, MA, pp. 86-91.
• Holmes, Alan (1986), <u>Slates from Abergynolwyn</u>, (The Story of Bryneglwys Slate Quarry), Gwynedd Archive Services, Caernarfon, Gwynedd, Wales, pp 113-114.
• Isherwood, Graham (1982); *Cwmorthin Slate Quarry*; Revised edition published by Adit Publications, Towerside, Pant-y-Buarth, Gwernaffield, Mold, Clwyd, CH7 5ER, UK.
• Isherwood, J. G., (1988); <u>Slate From Blaenau Ffestiniog</u>; AB Publishing, 33 Cannock St., Leicester LE4 7HR England.
• Lindsay, Jean, 1974, <u>A History of the North Wales Slate Industry</u>, David and Charles, Newton Abbot, London, pp.11-27, 246-61, 286-295, 324-325.
• Encyclopedia Britannica, (1965), Volume 23, p. 296 (population of Caernarvonshire and Merioneth in 1871).
• Lewis, M. J. T., and Denton, J. H., (1874), <u>Rhosydd Slate Quarry</u>, The Cottage Press, Shrewsbury, England, P. 98.
• McKinney, Margot, 1976, *"The Welsh Heritage of the Slatebelt,"* produced at Green Mountain College under a grant from HEW, Poultney, Vermont, Journal Press, Inc.
• Richards, Alan John (1994), <u>Slate Quarrying at Corris</u>; Gwasg Carreg Gwalch, Llanrwst, Gwynedd, Wales.
• Richards, Alan John (1995); <u>Slate Quarrying in Wales</u>; Gwasg Carreg Gwalch, Iard yr Orsaf, Llanrwst, Gwynedd, Wales.
• *The Slate Industry of North Wales, Unit 4;* A collection of historical documents available at the Llechwedd Slate Quarry historical site at Blaenau Ffestiniog, North Wales.
• Williams, Merfyn (1991); <u>The Slate Industry</u>, C. J. Thomas and Sons, (Haverfordwest) Ltd., Press Buildings, Merlins Bridge, Haverfordwest, Dyfed SA61 1XF, UK.
• Williams, M. C., and Lewis, M. J. T., 1989; <u>Gwydir Slate Quarries</u>, Snowdonia National Park Centre, Plas Tan y Bwlch, Maentwrog, Blaenau Ffestiniog, Gwynedd LL41 3YU, Wales.

CHAPTER 5 — VERMONT/NEW YORK

• Beers, F. W., (1969); <u>Atlas of Rutland County</u>, Charles E. Tuttle Publishing Co., Rutland, VT, pp. 111-114.
• Bowles, Oliver, (1934); <u>The Stone Industries</u>, 1st Ed., McGraw-Hill Book Co. Inc., New York and London, p. 278.
• Joslin, J., Fisbie, B. and Ruggles, F., (1875); <u>A History of the Town of Poultney, Vermont, From Its Settlement to the Year 1875</u>, Poultney: Journal Printing Office, pp. 181-187.
• Morrow, John A., (1970); <u>A Century of Hard Rock - The Story of Rising and Nelson Slate Company</u>, Grastorf Press, Granville, NY.
• *Ruggles Stone Machinery Catalog;* Ruggles Machine Co., Poultney, VT (Established 1828). This catalog provided the illustrations of the slate-working equipment in this chapter. Special thanks to Paul B. Boyce and the East Poultney Historical Society.
• Sharrow, Gregory (ed.), (1992); <u>Many Cultures, One People</u>; The Vermont Folk Life Center, Middlebury, VT 05753, pp. 232-235.
• Smith, H. P., and Rann, W. S. (ed.), (1886); <u>History of Rutland County, Vermont</u>; D. Mason and Co., Syracuse, NY, pp. 192-198.

CHAPTER 6 — PENNSYLVANIA

• Behre, Charles E., (1933); <u>Slate in Pennsylvania</u>; Pennsylvania Geological Survey, Fourth Series, Bulletin M 16, pp. 19-21, 112, 173-187.
• Bowles, Oliver, (1934); <u>The Stone Industries;</u> 1st ed., McGraw-Hill Book Co., Inc., New York and London, pp. 229-289.
• History of Trinity Evangelical Lutheran Church, 1878-1928; pp. 10-11.
• Pennsylvania State University, School of Mineral Industries; <u>Properties and New Uses of Pennsylvania Slate;</u> The Pennsylvania State College Bulletin, Volume XLI, No. 30, July 18, 1947, pp 7-25, 134-137.
• (Author Unknown), <u>Slate Roofs</u>, (1926); available from Vermont Structural Slate Co., Fair Haven, Vermont 05743; phone: 802-265-4933/34 or 800-343-1900, fax: 802-265-3865. A version is also available from Hilltop Slate Co., PO Box 201, Middle Granville, NY 12849; phone: 518-642-2270/642-1453, fax: 518-642-1220.

CHAPTER 7 — PEACH BOTTOM

• Behre, Charles E., (1933); <u>Slate in Pennsylvania</u>; Pennsylvania Geological Survey, Fourth Series, Bulletin M 16; pp.359-390.
• Eisenberg, H. O., *"The Story of Slate"*; Slate Centennial; pp.13-16. (Date unknown).
• Faill, Roger T., and Sevon, W. D. (Eds.), (1994); *"Guidebook for the 59th Annual Field Conference of Pennsylvania Geologists,"* specifically an article by Berkheiser, S.W., entitled *"Some Commercial Aspects of the Peach Bottom Slate: The Problem of Being Too Good"*; pp 143-145.
• Norris, John C., (1898); <u>History and Characteristics of the Peach Bottom Roofing Slate</u>; pp 4-16.

CHAPTER 8 — VIRGINIA/GEORGIA

• Bridgen, John; *"Turning Stone into Bread,"* article of unknown origin and date, pp. 10-14.
• Buckingham Slate Corporation BuyLine 0459; January, 1990, and other published information, Buckingham-Virginia Slate Corporation, One Main Street, Arvonia, VA 23004-0008 (1994).
• Chambers, S. Allen, Jr., (1989); *"Of the Best Quality - Buckingham Slate,"* article in "Virginia Cavalcade," Spring 1989, pp. 158-171.
• Lindsay, Jean, (1974); *A History of North Wales*; David and Charles, Newton Abbot, London; p. 254.
• Redden, J. A., (1961), *"Slate in Virginia,"* article in *"Mineral Industries Journal,"* Vol. VIII, No. 3, September, 1961, published by the Virginia Polytechnic Institute, Mineral Industries Departments, School of Engineering and Architecture, pp. 1-5.
• Terrell, Patricia, (1962); *"Virginia Slate Industry: An Old Reliable Shows New Vigor,"* article in *"The Commonwealth,"* December 1962, p. 110.
• Tucker, Beverley R. Jr., (date unknown); *"Slate, Past Present and Future,"* Buckingham-Virginia Slate Corporation, Richmond, Va.
• *Galite Lightweight Structural Concrete, A Basic Manual*; Georgia Lightweight Aggregate Company, Atlanta, GA.
• Geological Survey of Georgia, Bulletin 27, 1912.
• Knight, Lucian Lamar, (1917); <u>Georgia and Georgians</u>, Vol. II; Lewis Publishing Co., Chicago and New York.
• <u>Memoirs of Georgia</u>, Volume 1, (1895), The Southern Historical Association, Atlanta, GA, p. 206.
• Mintz, Leonora, *"Slate Discovery Brought Mines to Van Wert"* (newspaper article found in Rockmart, Georgia, Public Library; no date or source accompanied article). Other photocopied articles from the same library, kept in a manila file folder in a back room, also devoid of dates, page numbers or sources, were also used in the section on Georgia slate in this chapter.
• Sargent, Gordon D., and Jackson, Olin; "The Town that Stone Built"; *North Georgia Journal*; Summer, 1996; pp. 28-35.
• *Some Historical Facts About Rockmart's First Industry - Slate*; The Rockmart Journal, Thursday, July 20, 1967; p. 7-A.

CHAPTER 13 — INSTALLING

• Hoppen, Ewald A., and Wagner, Dr. Wolfgang; <u>Firschungen zur Modernisierung des Schieferbergbaus</u>, 1995.
• Lorenz-Burmann-Schule Eslohe, Uberbetriebliche Unterweisung; Dachdeckungen.

CHAPTER 15 — ROOF INSCRIPTIONS AND DESIGNS

• <u>The Slate Roofer</u> (1905), by Auld and Conger Co., Cleveland, Ohio.

CHAPTER 19 — FLASHING

• <u>Copper Flashings and Weatherings - A Practical Handbook</u> (1951); Published by the Copper Development Association, Kendals Hall, Radlett, Herts, England.
• <u>Slate Roofs</u> (author unknown), (1926); available from Vermont Structural Slate Co., Fair Haven, Vermont 05743; an abbreviated version is also available from Hilltop Slate Co., Middle Granville, NY 12849.
• <u>Architectural Sheet Metal Manual</u> (Fifth edition, 1993); Sheet Metal and Air Conditioning Contractors National Association, Inc, 4201 Lafayette

Center Drive, Chantilly, Virginia 22021.
- The NRCA Roofing and Waterproofing Manual (Third Edition - 1990); National Roofing Contractors Association, O'Hare International Center, 10255 W. Higgins Road, Suite 600, Rosemont, IL 60018-5607; Ph: 708-299-9070; Fax: 708-299-1183.
- Soldering Manual, second edition, revised (1978); American Welding Society, Inc., 2501 N. W. 7th Street, Miami, FL 33125.

CHAPTER 20 — VALLEYS

- Jose Luis Menendez Seigas, (1995); Architecture and Techniques of Slate Roofing; Centro Technologico de la Pizarra de Galicia.

CHAPTER 21 — TILE ROOFS AND MISC.

- Hobson, Vincent H., (2001), Historic and Obsolete Roofing Tiles; Remai Publishing Co., Inc., Evergreen Colorado, USA

ADDITIONAL BIBLIOGRAPHY

special thanks to Terry Hughes of the Stone Roof Association;
http://www.stoneroof.org.uk

- Ashurst, John, and Dimes, Francis G. London: Butterworth-Heinemann, 1990: pp 137-144. Conservation of Building and Decorative Stone - Volume 1.
- Bates, Robert L., New York: Dover Publications, Inc., 1969: pp 59-69., Geology of the Industrial Rocks and Minerals.
- Behre, Charles H. Jr., Harrisburg, PA: Department of Forests and Waters/Topographic and Geologic Survey, 1927: pp 1-308. Slate in Northampton County Pennsylvania — Pennsylvania, Geological Survey, Fourth Series, Bulletin M9.
- Behre, Charles H. Jr., Harrisburg, PA: Department of Internal Affairs/Topographic and Geologic Survey, 1933: pp 1-400. Slate in Pennsylvania — Pennsylvania Geological Survey, Fourth Series, Bulletin M16.
- Bowles, Oliver. New York: McGraw-Hill Book Company, Inc., 1934: pp 229-289. The Stone Industries.
- Bowles, Oliver. Washington, D.C.: United States Department of the Interior, June 1955: pp 1-12. Slate — Bureau of Mines Information Circular 7719.
- Brumbaugh, James E., New York: Macmillan Publishing Company, 1986: pp 410-432. Complete Roofing Handbook — Installation, Maintenance, Repair.
- Building Stone Board of Regents of the Smithsonian Institution (Annual Report 1886). Washington: Government Printing Office, 1889: pp 464-471, 488, 509, 510, 549, 551, 556, 567, 572, 573, 591, 595, 603, 606, 609, 616, 629. Report of the United States National Museum.
- Clifton-Taylor, Alec. London: Faber and Faber Limited, 1972: pp 158-175. The Pattern of English Building.
- Dale, T. Nelson et al., Washington: Government Printing Office, 1914: pp 1-220., Slate in the United States — Department of the Interior, United States Geological Survey, Bulletin 586.
- Day, David Talbot. Washington: Government Printing Office, 1892: pp 373-440. Mineral Resources of the United States, 1889 and 1890 (United States Geological Survey).
- Downing, Andrew Jackson. New York: Dover Publications, 1981 (reprint of 1873 edition). Victorian Cottage Residences.
- "Early Roofing Materials." 1970 (Vol. II, Nos. 1-2): pp 18-51. Bulletin of APT (The Association for Preservation Technology).
- Earney, Fillmore C. The Journal of Geography, Volume LXII, No. 7, October, 1963: pp 300-310. The Slate Industry of Vermont.
- Evergreen Slate Co. Inc., Granville, New York: Evergreen Slate Company., Slate Roofing — A Complete Guide to Roofing Slate and Application Techniques.
- Garcia-Guinea, J. "Spanish Roofing Slate Deposits." 1997 Dec. (V. 106): pp B205. Transactions (Section B, Applied Earth Science).
- Gove, Les. "Slate Roof Repairs." 1990. Journal of Light Construction, Stone World Magazine.
- Hance, Peter, and Hart, David McLauren. Laconia, NH: City, Planning Department, 1978: pp 1-130., Slates, Shingles, and Shakes.
- Harrisburg: Board of Commissioners for the Second Geological Survey, 1883: pp 83-160. The Geology of Lehigh and Northampton Counties - Second Geological Survey of PA: Report of Progress, D3, Vol. I.
- Hawkins, Judy, and MacDonald, Susan. "Roofs of England: Reviving A Lost Industry." July 1997 (N. 32): p 10. English Heritage Conservation Bulletin.
- Hebert, Tony. Roofs of England. "On the Slate (Welsh Quarries)." Aug. 1989 (V. 5, No. 11): pp 16-20. Traditional Homes.
- Hunt, B. "The Changing Face of Slate — How to Assess Roofing Slate in the Light of Some Recent Failures and A Proposed New European Standard." 1994 (V. 29, N. 4): pp 14. Stone Industries.
- Kimball, Fiske. New York: Dover Publications, Inc., 1966 (original publication 1922): pp 40-44. Domestic Architecture Of the American Colonies and of the Early Republic.
- Levine, Jeffrey S. "Slate from the Source." 1997 Aug. (V. 25, N. 4): pp 36-39. Old House Journal.
- "Living National Treasure." 16 May 1996 (Vol. 190, No. 2): pp 70-71. Country Life; An article exploring the task of riving (i.e. splitting) quarried slate by hand, featuring a profile of Rex Barrow in England's Lake District.
- Marshall, Philip C. "Polychromatic Roofing Slate of Vermont and New York."
- 1979 (Vol. XI, No. 3): pp 77-88. APT Bulletin.
- Marshall, Philip C. "Slate Roofs: Conserving A New England Resource." Apr 1979 (Volume 4, Number 1): pp 1-2.
- McKee, Harley J. "Slate Roofing." 1970 (Vol. II, Nos. 1-2): pp 77-84. Bulletin of APT (The Association for Preservation Technology).
- Merrill, George P., New York: John Wiley & Sons, 1897: pp 345-365, 382-385. Stones for Building and Decoration.
- Merriman, M. "Perspectives — The Strength and Weathering Qualities of Roofing Slate, A Reprint of an 1892 Article from the Transactions of the American Society of Civil Engineers." Fall/Winter 1985 (No. 9): pp 12+. Technology and Conservation.
- Merriman, Mansfield. "The Slate Regions of Pennsylvania." July 1898, Volume XVII, Number 2 (Indianapolis, Indiana: Stone Magazine Review Pub. Co, 1889): pp 77- 90.
- Morrison, Hugh. New York: Dover Publications, Inc., 1952: pp 36, 75, 170, 271, 514. Early American Architecture — From the First Colonial Settlements to the National Period.
- Morrow, John A. Granville, NY: Grastorf Press, 1970: pp 1-32. A Century of Hard Rock — the story of Rising and Nelson Slate Company.
- New York: Revere Copper Products, Inc., 1987 (7th ed.). Copper & Common Sense, To obtain a copy of this publication, contact: Revere Copper Products, Inc., Corporate Headquarters, One Revere Park, Rome, NY 13440-5561 Tel: 800-448-1776. Fax 315.338.2224. Web: www.reverecopper.com.
- Perkins, George H. Montpelier, VT: Argus and Patrick Printing House, 1904: pp 47-51. Report of the State Geologist on the Mineral Industries and Geology of Certain Areas of Vermont, 1903-1904 — Fourth of Series.
- Perkins, George H. Montpelier, VT: Argus and Patriot Press, 1906: pp 56-58, 220-231, 260-261. Report of the State Geologist on the Mineral Industries and Geology of Certain Areas of Vermont, 1905-1906 - Fifth of Series.
- Perkins, George H. Report of the State Geologist on the Mineral Industries and Geology of Vermont, 1911-1912 — Eighth of Series. Montpelier, VT: Capital City Press, 1912.
- Peterson, Charles E. (ed.). Mendham, NJ: The Astragal Press, 1976: pp 38-39, 138-141, 384-387. Building Early America.
- Richards, Koziol, and Stockbridge, Jerry. "Detecting Water Leaks in Slate and Tile Roofs." 1987 (Vol. 19, No. 2): pp 6-9. APT Bulletin.
- Rockhill, Dan. "Tudor Hardtop: Building A Slate Roofed Cottage on A Kansas Moor." 1988 June/July (V. 47): pp 56-60., Fine Homebuilding.
- School of Mineral Industries and Experiment Station. The Pennsylvania State College Bulletin, Volume XLI, July 18, 1947, No. 30: pp 1-168. Properties and New Uses of PA Slate — Bulletin #47: Utilization of Waste Slate.
- Stanier, Peter H. Aylesbury, Bucks, United Kingdom: Shire Publications Ltd, 1985: pp 23-28. Shire Album 134: Quarries and Quarrying.
- Stone, Ralph W. Harrisburg, PA: Dept. of Internal Affairs/Topographic and Geologic Survey, 1932: pp 6, 14, 239, 299. Building Stones of PA — Pennsylvania Geological Survey, Fourth Series, Bulletin M15.
- Sweetser, Sarah M. "Roofing for Historic Buildings." Washington, D.C.: U.S. Department of the Interior, Technical Preservation Services Division, 1975: pp 1-8. Preservation Briefs, No 4.
- Vaux, Calvert. New York: Da Capo Press, 1968 (reprint of original 1857 publication). Villas and Cottages.
- "Wales Covers Ireland." 1987 Jan. (V.18, N. 1): pp 24-25. Plan: Architecture and Interior Design in Ireland.
- Warseck, K. "Historic Roofing." 1989 June (V. 70): pp 98-103. Progressive Architecture.
- Weaver, Kenneth N. (Director). Baltimore, Maryland: State of Maryland, 1969: pp 39-45, 104-106. The Geology of Harford County, Maryland — State of Maryland, Department of Natural Resources, MD Geol. Survey.
- Williams, Mefyn. Buckinghamshire, United Kingdom: Shire Publications Ltd, 1991: pp 1-32. Shire Album 268: The Slate Industry.
- Woodward, George E. and Thompson, Edward G. New York: Dover Publications, Inc., 1988 (reprint of original 1869 edition). A Victorian Housebuilder's Guide — "Woodward's National Architect" of 1869.

Slate Roofing Industry Resource Guide

New and Used Roofing Slates, Ceramic Tiles, Tools, Equipment, Fasteners, and Flashing Metal

This listing is for informational purposes only and does not constitute an endorsement. For the latest updated industry resource listings see www.slateroofcentral.com.

Companies marked with an asterisk (*) are featured in a special ad section following this listing.

SOURCES OF NEW SLATE
AUSTRALIA
MINTARO SLATE QUARRIES PTY LTD, PO Box 8, Mintaro, South Australia 5415; Telephone: 61 8 8843 9077; Fax: 61 8 8843 9019; Location: Quarries and Administration at Mintaro, near Clare, South Australia

SPALDING SLATE & STONE, Hill River Rd, Spalding; (08) 8845 2191; Clare Quarry, Broughton Valley Rd, Spalding; (08) 8843 4250

BRAZIL
AARÃO & COHEN CORP., Rua Turim, 146 - Congonhas - MG BRAZIL; Ph 36.415-000; aaraoecohen@ieg.com.br

B2B INTERNACIONAL COMMERCIO E EMPREENDIMENTOS LTDA., R. Frei Gaspar, 931 sala 11, Centro Sao Vicente-SP, BRAZIL, CEP 11310-080; Ph 0055 13 3468 1212; Fx 0055 13 3468 1011; export@b2binternacional.com.br; info@b2binternacional.com.br

BRTRADE (Ian Orellana) – Ph 55-31-99857756; Fx 55-31-33780037; brtrade@yahoo.com

MUNDIAL STONES, Rodovia MG 060 - Km 151 - Papagaio - MG - BRAZIL; Ph 55-37-32741117; Fx 55-37-32741999; exporting.slate@nwm.com.br

OPEN MART COMERCIAL EXPORTADORA, Ph 00 55 31 3292-4692; www.openmart.com.br

STONE TRADE, Papagaio – Minas Gerais – BRAZIL; Ph/Fx 55 37 3274 0044; www.stone-trade.com.br

CANADA
(Many American suppliers also carry Canadian slates.)
GLENDYNE QUARRY, 396 rue Principale, St. Marc du Lac Long Quebec G9L 1T0 CANADA; Ph 418-893-7221; Fx 418-893-7346;

NORTH COUNTRY SLATE, 8800 Sheppard Ave. East, Toronto, Ontario CANADA M1B5R4; Ph 416-724-4666 or 800-975-2835; Fx 416-281-8842; info@ncslate.com; www.northcountryslate.com

NORTHERN ROOF TILES, 50 Dundas St. E., Dundas, ON CANADA L9H 7K6 Ph 905-689-4035, 888-678-6866; Fx 905-689-7099; sales@nothernrooftiles.com; www.northern-rooftiles.com

ROOF TILE MANAGEMENT, INC., 2535 Drew Rd., Mississauga Ontario CANADA L4T 1G1; Ph 905-672-9992; Fx 905-672-9902; www.rooftilemanagement.com

UNIVERSAL SLATE INTL., INC., 3821 9th St. SE, Calgary ALB CANADA T2G 3C7; Ph 403-287-7763 or 888-67-SLATE; Fx 403-287-7736; zimmer@universalslate.com; www.universalslate.com

CHINA
(Many American suppliers also carry Chinese slates.)
BEIJING ORIENT SLATE & STONE CO., Ph (86)10-6428-1976; root@orientslate.com; www.orientslate.com

BAODING VITIAN TRADING CO., LTD., Ph 0086-312-2275655; Fx 0086-312-3023123; baojiangyu@yahoo.com.cn

CENTRAL & WEST CHINA SLATE CO., Wuhan China 430077; Ph 86-27-8678-8771; Fx 86-8678-8771

CHINA BSS NATURAL SLATE; Ph 8610-64968538; Fx 8610-64974573; slate@starstone.sina.net; contact Ms Yan

GHY STONE CO., LTD., NO.3, Shenggu North Road Dongcheng District, Beijing, China 100029; Ph 8610-64440324; Fx 8610-64426709; sales@ghystone.com; www.ghystone.com

HEBEI IMPORT & EXPORT CORPORATION, No. 486 West Heping Road, Shijiazhuang, P.R. China; Ph 86-311-7813720 ext. 6407; Fx 86-311-7050674; www.stone.hebei.net.cn

IEL INTERNATIONAL LIMITED, 1301 Winsome House, 73, Wyndham Street, Central, Hong Kong; Ph 852- 2522-2405; Fx 852-2522-2834; info@iel.com.hk; www.iel.com.hk/slate

LIU SHAOYU, Fx 0086-29-6243506 6280322; shaoyu@public.xa.sn.cn

SHAANXI LEESTONE CO., LTD., Ph (86)029-7661183 7661189; Fx (86)029-7661184; slate-quarry@globalsources.com, www.naturalslate.com.cn/test

SHAANXI WEIYIDA TRADE & DEVELOPMENT CO. LTD., No.96 Xiying Road Xii¯an Shaanxi P.R. China 710054; Fx 86- 29 - 5514155/5543799; Ph 86- 29- 5522582/5513336; darsim@163.com

SHANGHAI LEIHUA SLATE CO., LTD., www.leihuaint.com; contact information: Tel:+86-21-6406-5640, Fax:+86-21-6406-0292, Mobile: +86-133-7915-0688, Email: info@leihuaint.com

SHENZHEN SEG GENERAL TRADING CO., LTD., 6/F West, Huafa Building, Huafa Rd. N. Shenzhen, China 518031; Ph: 86-755-326-3531; Fx: 86-755-323-4127

WEICHANG NATURAL STONE (SLATE) CO., LTD.; Hong Kong; Ph 0086-29-88193888, 8569906, 8562316; Fx 0086-29-88193889; Mobile 0086 1389 2888958, 1389-2888028 info@westones.com; www.westones.com

YUN TAN SLATE FACTORY OF JIANGXI, Ph0086-792-2596051; Fx 0086-792-2596051

Z. Z. INTERNATIONAL, INC., Ph 360-459-4093; Fx 775-402-7608; ding@zzinternational.com; zzinternational.com

FRANCE
ASPIGAL, ZI Les Jonceaux B.P. 131, Hendaye Cedex France 64701; Ph (33)-55-920-1757; Fx (33)-55-920-3916; aspigal@wanadoo.fr; www.aspigal.com

LARIVIERE, 36 bis rue Delaage, B.P. 446-49004 Angers Cedex 01, France; Ph 02 41 66 67 81; Fx 02 41 47 19 16; bmonnier@lariviere-sa.fr; www.lariviere-sa.fr

GERMANY
JOHANN & BACKES, Flurstr. 11, 55626 Bundenbach, Germany; Ph 0049-6544-9988-0; Fx 0049-6544-99-8850; info@johann-backes.de; www.naturschiefer.de

RATHSCHECK SCHIEFER, Postfach 17 52, Mayen-Katzenberg Germany D-56727; Ph (49)026519550; Fx (49)02651955100; info@rathscheck.de; www.rathscheck.com

SCHIEFERGRUBEN MAGOG GMBH & CO. KG, Alter Bahnhof 9, D-57392 Bad Fredeburg; Ph 0049-2974-9620-0; Fx 0049-2974-9620-20; info@magog.de; www.magog.de

INDIA
HARSHINI EXPORTS, harshiniexports@yahoo.com

HIMACHAL SLATE & STONE, Upper Julakari, Chamba, Himachal Pradesh 176310 India; Phone: 91-18-9922 4682; Cell: 91-18-1602 4682; Fax: 91-18-9922 2567; neerajnayar@hotmail.com U.S. Address and Contact: Dhiraj Nayar, 37 Exeter Rd, Short Hills, NJ 07078; Ph: 646 287 0047; 973 467 0169; Fax: 973 467 4743; dhiraj_nayar@yahoo.com

M/S SANKAR ANAND EXPORTS, 24-2-709, Rajagopalapuram, Dargamitta, NELLORE - 524 003 A.P., India; Ph +91 861 320518, 301934; Fx +91 861 322836; sanexpo@vsnl.com; www.slate-granite.com

NEERA J. NAYAR, Ph91-1899-224682,226123; Fx 91-1899-222567;

RAMESH REDDY, jagadeeswara_slates@yahoo.co.in

WORLDLINK EXIM OPERATIONS & SERVICES, 17/17, Inbarajapuram 1st Street, off Bajanai Koil Road, Choolaimedu, Chennai - 600 094, Tamil Nadu India; Ph+91-44-23614315; rch_gunjans@sancharnet.com

IRELAND
CAPCO ROOFING CENTER, Unit 47/48 Broomhill Close, Tallaght, Dublin 24 Ireland; Ph 01 462 0740; Fx 01 462 0741; or: Mount Tallant Avenue, Terenure, Dublin 6W Ireland; Ph 01 490 2755; Fx 01 490 1021; info@capco.ie; www.capco.ie

TEGRAL BUILDING PRODUCTS LTD., 6 South Leinster St., Dublin 2 Ireland; Ph 01 676 3974; Fx 01 676 2820; support@tegral.com; www.tegral.com

ITALY
EUROSLATE, 16040 Monteleone Di Cicagna, Orero Italy; Ph 39-0185-334-042; Fx 39-0185-334-233; www.euroslate.it

NORWAY
MINERA NORGE, AS, P.O. Box 68, 9501 Alta Norway; Ph 0047 784 35 333; Fx 0047 784 35 374; (Danmark-Finland-Iceland-Norway-Sweden); info@naturstein.no

SPAIN
(Many American suppliers also carry Spanish slates.)
CUPA PIZARRAS, Spanish HQ, La Medua, s/n, 32330, Sobradelo de Valdeorras (Orense), Ph: 00 34 988 335 580; Fax: 00 34 988 335 599; comercial@cupire.com

GESTIO D'ENDERROS INMOBLES S.L., C/ Galicia, 112 08223 Terrassa (Barcelona - SPAIN); Ph 34 937316515; Fx 34 937312455; gdenderrocs@cecot.es

PIZARRAS CASTRELOS, S.A.; Casaio s/n, 32337 Caballeda de Valdeorras (Orense), Spain; Ph 0034-988-324-760; Fx 0034-988-337-787; infocastrelos@castrelos.com; www.castrelos.com

PIZARRAS FRANVISA, c/o General Vives, 54, Ponferrada Leon, Spain; Ph 34-987-418-904; Fx 34-987-414-158; UK and USA office, 8 Avenue Mansions, St. Paul's Ave., London NW2 5UG UK; Ph 44-181-459-3857; Fx 44-181-830-4047; cubelos@mcmail.com

PIZARRAS SAMACA, S.A., E-32337 El Trigal Sobradelo de Valdeorras Orense, Spain; Ph 34-9-883-24-770; Fx 34-9-883-24-733

PIZARRAS VILLAR DEL REY, S.L., 06192 Villar Del Rey Badajoz, Spain; Ph 34-24-414-111 or 34-24-414-211; Fx 34-24-414-221

UK
(Some American suppliers also carry UK slates.)
ALFRED McALPINE SLATE LTD., Bethesda Bangor Gwynedd WALES; Ph 44 (0) 1248-600656; Fx 44 (0) 1248-601171; www.amslate.com

BURLINGTON SLATE LTD., Cavendish House Kirkby-in-Furness, Cumbria, WALES LA17 7UN; Ph (44)01229-889-661; Fx (44)01229-889-466; sales@burlingtonstone.co.uk; www.burlingtonstone.co.uk/roofing.htm

CUPA NATURAL SLATE, UK & Ireland office; 45 Moray Place, Edinburgh EH3 6BQ; Ph: 00 44

SOURCES OF NEW SLATE (CONTINUED)

131 22 53 111; Fax: 00 44 131 22 05 463; uk@cupirepadesa.com; www.cupa.es

CWT Y BUGAIL SLATE QUARRIES CO., LTD., Blaenau Ffestiniog Gwynedd LL413RG WALES; Ph 44-076-683-0204; Fx 44-076-683-1105

HONISTER SLATE MINE, Honister Pass, Borrowdale, Cumbria CA12 5XN, ENGLAND; Ph 017687 77230; info@honister-slate-mine.co.uk

SLATE AND STONE CONSULTANTS, Terry Hughes, Ceunant, Caernarfon, Gwynedd LL55 4SA, UK; Ph 44 (0) 1286 650402; terry@ slateroof.co.uk; www.stoneroof.org

SLATE WORLD, 158 Wandsworth Bridge Road, Fulham UK SW6 5UL; Ph (44)020-7384-9595; Fx (44)020-7384-9599; sales@slateworld.com www.slateworld.com

WELSH SLATE LTD., Unit 205, 52 Upper St., London N1 0QH UK; Ph 44(20)7354-0306; Fx 44(20)7354-8485; nblager@welshslate.com; www.welshslate.com

UNITED STATES
ALABAMA

EMACK SLATE COMPANY, INC., 9 Office Park Circle, Suite 120, Birmingham, AL 35223; Ph 205-879-3424; Fx 205-879-5420; jim@emackslate.com; www.emackslate.com

CALIFORNIA

*****AMERICAN SLATE CO.,** 1900 Olympic Blvd. Ste. 200, Walnut Creek, CA 94596; Ph 925-977-4880 or 800-553-5611; Fx 925-977-4885; slatexpert@americanslate.com; www.americanslate.com

ECHEGUREN SLATE, INC., 1495 Illinois St., San Francisco CA 94107; Ph 415-206-9343; Fx 415-206-9353; slate@echeguren.com; www.echeguren.com

COLORADO

SOURCE PRODUCTS GROUP INC., 16000 Huron St., Broomfield, CO 80020; Ph 303-280-9595; Fx 303-280-2600; www.petraslate.com; www.tileroofing.com

DELAWARE

SLATE INTERNATIONAL, INC., 3422 Old Capitol Trail, Ste. 1061, Wilmington, DE 19808; Ph 301-952-0120; Fx 301-952-0295; ardelis@comcast.net; www.slateinternational.com

FLORIDA

PREMIER ROOFING SPECIALISTS, INC., PO Box 2298, Lake City, FL 32056; Ph: 888-492-4789; Fax: 386-719-9905; ricardo@premierroofs.com

TEJAS BORJA USA; 401 Redland Rd., Homestead, FL 33030; Ph 305-594-4224 or 800-830-TILE; Fx 305-242-6595; adam@tejasborja-usa.com; www.tejasborja-usa.com

GEORGIA

*****BLACK DIAMOND SLATE LLC.,** P.O. 30957, Savannah, GA 31410; Ph 877-229-9277; Fx 912-898-2339; ken@blackdiamondslate.com; www.blackdiamondslate.com

CLASSIC SLATE AND TILE, 80 W. Wieuca Rd., Suite 204, Box 25, Atlanta, GA 30342; Ph: 404-847-0188; Fax: 404-847-0166; clastile@bellsouth.net; Contact: Steve Yoder

JGA SOUTHERN ROOF CENTER, 2200 Cook Dr., Atlanta, GA 30340; Ph 770-447-6466; Fx 770-840-9001; jimo@jgacorp.com; www.jgacorp.com

ILLINOIS

MORTENSON ROOFING CO., 9505 Corsair Rd., Frankfort, IL 60423; Ph 815-464-7300; Fx815-464-7850; www.mortensonroofing.com

*****RENAISSANCE ROOFING, INC.,** PO Box 5024, Rockford, IL 61125; Ph 815-547-1725 or 815-874-5695 or 800-699-5695; Fx 815-547-1425; info@claytileroof.com; www.claytileroof.com

*****TILE ROOFS, INC.,** 9505 Corsair Rd., Frankfort, IL 60423; Ph 708-479-4366 or 888-708-TILE; Fx 708-479-7865; tileroofs@aol.com; www.tileroofs.com

MARYLAND

ROOF CENTER, THE, 9055 Comprint Ct., Ste. 300, Gaithersburg, MD 20877; Ph 301-548-0548 or 800-503-5500; Fx 301-548-0828; roof@roofcenter.com; www.roofcenter.com

MASSACHUSSETS

*****MAHAN SLATE ROOFING COMPANY,** PO Box 2860, Springfield, MA 01101; Ph: 413-788-9529; Fax: 413-467-2177, Email

MICHIGAN

OLD WORLD DISTRIBUTORS, 1601 West KL Ave. Ste. 2, Kalamazoo, MI 49009; Ph 269-372-3916; Fx 269-372-9852; owdist@net-link.net; www.oldworlddistributors.com

NEW YORK

*****EVERGREEN SLATE CO., INC.,** 68 E. Potter Ave., Granville, NY 12832; Ph 518-642-2530; Fx 518-642-9313; slate@evergreenslate.com; www.evergreenslate.com

HILLTOP SLATE, INC., PO Box 201, Rt. 22A, Middle Granville, NY 12849; Ph 518-642-2270; Fx 518-642-1220; hilltopslate@aol.com; www.hilltopslate.com

NORTHEAST SLATE, 911 Central Ave., #152, Albany, NY 12206; Contact: Daniel Boone, ph: 518-339-1818; cell: 518-265-2766; fax: 518-391-2831; sales@northeastslate.com; http://www.northeastslate.com

*****RISING AND NELSON SLATE CO.,** P.O. Box 336, 2027 County Rte. 23, Middle Granville, NY 12849; Ph 518-642-3333; Fx 518-642-1819; slaterisingandnelson@adelphia.net

*****SHELDON SLATE PRODUCTS,** PO Box 199, Fox Road, Middle Granville, NY 12849; Ph: 518-642-1280; Fax: 518-642-9085

TATKO STONE PRODUCTS, 50 Columbus Street, Granville, NY 12832; Contact: Robert Tatko, Ph: 518-642-1702; Fax: 518-642-3255; Cell: 518-642-1733; http://www.tatkostone.com; tatkostone@joimail.com

NORTH CAROLINA

JOHN KING, 133 Hayfield Ct., Wilmington, NC 28411; Ph 910-686-9394; Fx 910-686-3812; jkingco1@earthlink.net

THE TILE MAN, INC., 520 Vaiden Rd., Louisberg NC 27549; Ph: 919-853-6923 or 888-263-0077; Fax: (919) 853-6634; info@thetileman.com; www.thetileman.com

OHIO

DURABLE SLATE CO., 1050 N. Fourth St., Columbus, OH 43201; Ph 614-299-5522 or 800-666-7445; Fx 614-299-7100; tile@durableslate.com; www.durableslate.com

PENNSYLVANIA

DALLY SLATE CO., 500 Railroad Ave., Pen Argyl, PA 18072; Ph 610-863-4172; Fx: 610-863-8388; www.dallyslate.com

*****JOSEPH JENKINS, INC.,** 143 Forest Lane, Grove City, PA 16127; Ph 814-786-9085 or 866-641-7141; Fx 814-786-8209; mail@josephjenkins.com; www.slateroofcentral.com

*****PENN BIG BED,** PO Box 184, 8450 Brown St., Slatington, PA 18080; Ph 610-767-4601; Fx 610-767-9252; pbbslate@ptd.net; www.pennbigbedslate.com

STRUCTURAL SLATE CO., 222 East Main St., Pen Argyl, PA 18072; Ph 610-863-4141 or 800-677-5283; Fx 610-863-7016; sscol@ptd.net; www.structuralslate.com

WILLIAMS AND SONS SLATE AND TILE, 6596 Sullivan Trail, Wind Gap PA 18091; Ph: 610-863-4161; Fax: 610-863-8128; www.williamsslate.com/index.htm; wmsslate@enter.net

TEXAS

BURLINGTON NATSTONE INC., 2701C W. 15th St., Ste. 505, Plano, TX 75075; Ph 972-985-9182; Fx 972-612-0847

ROOF TILE AND SLATE COMPANY, THE, 1209 Carroll St., Carrollton, TX 75006; Ph 972-446-0005 or 800-446-0220; Fx 972-242-1923; rtscdow@aol.com; www.claytile.com

TILESEARCH, INC., 216 James St., Roanoke, TX 76262; Ph 817-491-2444; Fx 817-491-2457; ts@tilesearch.net; www.tilesearch.net

VERMONT

*****CAMARA SLATE PRODUCTS, INC.,** P.O. Box 8, 963 S. Main St., Fair Haven, VT 05743; Ph 802-265-3200; Fx 802-265-2211; info@camaraslate.com, www.camaraslate.com

*****GREENSTONE SLATE,** 325 Upper Rd., PO Box 134, Poultney, VT 05764; Ph 802-287-4333; Fx 802-287-5720; info@greenstoneslate.com; www.greenstoneslate.com

JUST SLATE, 208 Frog Hollow Rd., Brandon, VT 05733; Ph/Fx 802-247-8145; justslate1@aol.com

*****NEW ENGLAND SLATE CO.,** 1385 US Rt 7, Pittsford, VT 05763; Ph 802-247-8809 or 1-888-NE-SLATE; Fx 802-247-0089; slate@neslate.com; www.neslate.com

TACONIC STONE LLC, 5 Brooklyn Heights, Fair Haven, VT 05743; Ph 802-265-8163; taconicstone@aol.com

TARAN BROTHERS SLATE COMPANY, 2522 Vermont Route 30 N, North Poultney, VT 05764; 802-265-3220

VERMONT SPECIALTY SLATE, INC., PO Box 4, Brandon, VT. 05733; store 855 North Street, Forestdale, VT 05745; Ph 1-866-US-SLATE; Fx 1-802-247-4209; info@vtslate.com; www.vtslate.com

*****VERMONT STRUCTURAL SLATE CO., INC.,** Box 98, 3 Prospect St., Fair Haven, VT 05743; Ph 802-265-4933 or 800-343-1900; Fx 802-265-3865; info@vermontstructuralslate.com; www.vermontstructuralslate.com

VIRGINIA

BUCKINGHAM-VIRGINIA SLATE CORP., 1 Main Street, PO Box 8, Arvonia, VA 23004; Ph 434-581-1131 or 800-235-8921; Fx 434-581-1130; bvslate@ceva.net; www.bvslate.com

*****VIRGINIA SLATE,** 100 East Main St., Richmond VA 23219; Ph: 804-282-7929 or 888-VA-SLATE; Fax: 804-285-4442; sales@virginiaslate.com; www.virginiaslate.com

WISCONSIN

ENCHANTED IMPORTS, INC., PO Box 266, Land O' Lakes, WI 54540; Ph 715-547-8000; Fx 715-547-8001, globalgatherings@aol.com; www.enchantedforestimports.com

SALVAGED ROOFING SLATE

*****ALLUVIUM CONSTRUCTION,** 200 Lake Shore Dr., Marlton, NJ 08053; Ph 856-767-2700; Fx 856-768-7766; alluviumconstruction@comcast.net; www.historicroofs.com; www.thesteeplepeople.com.

*****BLACK DIAMOND SLATE LLC.,** P.O. 30957, Savannah, GA 31410; Ph 877-229-9277; Fx 912-898-2339; ken@blackdiamondslate.c- om; www.blackdiamondslate.com

CUSTOM TILE ROOFING, INC., 2875 West Hampden Ave., Englewood, CO 80110; Ph 303-761-3831; Fx 303-761-3839; ctrvince@qwest.net; www.customtileroofing.com

DURABLE SLATE CO., 1050 N. Fourth St., Columbus OH 43201; Ph 614-299-5522 or 800-666-7445; Fx 614-299-7100; tile@durableslate.com; www.durableslate.com

ECHEGUREN SLATE, INC., 1495 Illinois St., San Francisco, CA 94107; Ph 415-206-9343; Fx 415-206-9353; slate@echeguren.com; www.echeguren.com

EMACK SLATE COMPANY, INC., 9 Office Park Circle, Suite 120, Birmingham, AL 35223; Ph 205-879-3424; Fx 205-879-5420; jim@emackslate.com; www.emackslate.com

GENUINE SLATE, P.O. Box 235, Whitehall, NY 12887; store 1209 Prospect St., Fair Haven, VT 05743; Ph 802-265-8300; info@genuineslate.com; www.genuineslate.com

GILBERT & BECKER CO., INC., 16-24 Clapp St., Dorchester, MA 02125; Ph 617-265-4343; Fx 617-265-0936; info@gilbertandbecker.com; www.gilbertandbecker.com

★**GREENSTONE SLATE**, PO Box 134, 325 Upper Rd., Poultney, VT 05764, Ph 802-287-4333; Fx 802-287-5720; info@greenstoneslate.com; www.greenstoneslate.com

★**JOSEPH JENKINS INC.**, 143 Forest Lane, Grove City, PA 16127; Ph 814-786-9085 or 866-641-7141; Fx 814-786-8209; mail@joseph-jenkins.com; www.slateroofcentral.com

JUST SLATE, 208 Frog Hollow Rd., Brandon, VT 05733; Ph/Fx 802-247-8145; justslate1@aol.com

JOHN KING, 133 Hayfield Ct., Wilmington, NC 28411; Ph 910-686-9394; Fx 910-853-3812; jkingco1@earthlink.net

LANTZ SLATE ROOF REPAIR, 2431 Creekhill Rd., Lancaster, PA 17601; Ph 717-656-2620; Fx 717-656-7727

MORTENSON ROOFING CO., 9505 Corsair Rd., Frankfort, IL 60423; Ph 815-464-7300; Fx 815-464-7850; www.mortensonroofing.com

NORTH AMERICAN BÖCKER, 302 West Lane St.., Raleigh, NC 27603; Ph 800-624-8076; Fx 919-832-8439; info@nabocker.com; www.nabocker.com

NOSAK IMPROVEMENTS, INC., 2121 E. Ute St. Tulsa, OK 74110; Ph 918-230-0005, 918-599-0368; Fx 918-599-0277; paulnosak13@aol.com

OLD WORLD DISTRIBUTORS, 6101 West KL Ave., Kalamazoo, MI 49009; Ph 269-372-3916; Fx 269-372-9852; owdist@net-link.net; www.oldworld-didtributors.com

★**RENAISSANCE ROOFING, INC.**, PO Box 5024, Rockford, IL 61125; Ph 815-547-1725 or 815-874-5695 or 800-699-5695; Fx 815-547-1425; info@claytileroof.com; www.claytileroof.com

★**RECLAIMED ROOFS, INC.**, 7454 Lancaster Pike #328, Hockessin, DE 19707; Ph (302) 369-9187; Fx (302) 397-2742; Mobile (302) 388-1155; doug@reclaimedroofs.com; www.reclaime-droofs.com

ROOF CENTER, THE, 9055 Comprint Ct., Ste. 300, Gaithersburg, MD 20877; Ph 301-548-0548 or 800-503-5500; Fx 301-548-0828; roof@roofcenter.com; www.roofcenter.com

ROOF TILE AND SLATE COMPANY, THE, 1209 Carroll St., Carrollton, TX 75006; Ph 972-446-0005 or 800-446-0220; Fx 972-242-1923; rtsc-dow@aol.com; www.claytile.com

RYAN CONTRACTING & ROOFING, 360 Merrimack Street, Bldg #5, Lawrence, MA 01843; Ph 888-ROOF-SOS or 978-557-9413

SLATE & COPPER CO., 1024 W. Fourth St., Erie, PA 16057; Ph 814-455-7430; Fx 267-200-0800; slatecut@aol.com; www.slateandcopper.com

SLATEWORKS ROOFING, 117 Elizabeth Street, Evans City, PA 16033; Contact: Ron Kugel; Ph: 724-538-3538; Cell: 724-316-7702; Fax: 724-538-3538

TARAN BROTHERS SLATE COMPANY, 2522 Vermont Route 30 N, North Poultney, VT 05764; 802-265-3220

TILE MAN, INC., THE, 520 Vaiden Rd., Louisberg NC 27549; Ph 919-853-6923 or 888-263-0077; Fx 919-853-6634; info@thetileman.com; www.thetileman.com

★**TILE ROOFS, INC.**, 9505 Corsair Rd., Frankfort, IL 60423; Ph 708-479-4366 or 888-708-TILE; Fx 708-479-7865; tileroofs@aol.com; www.tileroofs.com

TILESEARCH, INC., 216 James St., Roanoke, TX 76262; Ph 817-491-2444; Fx 817-491-2457;

ts@tilesearch.net; www.tilesearch.net

★**VERMONT RECYCLED SLATE**; PO Box 71; Fair Haven, VT 05743; Ph 802-265-4506; www.usedslate.com;

★**VERMONT SPECIALTY SLATE, INC.**, PO Box 4, Brandon, VT 05733; Store 855 North Street, Forestdale, VT 05745; Ph 1-866-US-SLATE; Fx 802-247-4209; info@vtslate.com; www.vtslate.com

VINTAGE SLATE, 265 Furnace St, Poultney, VT 05764; Phone: 802-287-2559; Cell: 802-342-0915; info@vintageslate.com; http://www.vintageslate.com

NEW CERAMIC ROOFING TILE

★**ALLUVIUM CONSTRUCTION**, 200 Lake Shore Dr., Marlton, NJ 08053; Ph 856-767-2700; Fx 856-768-7766; alluviumconstruction@comcast.net; www.historicroofs.com;

★**AMERICAN SLATE CO.**, 1900 Olympic Blvd., Ste. 200 Walnut Creek, CA 94596; Ph 925-977-4880 or 800-553-5611; Fx 925-977-4885; slatexpert@americanslate.com; www.americanslate.com

EMACK SLATE COMPANY, INC., 9 Office Park Circle, Suite 120, Birmingham, AL 35223; Ph 205-879-3424; Fx 205-879-5420; jim@emack-slate.com; www.emackslate.com

GENUINE SLATE, P.O. Box 235, Whitehall, NY 12887; store 1209 Prospect St., Fair Haven, VT 05743; Ph 802-265-8300; info@genuineslate.com; www.genuineslate.com

JGA SOUTHERN ROOF CENTER, 2200 Cook Dr., Atlanta GA 30340; Ph 770-447-6466 or 800-763-0118; Fx 770-840-8941; jimo@jgacorp.com; www.jgacorp.com

LUDOWICI ROOF TILE, PO Box 69; New Lexington, OH 43764; Ph 800-917-8998; teresa.spencer@ludowici.com; www.ludowici.com

MORTENSON ROOFING CO., 9505 Corsair Rd., Frankfort, IL 60423; Ph 815-464-7300; Fx 815-464-7850; www.mortensonroofing.com

NORTHERN ROOF TILES, 50 Dundas St. E., Dundas, ON Canada L9H 7K6 Ph 905-689-4035, 888-678-6866; Fx 905-689-7099; sales@nothern-rooftiles.com; www.northernrooftiles.com

OLD WORLD DISTRIBUTORS, 6101 West KL Ave. Ste. 2, Kalamazoo, MI 49009; Ph 269-372-3916; Fx 269-372-9852; owdist@net-link.net; www.oldworlddistributors.com

★**RENAISSANCE ROOFING, INC.**, PO Box 5024, Rockford IL 61125; Ph 815-547-1725 or 815-874-5695 or 800-699-5695; Fx 815-547-1425; info@claytileroof.com; www.claytileroof.com

ROOF CENTER, THE, 9055 Comprint Ct., Ste. 300, Gaithersburg, MD 20877; Ph 301-548-0548; or 800-503-5500; Fx 301-548-0828; roof@roofcenter.com; www.roofcenter.com

ROOF TILE AND SLATE COMPANY, THE, 1209 Carroll Ave., Carrollton, TX 75006; Ph 972-446-0005 or 800-446-0220; Fx 972-242-1923; rtsc-dow@aol.com; www.claytile.com

ROOF TILE MANAGEMENT, INC., 2535 Drew Rd., Mississauga Ontario Canada L4T 1G1; Ph 905-672-9992; Fx 905-672-9902; www.rooftilemanagement.com

SOURCE PRODUCTS GROUP INC., 16000 Huron St., Broomfield, CO 80020; Ph 303-280-9595; Fx 303-280-2600; www.petraslate.com; www.tileroofing.com

TEJAS BORJA USA; 401 Redland Rd., Homestead, FL 33030; Ph 305-594-4224 or 800-830-TILE; Fx 305-242-6595; adam@tejasborja-usa.com; www.tejasborja-usa.com

TILESEARCH, INC., 216 James St., Roanoke TX 76262; Ph 817-491-2444; Fx 817-491-2457; ts@tilesearch.net; www.tilesearch.net

UNIVERSAL SLATE INTL., INC., 3821 9th St. SE,

Calgary ALB Canada T2G 3C7; Ph 403-287-7763 or 888-67-SLATE; Fx 403-287-7736; zimmer@universalslate.com; www.universalslate.com

USED CERAMIC ROOF TILE

★**ALLUVIUM CONSTRUCTION**, 200 Lake Shore Drive Marlton, NJ 08053; Ph 856-767-2700; Fx 856-768-7766; alluviumconstruction@comcast.net; www.historicroofs.com, www.thesteeplepeople.com

DURABLE SLATE CO., 1050 N. Fourth St., Columbus OH 43201; Ph 614-299-5522 or 800-666-7445; Fx 614-299-7100; tile@durableslate.com; www.durableslate.com

EMACK SLATE COMPANY, INC., 9 Office Park Circle, Suite 120, Birmingham, AL 35223; Ph 205-879-3424; Fx 205-879-5420; jim@emack-slate.com; www.emackslate.com

GENUINE SLATE, P.O. Box 235, Whitehall, NY 12887; store 1209 Prospect St., Fair Haven, VT 05743; Ph 802-265-8300; info@genuineslate.com; www.genuineslate.com

GILBERT & BECKER CO., INC., 16-24 Clapp St., Dorchester, MA 02125; Ph 617-265-4343; Fx 617-265-0936; info@gilbertandbecker.com; www.gilbertandbecker.com

JOHN KING, 133 Hayfield Ct., Wilmington, NC 28411; Ph 910-686-9394; Fx 910-853-3812; jkingco1@earthlink.net

MORTENSON ROOFING CO., 9505 Corsair Rd., Frankfort, IL 60423; Ph 815-464-7300; Fx 815-464-7850; www.mortensonroofing.com

NOSAK IMPROVEMENTS, INC., 2121 E. Ute St. Tulsa, OK 74110; Ph 918-230-0005, 918-599-0368; Fx 918-599-0277; paulnosak13@aol.com

OLD WORLD DISTRIBUTORS, 6101 West KL Ave. Ste. 2, Kalamazoo, MI 49009; Ph 269-372-3916; Fx 269-372-9852; owdist@net-link.net; www.oldworlddistributors.com

★**RENAISSANCE ROOFING, INC.**, PO Box 5024, Rockford, IL 61125; Ph 815-547-1725 or 815-874-5695 or 800-699-5695; Fx 815-547-1425; info@claytileroof.com; www.claytileroof.com

ROOF CENTER, THE, 9055 Comprint Ct., Ste. 300, Gaithersburg, MD 20877; Ph 301- 548-0548 or 800-503-5500; Fx 301-548-0828; roof@roofcenter.com; www.roofcenter.com

ROOF TILE AND SLATE COMPANY, THE, 1209 Carroll Ave., Carrollton, TX 75006; Ph 972-446-0005 or 800-446-0220; Fx 972-242-1923; rtsc-dow@aol.com; www.claytile.com

RYAN CONTRACTING & ROOFING, 360 Merrimack Street, Bldg #5, Lawrence, MA 01843; Ph 888-ROOF-SOS or 978-557-9413

TILE MAN, INC., THE, 520 Vaiden Rd., Louisberg, NC 27549; Ph 919-853-6923 or 888-263-0077; Fx 919- 853-6634; info@thetileman.com; www.thetileman.com

★**TILE ROOFS, INC.**, 9505 Corsair Rd., Frankfort, IL 60423; Ph 708-479-4366 or 888-708-TILE; Fx 708-479-7865; tileroofs@aol.com; www.tileroofs.com

TILESEARCH, INC., 216 James St., Roanoke, TX 76262; Ph 817-491-2444; Fx 817-491-2457; ts@tilesearch.net; www.tilesearch.net

SLATE TOOLS & EQUIPMENT

("T" = Tools; "E" = Equip.; "F" = Flashing)

ABC SUPPLY, One ABC Pkwy, Beloit, WI 53511; Ph: 608-362-7777 or 800-786-1210; Fx 608-362-6215; www.abcsupply.com; (T, E, F)

ACRO BUILDING SYSTEMS, 2200 W. Cornell St., Milwaukee, WI 53209; Ph 414-445-8787 or 800-267-3807; Fx 414-445-8792; info@acrobuild-ingsystems.com; www.acrobuildingsystems.com; (E - roof brackets, ladder accessories)

AJC HATCHET CO., 1227 Norton Rd., Hudson, OH 44236; Ph 330-655-2851 or 800-428-2438; Fx

330-650-1000; info@ajctools.com; www.ajctools.com; (*T* - slate ripper)

BENO J. GUNDLACH CO., 211 North 21st St., Belleville, IL 62226; Ph 618-233-1781; Fx 618-233-3636; www.benojgundlachco.com; contact.us@ benojgundlachco.com; (*T* - cutter)

*****BERGER BROS. CO.**, 805 Pennsylvania Blvd., Feasterville, PA 19053; Ph 215-355-1200 or 800-523-8852; Fx 215-355-7738; berger@berger-brothers.com; www.bergerbrothers.com or www.snowbrakes.com (*E, F*)

CARL KAMMERLING AND CO., PO Box 10 02 40, D-42002 Wuppertal, Bendahler Str. 110, Wuppertal(barmen), Germany 42285; Ph (49)0202-8903-0; Fx (49)0202-8071-5 (*T*)

CASSADY-PIERCE CO., 2295 Preble Ave., Pgh., PA 15233; Ph 412-321-8987 or 800-227-7239; Fx 412-321-4076; www.cassadypierce.com (*T, E, F*)

CEKA WORKS, LTD., Pwllheli LL53 5LH North Wales; Ph (44)01758-70-10-70; Fx (44)1758-70-10-90; sales@ck-tools.com; www.ck-tools.com (*T*)

ECHEGUREN SLATE, INC., 1495 Illinois St., San Francisco, CA 94107; Ph 415-206-9343; Fx 415-206-9353; slate@echeguren.com; www.eche-guren.com (*T*)

ESTWING MANUFACTURING CO., 2647 Eighth St., Rockford, IL 61109; Ph 815-397-9558; Fx 815-397-8665; sales@estwing.com; www.estwing.com (*T* -hammers, prybars, chisels)

*****EVERGREEN SLATE CO., LLC.**, 68 E. Potter Ave., Granville, NY 12832; Ph 518-642-2530; Fx 518-642-9313; slate@evergreenslate.com; www.ever-greenslate.com (*T*)

FLAME ENGINEERING, INC., PO Box 577 Lacrosse, KS 67548; Ph 785-222-2873, 800-255-2469; Fx 785-222-3619; flame@flameengineering.com ; www.flameengineering.com (*T* - torches, dragon wagons)

FULTON CORP., 303 8th Ave., Fulton, IL 61252; Ph 800-252-0002; Fx 815-589-4433; www.fulton-corp.com (*E* - ladder hooks)

GILBERT & BECKER CO., INC., 16-24 Clapp St., Dorchester, MA 02125; Ph 617-265-4343; Fx 617-265-0936; info@gilbertandbecker.com; www.gilbertandbecker.com (*T* - hammer, stakes, ripper)

*****GREENSTONE SLATE**, PO Box 134, 325 Upper Rd., Poultney, VT 05764; Ph 802-287-4333; Fx 802-287-5720; info@greenstoneslate.com; www.greenstoneslate.com (*T*)

GT PRODUCTS, INC., 86 Union Street, Mineola, NY 11501; Ph 516-625-1870; Fx 516-625-1226; gtproduct@aol.com; www.geotechproduct.com (*T*- hammer, cutters, rippers)

INTERSTATE MANUFACTURING & SUPPLY, 6363 Highway #7, St. Louis Park, MN 55416; Ph 800-328-6766; Fx 952-926-2313 (*E, F*)

*****JOSEPH JENKINS, INC.**, 143 Forest Lane, Grove City, PA 16127; Ph 814-786-9085 or 866-641-7141; Fx 814-786-8209; mail@joseph-jenkins.com; www.slateroofcentral.com (*T, E,F*)

JGA SOUTHERN ROOF CENTER, 2200 Cook Dr., Atlanta, GA 30340; Ph 770-447-6466 or 800-763-0118; Fx 770-840-8941; jimo@jgacorp.com; www.jgacorp.com (*T*)

JUST SLATE, 208 Frog Hollow Rd., Brandon, VT 05733; Ph/Fx 802-247-8145; justslate1@aol.com (*T*)

*****NEW ENGLAND SLATE CO.**, 1385 US Rt. 7, Pittsford, VT 05763; Ph 802-247-8809 or 888-NE-SLATE; Fx 802-247-0089; slate@neslate.com; www.neslate.com (*T*)

NORTH AMERICAN BÖCKER, 302 West Lane St., Raleigh, NC 27603; Ph 800-624-8076; Fx 919-832-8493; info@nabocker.com; www.nabocker.com (*T, E*)

OLD WORLD DISTRIBUTORS, 6101 West KL Ave. Ste. 2, Kalamazoo, MI 49009; Ph 269-372-3916; Fx 269-372-9852; owdist@net-link.net; www.old-

worlddistributors.com (*T, E, F*)

P. F. FREUND & CIE. GMBH., Hahnerberger Str. 94-96, Postfach 150 125, Wuppertal, Germany D-42349; Ph (49) 0202-409-29-0; Fx (49) 0202-409-2929; info@freund-cie.com; www.freund-cie.com (*T*)

*****PENN BIG BED**, PO Box 184, 8450 Brown St., Slatington, PA 18080; Ph 610-767-4601; Fx 610-767-9252; www.pennbigbedslate.com (*T*)

REIMANN & GEORGER CORP., PO Box 681, Buffalo, NY 14240; Ph 716-895-1156; Fx 716-895-1547; sales@rgcproducts.com; www.rgcproducts.com (*E* - ladder hooks)

*****RENAISSANCE ROOFING, INC.**, PO Box 5024, Rockford, IL, 61125; Ph 815-547-1725 or 815-874-5695 or 800-699-5695; Fx 815-547-1425; info@claytileroof.com; www.claytileroof.com (*T*)

ROOF CENTER, THE, 9055 Comprint Ct., Ste. 300, Gaithersburg, MD 20877; Ph 301-548-0548 or 800-503-5500; Fx 301-548-0828; roof@roofcenter.com; www.roofcenter.com (*T, E, F*)

ROOF TILE AND SLATE COMPANY, THE, 1209 Carroll Ave., Carrollton, TX 75006; Ph 972-446-0005 or 800-446-0220; Fx 972-242-1923; rtsc-dow@aol.com; www.claytile.com (*T*)

SIEVERT INDUSTRIES, INC., 5255 Zenith Parkway, Loves Park, IL 61111; Ph 877-639-1319; Fx 815-639-1320; mmelito@sievertindustries.com; www.sievertindustries.com (soldering equipment)

*****JOHN STORTZ AND SON, INC.**, 210 Vine St., Philadelphia, PA 19106; Ph 215-627-3855 or 888-847-3456; Fx 215-627-6306; john@stortz.com; www.stortz.com (*T, E* - hammers, rippers, cutters, tongs, etc.)

STRUCTURAL SLATE CO., 222 East Main St., Pen Argyl, PA 18072; Ph 610-863-4141 or 800-677-5283; Fx 610-863-7016; ssco1@ptd.net; www.structuralslate.com (*T*)

TILESEARCH, INC., 216 James St., Roanoke, TX 76262; Ph 817-491-2444; Fx 817-491-2457; ts@tilesearch.net; www.tilesearch.net (*T*)

VERMONT SPECIALTY SLATE, PO Box 4, Brandon, VT 05733; store 855 North St., Forestdale, VT 05745; Ph 1-866-US SLATE or 802-247-6615; Fx 802-247-4209; info@vtslate.com; www.vtslate.com (*T*)

*****VERMONT STRUCTURAL SLATE CO., INC.**, Box 98, 3 Prospect Ave., Fair Haven, VT 05743; Ph 802-265-4933 or 800-343-1900; Fx 802-265-3865; info@vermontstructuralslate.com; www.ver-montstructuralslate.com (*T*)

WILLIAMS AND SONS SLATE AND TILE, 6596 Sullivan Trail, Wind Gap, PA 18091; Ph 610-863-4161; Fx 610-863-8128; wmsslate@enter.net (*T*)

FLASHING METAL

ABC SUPPLY, One ABC Pkwy, Beloit, WI 53511; Ph: 608-362-7777 or 800-786-1210; Fx 608-362-6215; www.abcsupply.com

*****BERGER BROS. CO.**, 805 Pennsylvania Blvd., Feasterville, PA 19053; Ph. 215-355-1200 or 800-523-8852; Fx. 215-355-7738; berger@berger-brothers.com; www.bergerbrothers.com or www.snowbrakes.com

CAMBRIDGE-LEE INDUSTRIES INC., 475 Jersey Ave., New Brunswick, NJ 08903; Ph 800-852-2885; Fx 732-846-8476

CASSADY-PIERCE CO., 2295 Preble Ave., Pgh., PA 15233; Ph 412-321-8987 or 800-227-7239; Fx 412-321-4076; www.cassadypierce.com

COPPER SALES (UNA-CLAD), 1001 Lund Blvd., Anoka, MN 55303; Ph 763-576-9595 or 800-426-7737; Fx 763-576-9596; www.unaclad.com

FLORIDA METAL PRODUCTS, PO Box 6310, Jacksonville, FL 32236; Ph 904-783-8400 or 800-634-3937; Fx 904-783-8403; www.flamco.com; flamco@flamco.com (ridge iron and copper)

FOLLANSBEE STEEL, PO Box 610, Follansbee, WV 26037; 800-624-6906; folrfg@lbcorp.com;

www.follansbeeroofing.com

INTERSTATE MANUFACTURING AND SUPPLY, 6363 Highway #7, St. Louis Park, MN 55416; Ph 800-328-6766 or 952-926-2611; Fx 952-926-2313

*****JOSEPH JENKINS, INC.**, 143 Forest Lane, Grove City, PA 16127; Ph 814-786-9085 or 866-641-7141; Fx 814-786-8209; mail@joseph-jenkins.com; www.slateroofcentral.com

OLD WORLD DISTRIBUTORS, 6101 West KL Ave., Kalamazoo, MI 49009; Ph 269-372-3916; Fx 269-372-9852; owdist@net-link.net; www.oldworld-distributors.com

ROOF CENTER, THE, 9055 Comprint Ct., Ste. 300, Gaithersburg, MD 20877; Ph 301-548-0548 or 800-503-5500; Fx 301-548-0828; roof@roofcenter.com; www.roofcenter.com

ROOF TILE MANAGEMENT, INC., 2535 Drew Rd., Mississauga, Ontario CANADA L4T 1G1; Ph 905-672-9992; Fx 905-672-9902; www.rooftilem-anagement.com

SLATE & COPPER SALES CO., 201 German St., Erie, PA 16507; Ph 814-455-7430; Fx 267-200-0800; sales@slateandcopper.com; www.slateand-copper.com

UNIMET METAL SUPPLY, 557 Main Street, Orange, NJ 07050; Ph 800-526-4004; Fx 973-673-6477

WILLIAM METALS AND WELDING., 8052 State Street, Garrettsville, OH 44231-1023; Ph 800-842-3762; Fx 330-527-2748

NAILS, RIVETS, SLATE HOOKS

DORR WHOLESALE SUPPLY, 209 Riverside Heights, Manchester Center, VT 05255; Ph 802-362-2344 (slate hooks)

INTERSTATE MANUFACTURING AND SUPPLY, 6363 Highway #7, St. Louis Park, MN 55416; Ph 800-328-6766 or 952-926-2611; Fx 952-926-2313 (nails, rivets)

*****JOSEPH JENKINS, INC.**, 143 Forest Lane, Grove City, PA 16127; Ph 814-786-9085 or 866-641-7141; Fx 814-786-8209; mail@joseph-jenkins.com; www.slateroofcentral.com (nails, rivets, slate hooks)

*****NEW ENGLAND SLATE CO.**, 1385 US Rt 7, Pittsford, VT 05763; Ph 802-247-8809 or 1-888-NE-SLATE; Fx 802-247-0089; slate@neslate.com; www.neslate.com (nails, slate hooks)

OLD WORLD DISTRIBUTORS, 6101 West KL Ave. Ste. 2s, Kalamazoo, MI 49009; Ph 269-372-3916; Fx 269-372-9852; owdist@net-link.net; www.old-worlddistributors.com (nails, slate hooks)

*****RENAISSANCE ROOFING, INC.**, PO Box 5024, Rockford, IL 61125; Ph 815-547-1725 or 815-874-5695 or 800-699-5695; Fx 815-547-1425; info@claytileroof.com; www.claytileroof.com (nails, slate hooks)

*****SLATE ROOF SPECIALTIES**, PO Box 362, Stowe, VT 05672; 802-498-4158 (slate hooks)

SWAN SECURE PRODUCTS, 7525 Perryman Court, Baltimore, MD 21226; Ph 410-360-9100; Fx 410-360-2288; www.swansecure.com (nails)

WIRE WORKS, INC., 2910 E. Commercial, Pahrump, NV 89048; Ph 800-341-8828, 775-751-5555; Fx 775-751-9250; staff@wire-works-inc.com; www.wire-works-inc.com (nails, slate hooks)

* see ad following this section

For the latest up-to-date listings visit slateroofcentral.com

INDUSTRY SPOTLIGHT
CAMARA SLATE

Reprinted from *Traditional Roofing*, Spring, 2002 (www.traditionalroofing.com)
by Joseph Jenkins

The "Slate Valley" of western Vermont is well-known world wide for its high quality, multi-colored roofing slates — some of the best slates in the world come from here. Stretching for 25 miles from north to south, the valley is dotted by numerous quarries of stone that glistens purple, red, green, gray, or black; stone that has been historically proven to withstand the test of time on the roofs of buildings. Many rugged men work these quarries, wrestling from the Earth the massive stone slabs that will be skillfully worked by hand into individual roofing shingles.

Dave Camara is one of those workers. President and founder of Camara Slate Inc., Dave and four of his sons now operate three working quarries spread throughout the valley. Camara Slate also owns an additional dozen or so area quarries for possible future development.

Dave started in the business by salvaging slates from old buildings in the early 80s, trucking the salvaged slates around the nation, and back-hauling steel building components. His slate salvaging business expanded rapidly, allowing, from year to year, the purchase of bigger trucks and more slates, and finally his first slate quarry. Truly a tale of hard work, determination, and ingenuity, Dave Camara, with the help of his family, has risen to the top of the field in the slate roofing manufacturing business. Camara Slate now offers for sale new slates in a rainbow of colors: gray, unfading red, Spanish black, unfading green, unfading mottled purple, "sea green," purple, and Vermont gray black. They also produce slate flooring, flagstones, cladding, sills and copings, treads and risers, countertops, and structural slate in a variety of colors. Pallets of salvaged slates can still be found in the Camara stockyard on Route 22A just outside Fair Haven, although these are now dwarfed by the extensive inventory of new slates that fill the yard.

Dave Camara and a supply of "sea green" roofing slates awaiting shipment at the Camara Slate Co., stockyard on Rt 22A just outside Fair Haven, Vermont.

Camara and his sons express a strong pride in their products; their attention to detail and concern for quality and reputation is remarkable. Shawn Camara (age 30) runs the Blissville Quarry at the northern end of the valley, working alongside the other men there, splitting slates and keeping an eye on quality control. Here they produce the unfading slates: mottled green and purple, unfading gray and unfading green. Dave Camara Jr. (age 33) works in the pit at the Blissville quarry — an experienced "rockman," responsible for selecting the high-quality stone that is needed for splitting into roofing shingles.

Mike Camara (age 32) runs the West Pawlet quarry at the southern end of the valley, while Danny (age 26) is the rockman there. Here they produce Vermont black slates, semi-weathering gray, and semi-weathering green slates. From their Wells quarry, in the center of the valley, also come sea green and semi-weathering gray slates.

Camara slates are shipped throughout the United States including Hawaii, as well as to Canada. Their roofing slates have the traditional punched nail holes as opposed to non-countersunk drilled holes that are found on lower quality slates. The holes are punched to allow for either a three inch or four inch headlap. Camara's slate prices are very competitive; their product quality appears to be quite high; their attention to detail and concern for customer satisfaction is genuine, and they offer information that is no-nonsense and straightforward. As a result, Camara Slate is gaining an impressive reputation among roofing contractors in the United States. They're certainly worth a look when considering the purchase of virtually any slate product.

Shawn Camara splitting a slate block into roofing slates at the Blissville Quarry.

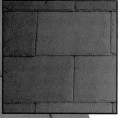
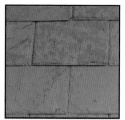

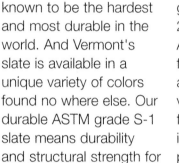

The Twelfth Century
Slate Roofing Company

Douglas L. Raboin

Slate Roof Specialist

10 Spring Valley Road
Burlington, Massachusetts 01803
(617) 666-3888

MA Reg.# 100693

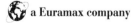
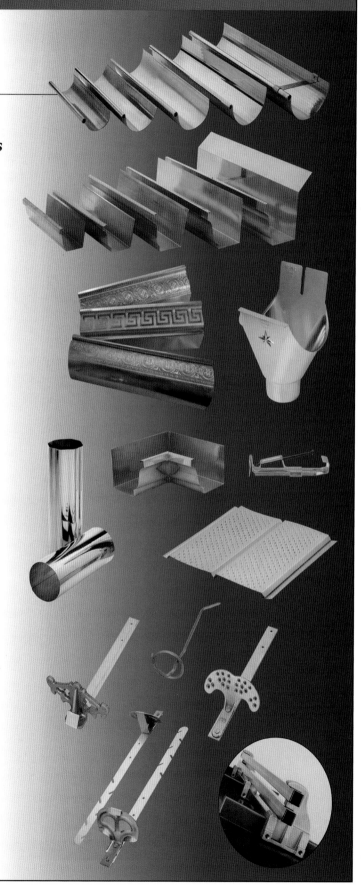

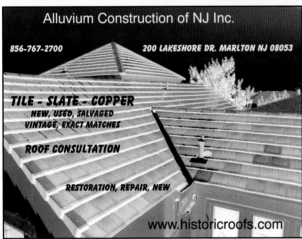

Index